OXFORD ENGINEERING SCIENCE SERIES

GENERAL EDITORS

THE OXFORD ENGINEERING SCIENCE SERIES

7. K. H. HUNT: *Kinematic geometry of mechanism*
10. P. HAGEDORN: *Non-linear oscillations* (Second edition)
13. N. W. MURRAY: *Introduction to the theory of thin-walled structures*
14. R. I. TANNER: *Engineering rheology*
15. M. F. KANNINEN and C. H. POPELAR: *Advanced fracture mechanics*
16. R. H. T. BATES and M. J. McDONNELL: *Image restoration and reconstruction*
19. R. N. BRACEWELL: *The Hartley transform*
20. J. WESSON: *Tokamaks*
22. C. SAMSON, M. LeBORGNE, and B. ESPIAU: *Robot control: the task function approach*
23. H. J. RAMM: *Fluid dynamics for the study of transonic flow*
24. R. R. A. SYMS: *Practical volume holography*
25. W. D. McCOMB: *The physics of fluid turbulence*
26. Z. P. BAZANT and L. CEDOLIN: *Stability of structures: elastic, inelastic, fracture, and damage theories*
27. J. D. THORNTON: *Science and practice of liquid–liquid extraction* (Two volumes)
28. J. VAN BLADEL: *Singular electromagnetic fields and sources*
29. M. O. TOKHI and R. R. LEITCH: *Active noise control*
30. I. V. LINDELL: *Methods for electromagnetic field analysis*
31. J. A. C. KENTFIELD: *Nonsteady, one-dimensional, internal, compressible flows*
32. W. F. HOSFORD: *Mechanics of crystals and polycrystals*
33. G. S. H. LOCK: *The tubular thermosyphon: variations on a theme*
34. A. LIÑÁN and F. A. WILLIAMS: *Fundamental aspects of combustion*
35. N. FACHE, D. DE ZUTTER, and F. OLYSLAGER: *Electromagnetic and circuit modelling of multiconductor transmission lines*
36. A. N. BERIS and B. J. EDWARDS: *Thermodynamics of flowing systems: with internal microstructure*
37. K. KANATANI: *Geometric computation for machine vision*
38. J. G. COLLIER and J. R. THOME: *Convective boiling and condensation* (Third edition)
39. I. I. GLASS and J. P. SISLIAN: *Nonstationary flows and shock waves*
40. D. S. JONES: *Methods in electromagnetic wave propagation* (Second edition)
42. G. A. BIRD: *Molecular gas dynamics and the direct simulation of gas flows*
43. G. S. H. LOCK: *Latent heat transfer: an introduction to fundamentals*

Latent Heat Transfer

An Introduction to Fundamentals

G.S.H. LOCK

Department of Mechanical Engineering
University of Alberta, Canada

Oxford New York Toronto
OXFORD UNIVERSITY PRESS
1994

Oxford University Press, Walton Street, Oxford OX2 6DP
Oxford New York Toronto
Delhi Bombay Calcutta Madras Karachi
Kuala Lumpur Singapore Hong Kong Tokyo
Nairobi Dar es Salaam Cape Town
Melbourne Auckland Madrid
and associated companies in
Berlin Ibadan

Oxford is a trade mark of Oxford University Press

Published in the United States
by Oxford University Press Inc., New York

A catalogue record for this book is available from the British Library

Library of Congress Cataloging in Publication Data
Lock, G. S. H.
Latent heat transfer : an introduction to fundamentals /
G. S. H. Lock. – 1st ed.
(Oxford engineering science series ; 43)
Includes bibliographical references and index.
1. Heat – Transmission. 2. Fusion, Latent heat of.
3. Evaporation, Latent heat of. I. Title. II. Series.
TJ260.L59 1994 621.402'2 – dc20 94-11638
ISBN 0 19 856285 3

Printed in Great Britain on acid-free paper by
Bookcraft Ltd., Midsomer Norton, Avon

To the memory of

JOSEPH BLACK

scholar and gentleman

PREFACE

Heat is subtle. It has no shape, no colour, and is not detectable through taste, touch, or smell. It makes so sound. We know it only as an abstract idea, as the transfer of thermal energy in the absence of a mass transfer. We infer its existence from its effects; most notably by the changes in temperature it produces. Despite these many and widespread effects, ranging from the boiling of a pan of water to the evolution of the universe, heat has been defined and understood only during the past 150 years. In this period, the thermometer has been the principal tell-tale of heat transfer or, more precisely, of *sensible* heat transfer.

Hidden things, by definition, are not easily or quickly seen. It is not surprising, therefore, that latent heat should present us with a particular difficulty. Its existence cannot be inferred with a thermometer. Latent heat is both hidden and subtle. For these reasons, the subject of latent heat transfer is difficult to present to undergraduates, many of whom struggle enough with the concepts of thermodynamics and the complexities of fluid mechanics. Yet latent heat transfer is so important and pervasive that a unified treatment has become essential: firstly, to make up the deficiency found in most undergraduate heat transfer courses; and secondly, to delineate the common ground of industrial engineers and environmentalists who face a growing number of problems in latent heat transfer.

In a university context, perhaps the best solution is to provide a comprehensive, graduate-level course. This book appears to be the first to bring together suitable material for such a course. It has been designed as a junior graduate level text, principally for students in mechanical and chemical engineering. Its main purpose is twofold: to provide a unified, general coverage of latent heat transfer; and to bridge the gaps between a cursory undergraduate treatment and the advanced graduate treatment of particular topics, e.g. boiling. A secondary purpose is to broaden the student's exposure to important heat transfer problems which are not broached in the traditional curriculum. The book places deliberate emphasis on the fundamental thinking so necessary for engineers who must be flexible and adaptable in an innovative environment where rapid shifts in technology are commonplace. To that extent the book is multidisciplinary and forward looking.

The material contained herein has emerged from a junior graduate course at the University of Alberta where successive classes have helped me shape the topics and the treatment to meet the needs of most engineering students. In this regard, I have tried to take into account the characteristics and the expectations of the typical North American graduate student: his or her undergraduate preparation in heat transfer; her or his plans for advanced/industrial work in the near future; and the rapidly changing needs of industry. My own students have taught me many things. From a course questionnaire they completed, for example, I learned that they place the same weight on the practice of technique as on the use of illustration and exemplification; however, they consider neither of these to be half as important as insight and understanding. This finding alone would justify the emphasis on fundamentals in this book.

I also recognize that graduate students have other significant differences from undergraduate students. In the use of illustrative practical material, for example, they prefer a balance between quantitative, idealized problems and qualitative, real problems. Likewise, in the employment of technique, they claim to benefit as much from analytical technique (here the use of scale analysis) as from computational routine. And in searching for insight and perspective they place more emphasis on physico-mathematical models than on experimental or empirical data. These findings have also influenced my thinking.

After forty years under the spell of Socrates, I have come to realize that teaching and learning in a university setting are twins, and their essence is very simple. It is love: specifically, the twinning of the love of knowledge and the love of humankind. When the student earnestly wishes to learn, and the scholar wishes to learn more, we have the basis of a symbiotic relationship which both requires and develops the attributes of trust, confidence, and mutual respect. Such a shared experience leads not only to the meaningful acquisition of specific knowledge, but to the flowering of the mind and, ultimately, to a sense of perspective commonly called wisdom.

Typically, the sharing of this knowledge begins in the lecture theatre with mind acting upon mind. However, it quickly extends to the library or private study area where the student relies heavily on books. As a surrogate for the professor, a book must keep faith with the ideals of teaching as they are characteristically expressed through the sequential logic of a given lecture. But it must also recognize the multi-facetted process by which humans actually learn: through observation, practice, and exploration. Accordingly, the textual (lecture) material provided herein has been buttressed by examples, exercises, and projects.

Chapter 8 is devoted exclusively to worked examples which illustrate the development of theoretical ideas and apply them to real situations; the student may thus become familiar with computational routine while developing a sense of the physical magnitudes encountered in practice. The exercises at the end of each chapter help the student check and reinforce his or her understanding of the fundamentals while becoming more familiar with key variables. The projects listed at the end of the four main chapters are each a major exercise in the exploration of open-ended topics; many of these are multidisciplinary and almost all of them will give the student a healthy appreciation of the current limits of our knowledge. The exercises and worked examples are designed for private study but the projects, upon completion, are best shared in a class discussion led by a particular student or group. Taken together with the lectures, these adjunct activities constitute a full learning experience: based on observation, reinforced by practice, and expanded through exploration.

Moreover, this experience may be tailored to the specific needs and interests of individual students. There are no examples within the text, except for the purpose of conveying typical magnitudes at a point where a student might naturally wonder about them. This omission is deliberate. It is my belief that the mind set required during the introduction of new ideas and concepts is different from that needed when working through a numerical illustration, step by step. Accordingly, the two forms of presentations have been separated, leaving the student to decide if and when an exercise or a worked example would be a helpful elaboration or an unnecessary digression. Students with an immediate grasp of the textual material may thus proceed more rapidly, while others can take the time to confirm their understanding and develop their confidence. The professorial role in setting the pace and in seeking out appropriate additional illustrative material is crucial.

Before outlining the structure and use of the book it might be worthwhile explaining what it is not intended to do. At the outset, I decided not to deal extensively with industrial design methodology. This would have required a greater familiarity with industrial circumstances than most students possess, and it restricts the end use of the material. The projects and worked examples provide sufficient exposure for most students. In line with this decision I have limited the use of numerical analysis to straightforward computational schemes involving little more than simple iteration. It is my belief that numerical analysis is too important and too time-consuming to incorporate as a mere feature. Furthermore, it is better handled in a separate course. Thirdly, I decided to treat thermophysical properties as constants, again to avoid computational complexity. These restrictions freed me to focus entirely on the fundamentals of latent heat transfer. The book thus enables a student with limited industrial experience and numerical skills to complete a comprehensive introduction to latent heat transfer in one term, and thus establish a foundation from which to launch more advanced study or begin industrial practice.

In Chapter 1, the origins and significance of latent heat transfer are traced up to the present day. Both industrial and environmental contexts are considered. In addition, the technique of scale analysis is introduced and treated in detail. It is used repeatedly in subsequent chapters. Having used the technique in teaching for almost three decades, I have often been puzzled by its absence in heat transfer texts. Recent books by Bejan have finally remedied the deficiency.

Chapters 2 and 3 are preparatory. They assume that the student has a thorough knowledge of undergraduate thermodynamics and fluid mechanics, but in the graduate school, with a significant proportion of foreign students, the level of preparedness may vary a great deal. Some students, for example, are already familiar with tensor analysis while others have not seen vector analysis. As a compromise, the treatment here is limited to vector calculus with the component forms of the governing partial differential equations being introduced as soon as possible. Each of the preparatory chapters provides a concise review of undergraduate material but go on to introduce new concepts and additional phenomena. At this early stage, it is often sufficient to follow the gist of the arguments, although the student will be repeatedly asked to recall this material, and the equations which stem from it, in later chapters.

This brings us to the four main chapters (4–7). In each of these, considerable emphasis is placed on the development of physical models and their mathematical description, i.e. on the phrasing and formulation of a problem. Scale analysis is shown to be a powerful and widely applicable tool which quickly establishes an economical relation between the central variables and parameters in a given situation. Typically, the situation chosen is the simplest example of the essential physics possible, but the principal variables, once revealed, may be used in more complex situations. The Reynolds number and Jakob number first uncovered in laminar film condensation, for example, are by no means restricted to this problem. Another important feature of these chapters, and indeed the entire book, is the recognition that most practical problems in latent heat transfer entail metastable departures from saturated (equilibrium) conditions. Should these departures exceed a stability threshold they lead to a dramatic change of behaviour illustrated, for example, by grain growth in metals or hysteresis in boiling.

Finally, a few words on the use of the book, beginning with the observation that it contains more material than can be digested in a typical one-term course containing 36–40 hours of lectures. The material may therefore be used selectively to suit the needs of a particular student body. For those who need to spend more time strengthening and deepening their knowledge of thermodynamics and fluid mechanics, the course may consist of Chapters 2 and 3 followed by Chapters 4, 5, and 6 or, if that is too much, the coverage of latent heat transfer problems could be limited to fluids only, i.e. Chapters 5 and 6. This would certainly be sufficient to round off the heat transfer knowledge of a typical engineer. On the other hand, students with a strong background in thermodynamics and heat transfer could proceed directly to Chapters 4, 5, 6, and 7. This would place them in a strong position for more advanced study, in course work or in research.

Clearly, the particular choice of material, and the order in which it is presented, must be tailored to the particular student body: their previous course work, their interests, and their future plans. A similar comment applies to the use of the exercises, examples, and projects. My experience has taught me that many students benefit from the exercises, but most prefer the challenge of a self-study project of their own choosing. The opportunity to make an oral presentation also benefits most graduate students despite their initial trepidation. For this reason, I usually schedule oral rather than written examinations. Apart from the advantages I gain in assessing the student's depth of knowledge more accurately, I find that the students themselves prefer orals once they have overcome their initial apprehension.

This book completes a trilogy on heat transfer topics featuring change of phase. *The growth and decay of ice* was published by Cambridge University Press in 1990 while *The tubular thermosyphon* was published by Oxford University Press in 1992. In writing both of these earlier books I benefitted greatly from the encouragement and friendly criticism of many colleagues whose contributions I could not have done without. The present work is not an exception. I would therefore like to record my gratitude to colleagues, within my own university and beyond, who have been kind enough to read individual chapters in draft form. Among those who deserve my thanks are: Dr S. Banerjee, Dr A. Bar-Cohen, Dr J. C. Chen, Dr Vijay Dhir, Dr W. J. Garland, Dr P. Griffith, Dr H. Henein, Dr H. R. Jacobs, Dr E. P. Lozowski, Dr A. E. Mather, Dr R. B. Mesler, Dr A. F. Mills, Dr A. C. M. Sousa, Dr E. M. Sparrow, Dr D. J. Steigmann, Dr R. Viskanta, Dr M. E. Weber, and Dr L. C. Witte. I have learned much from their incisive comments which have led to many improvements in the text and illustrations. Equally, I am indebted to my students whose questions not only caused me to clarify my own explanations, but led me to ask new questions. Some of these students later assisted in the development of the exercises and examples. The illustrations were prepared by Mr Hiroshi Yokota in the Graphics Division of the University of Alberta, while the typing of successive drafts was undertaken with care and patience by Helen Wozniuk, Gail Anderson, and Betty-Ann Bloedorn. To all of these I am deeply grateful.

Edmonton G. S. H. L.
April 1994

ACKNOWLEDGEMENTS

Thanks are extended to the following organizations and individuals for their courtesy and kind permission to publish, in whole or in part, diagrams and photographs identified in the text and more fully in the bibliography:

Academic Press and R. Clift; American Institute of Chemical Engineers; Baltimore Aircoil InterAmerican Corporation; American Society of Mechanical Engineers, A. E. Bergles and E. M. Sparrow; H. R. Bardarson; J. G. Collier; Cambridge University Press; Dover Publications Inc; Lucile H. Faires; Institution of Mechanical Engineers and K. T. Disley; International Journal of Heat and Mass Transfer, M. Cerza, D. Hasson, I. A. Kopchikov, and E. M. Sparrow; Lasse Makkonen; McGraw–Hill Inc and D. E. Gray; Taylor and Francis Ltd, G. Hetsroni, and S. L. Chen.

CONTENTS

NOMENCLATURE xix

1. INTRODUCTION 1
   - 1.1 Early history of latent heat 1
   - 1.2 Calorimetry: the measure of heat 3
   - 1.3 Enthalpy, energy, and structure 5
   - 1.4 Interfaces 9
   - 1.5 Origin and significance of latent heat transfer problems 10
     - 1.5.1 The environmental context 10
     - 1.5.2 The industrial context 12
   - 1.6 Normalization and scale analysis 15
     - 1.6.1 Normalization 16
     - 1.6.2 Newtonian cooling 18
     - 1.6.3 Transient conduction 19
     - 1.6.4 Blasius flow 20
     - 1.6.5 Boundary layer convection 22
     - 1.6.6 Summary 23
   - 1.7 Purpose and scope 24
   - Selected bibliography 25

2. THERMODYNAMICS 27
   - 2.1 The behaviour of continua 27
     - 2.1.1 The energy equation 27
     - 2.1.2 The equations of motion, continuity, and diffusion 29
   - 2.2 Surface energy, surface tension, and wetting 31
     - 2.2.1 The transitional zone 31
     - 2.2.2 Wetting and spreading 32
   - 2.3 Growth at the interface 34
     - 2.3.1 Phase and interface motion 34
     - 2.3.2 The interface equation 35
     - 2.3.3 Interfacial resistance 36
   - 2.4 Thermodynamic surfaces 37
   - 2.5 The Clapeyron equation 41
   - 2.6 Systemic stability 44
     - 2.6.1 Equilibrium and stability 44
     - 2.6.2 Metastable and unstable equilibrium 45

2.7 Nucleation 46
Selected bibliography 51
Exercises 51

3. THE DYNAMICS OF HETEROGENEOUS FLUIDS 53
3.1 Smooth laminar films 54
3.1.1 The planar falling film 54
3.1.2 The annular film 56
3.2 Smooth turbulent films 58
3.2.1 The planar turbulent film 58
3.2.2 The annular film 60
3.3 Interfacial stability 60
3.3.1 Travelling waves on a horizontal interface 61
3.3.2 Travelling waves on a vertical interface 62
3.3.3 Stationary waves on a horizontal interface 63
3.3.4 Rivulet formation 64
3.4 The formation and motion of drops and bubbles 66
3.4.1 The origin of drops 66
3.4.2 The origin of bubbles 68
3.4.3 The motion of single bubbles or drops 69
3.5 Collective behaviour of bubbles and drops 72
3.5.1 Swarms and sprays 72
3.5.2 Accretion on drops and surfaces 72
3.6 The effects of confining walls 74
Selected bibliography 78
Exercises 79

4. SOLIDIFICATION AND FLUIDIFICATION 81
4.1 Scale analysis and the Stefan problem 81
4.1.1 Scales 81
4.1.2 The Stefan number 84
4.2 Simple Stefan problems 85
4.2.1 Simple planar problems 85
4.2.2 Simple problems in other geometries 88
4.3 The Neumann method 90
4.3.1 The classical solution and its limiting form 90
4.3.2 The Neumann method applied to a supercooled liquid 92
4.4 Supercooling and its effects 94
4.4.1 Pure supercooling 94
4.4.2 Constitutional supercooling 95
4.5 Thermal fabrication: casting and welding 98
4.5.1 Freezing of alloys 98

4.5.2 Casting 102
4.5.3 Soldering, brazing, and welding 104
4.6 Ablation, deposition, and accretion 106
4.6.1 Sublimation and deposition 106
4.6.2 Ablation and accretion 109
Selected bibliography 112
Exercises 112
Projects 114

5. CONDENSATION 115
5.1 The formation of drops and films 115
5.1.1 Free drops 115
5.1.2 Drops on a substrate 116
5.2 Scale analysis and the Nusselt problem 119
5.2.1 Scales 119
5.2.2 The Nusselt problem 121
5.3 External condensation on smooth films 123
5.3.1 The effects of Jakob number and Prandtl number 123
5.3.2 The effects of surface inclination and geometry 124
5.3.3 Transitional and turbulent flow 127
5.4 The role of the vapour 130
5.4.1 The effects of superheat in a pure vapour 130
5.4.2 Vapour mixtures: the basic features 131
5.4.3 The effects of vapour concentration and velocity 133
5.5 The role of the interface 136
5.5.1 Interfacial architecture 136
5.5.2 The effects of waves 137
5.6 Condensation in tubes 139
5.6.1 The vertical tube 140
5.6.2 The horizontal tube 143
Selected bibliography 147
Exercises 147
Projects 149

6. EVAPORATION 150
6.1 Inverse Nusselt problems 150
6.1.1 Film evaporation 150
6.1.2 Surface spray cooling 152
6.2 Pool boiling 155
6.2.1 Bubble formation 156
6.2.2 Bubble detachment and subcooled boiling 158
6.2.3 Saturated, supersaturated, and pre-transitional boiling 161

6.2.4 Film boiling 166
6.2.5 Transitional boiling 166
6.3 Boiling in liquid films 168
6.3.1 Boiling in a horizontal film 168
6.3.2 Boiling in a falling film 171
6.3.3 Quenching 172
6.4 Boiling in and around tubes 173
6.4.1 Incipient flow boiling 174
6.4.2 Flow boiling in a vertical tube 174
6.4.3 Flow boiling in a horizontal tube 178
6.4.4 Heat transfer rates 179
6.4.5 Performance limits 182
6.4.6 Boiling in tube bundles 184
Selected bibliography 185
Exercises 186
Projects 188

7. DIRECT CONTACT PROCESSES 190
7.1 Condensation and evaporation domains 190
7.2 Direct condensation 192
7.2.1 Condensation on drops 193
7.2.2 Condensation on jets 196
7.2.3 Other condensing systems 198
7.3 Direct evaporation 201
7.3.1 Evaporation from a drop surrounded by vapour 202
7.3.2 Evaporation from a jet surrounded by vapour 206
7.3.3 Evaporation from a drop surrounded by liquid 207
7.4 Direct crystallization 210
7.4.1 Growth from the vapour phase 210
7.4.2 Growth from the liquid phase 212
7.4.3 Re-crystallization 214
7.5 Combustion of condensed phases 215
7.5.1 Combustion near the surface of condensed phases 215
7.5.2 Combustion of liquid fuels 217
Selected bibliography 222
Exercises 223
Projects 224

8. WORKED EXAMPLES 225

INDEX 283

NOMENCLATURE

SI units are used throughout

a	amplitude ratio, coefficient
A	area, amplitude, constant
Ar	Archimedes number
B	constant, blowing parameter, buoyant force
Bi	Biot number
Bl	Black number
Bo	Bond number
c	specific heat
C	coefficient, capacity
d	diameter
D	diameter
$\mathfrak{D}$	diffusion coefficient
e	energy
E	total energy
f	friction factor, frequency
F	intensive work variable, force
Fo	Fourier number
Fr	Froude number
$\mathfrak{F}$	body force per unit mass
g	gravitational acceleration, specific Gibbs function
$\mathcal{G}$	Gibbs function
G	mass flux density
Gr	Grashof number
Gz	Graetz number
h	specific enthalpy, heat (or mass) transfer coefficient
H	total enthalpy, height
$\mathcal{H}$	enthalpy of reaction
i	vector
I	inertial force
j	diffusive flux density
J	formation rate per unit volume
Ja	Jakob number
k	thermal conductivity, Boltzmann constant
K	coefficient, conductance
Ku	Kutateladze number
l	length
L	length
LWC	liquid water content
m	mass, exponent
$\mathring{m}$	mass flux density
$\mathfrak{m}$	wetting coefficient, mass fraction
M	mass
Mo	Morton number
$\mathcal{M}$	molecular weight
n	number, normal displacement
Nu	Nusselt number
$O(\ )$	order of magnitude ( )
P	absolute pressure
Pe	Peclet number
Pr	Prandtl number
q	heat per unit mass
$\mathring{q}$	thermal energy flux density
Q	heat
r	radius, oxygen/fuel ratio
R	radius, gas constant, resistance
Re	Reynolds number
s	specific entropy, source
S	total entropy, spreading coefficient, supersaturation ratio
Sc	Schmidt number
Sh	Sherwood number
Ste	Stefan number
Su	Suratman number
t	absolute time, thickness
T	absolute temperature, surface tension force
u	specific internal energy, velocity
U	total internal energy, velocity
v	fractional volume, molar volume, velocity

$\mathcal{v}$	specific volume
V	velocity
$\mathcal{V}$	total volume
w	work, velocity, water content
$\mathring{w}$	work flux density
W	total work, velocity, width
We	Weber number
x	displacement, quality, concentration, fraction
X	displacement, extensive work variable
y	displacement
Y	displacement
z	displacement
Z	displacement

Subscripts

a	air
act	activation
A	ambient fluid, annulus, area
b	bubble, bulk, base, buoyancy
B	boiling
c	condensate, condensed, coolant
cap	capillary
C	critical, cold
d	diameter d, drop
db	dry bulb
dew	dew point
D	diameter D, drag
DIL	diluent
E	environment, eutectic
f	film, final
FL	flame
FC	forced convection
FU	fuel
g	gas, growing, generation
h	horizontal, hoop
H	hot
i	inertial, initial, ice, in, incipient
I	interface
k	component k
L	liquid, length
LV	liquid–vapour
m	mushy, mean, mixture
max	maximum
min	minimum
mo	mould
M	mass
n	normal, nth
N	nucleation point
NB	nucleate boiling
NC	natural convection
o	out
OX	oxidant
p	parent, isobaric
P	product
Q	heat
s	surface, superficial, substrate
S	solid, static
SL	solid–liquid
SV	solid–vapour
sat	saturation
sup	superheated
t	tangential, transitional
T	triple point, thermal, terminal, total
v	viscous, vertical
$\mathcal{v}$	per unit volume
V	vapour
w	water, wall
wb	wet bulb
W	work, width W
Λ	wavelength Λ
∞	distant
0	Y = 0, substrate, surface, entrance, upstream, initial

Superscripts

$\cdot$	per unit time
*	threshold
$\sim$	modified, molar
$-$	average
$'$	fluctuation, modified
$+$	non-dimensional
a	after
b	beginning
B	bulk
c	characteristic
E	eutectic
e	equilibrated
i	initial, ice
M	momentum
O	reference

s	superficial
T	thermal

Greek

α	ratio, inclination
β	constant, coefficient, driving force per unit volume
Γ	gamma function, mass flux per unit width, exchange coefficient
δ	increment, thickness
Δ	increment, thickness
ε	eddy diffusivity, void fraction
η	collection efficiency, similarity variable
θ	contact angle, temperature difference, enthalpy difference
Θ	temperature difference
κ	thermal diffusivity
λ	latent heat per unit mass
Λ	wavelength
μ	chemical potential, kinematic viscosity
ν	momentum diffusivity
ξ	displacement
π	stress tensor
$\boldsymbol{\Pi}$	stress vector
ρ	density
σ	surface energy density, surface tension, stress, Stefan–Boltzmann constant
τ	shear stress, time
ϕ	temperature difference, enthalpy difference
Φ	dissipation function

1
INTRODUCTION

1.1 Early history of latent heat

For thousands of years, humankind has been aware of changes in phase but has not always understood them. In prehistoric times, primitive people watched lava solidify, ice melt, water evaporate, and rain fall; with the dawn of civilization, the extensive use of fire provided the opportunity to observe and experiment with many boiling and melting phenomena. Most significant, perhaps, was the discovery of metals whose value and widespread use, in coinage and weaponry, led to the inevitable observation that different metals require a different 'heat' to melt them. But since the idea of 'heat' was (and remained) obscure and confused, and the scientific method itself was not widely recognized until the seventeenth century, it is not surprising that terms such as 'melting point' and 'latent heat' were unknown until modern times.

By the 1750s, the caloric theory of heat prevailed in an era when the equivalence of work and heat was not yet established; the laws of thermodynamics were unknown. Even the difference between heat, as thermal energy, and degree of heat, as temperature, was frequently ignored and not well understood. In this climate of uncertainty, Dr Joseph Black, a professor of medicine, began lecturing in chemistry at the University of Glasgow. He was an unusual academic for his time with great misgivings about theoretical constructs which he regarded as 'a mere waste of time'. In particular, he was not persuaded by the caloric theory which did not explain his own observations on heat.

In 1758, Black found himself at odds with the conventional wisdom on melting and solidification. This held that a solid heated to its melting point required only a little extra heating to bring about complete melting; likewise, a small amount of extra cooling was all that was believed necessary to completely freeze a liquid at its freezing point. Moreover, it was asserted that no further heat was transferred during these phase changes, other than that 'small amount' suggested by changes in thermometer readings. To demonstrate that these beliefs were inconsistent with the facts, Black conducted an historic experiment which is both elegant and simple. Reasoning that the rate of heat transferred to a glass of chilled water in a large, warm room would be almost the same as that transferred to a glass of iced water, he took two identical glasses and suspended them by wires in mid air. In one glass he put five ounces of water chilled to the freezing point; in the other was ice previously formed by pouring in the same amount of water and allowing the glass to rest in a snow-salt mixture. Using a thermometer, in the water-filled glass, he was able to determine the rate of (sensible) heat gain from the atmosphere. This rate, applied over the time required to completely melt the ice in the second glass, thus provided an estimate of the heat of melting in conditions where the temperature of the melting mixture was observed to vary very little.

Black went on to improve his experimental technique and demonstrate that similar behaviour may be observed during freezing. Not content with this, he extended his work to evaporation and condensation, the latter presenting a crucial problem to Scottish whisky distillers struggling with the difficulties of overheating, implosion, and flammability. By 1762, Black had demonstrated conclusively that the conventional wisdom was wrong. The phase change processes he had studied could not be explained in terms of sensible heat, especially in 'small amounts'. He said of the heat transferred during a phase change: 'it is concealed, or latent, and I gave it the name LATENT HEAT'.

It is remarkable that Black should have discovered latent heat with little more than two basic concepts: sensible heat capacity and Newton's law of cooling with its implicit scale of temperature. The caloric theory of heat played no part in this discovery except in the general notion that heat flows from hot regions to cold regions. It is no less remarkable that Black sought no benefit from his discovery, being content to continue his work and go about his professorial duties; he was evidently happy in the thought that his efforts might benefit the public. Among those who did benefit was James Watt, who worked for a time as Black's assistant. Watt concentrated on the heats of evaporation and condensation, demonstrating that, within experimental error, they had exactly the same value. Using a vacuum pump he also found that the latent heat increased as the pressure decreased. Thus we see the beginnings of the saturation curve.

Watt was an engineer and quick to exploit Dr Black's 'doctrine of latent heat' in the development of the reciprocating steam engine, which at the time generated much needed pumping power by first admitting and then condensing the steam inside the same cylinder. This wasted fuel. Watt contrived to remove the steam from the cylinder at the end of each cycle before condensing it externally. Having measured the latent heat to be about one thousand times greater than the specific heat, Watt knew that the fuel savings would be substantial. He was able to improve the thermal efficiency by over 500 per cent! This improvement, perhaps more than any other, ushered in a new era of industrial development and has been described as the most important technical advance of the nineteenth century.

While Joseph Black was the first to identify the true nature of latent heat, he was not alone in exploring its significance. Around the same time, Johann Leidenfrost made several insightful observations on the behaviour of liquid drops brought into contact with a high-temperature substrate:

specifically, 'an iron spoon ... heated over glowing coals ...'. Such an experiment, and such language, would not survive the editorial review of today's international journals, but it is a lesson in humility to follow the keenness of his mind. From these simple observations he gives us the concept of film boiling and, of course, the Leidenfrost temperature marking its lower extremity. For good measure he warns us about the violence of the transition, notes the existence of surface fouling, and introduces the consequences of evaporation on a binary solution.

Interest in phase change phenomena continued in the nineteenth century, especially during the Victorian era. In Britain, the work of James Joule on thermodynamics in general, and heat in particular, included a presentation to the Royal Society of London entitled 'On the surface condensation of steam', a topic of both theoretical and practical interest during the late stages of the industrial revolution which, as noted above, was driven largely by steam power. Within a decade, Lord Kelvin had introduced a theory of nucleation. Not much later, in Europe, Rudolf Clausius established the character of the saturation curves and Josef Stefan published his seminal analysis on the freezing of the seas. We continue to describe solidification, and some aspects of vapour diffusion, as Stefan problems.

Many of the great physicists of the day directed their attention toward what we might call thermoscience. Not least among these was Lord Rayleigh, whose contributions include descriptions of the instabilities at liquip–vapour surfaces, the break up of jets in particular. His work takes us into the twentieth century when his paper 'On the pressure developed during the collapse of a spherical cavity' appeared. This laid the groundwork for future studies of vapour bubble growth and collapse. About the same time, Wilhelm Nusselt published his classical analysis of film condensation which, to this day, provides most engineering undergraduates with their introduction to phase change heat transfer. Less than twenty years later, in Japan, Shiro Nukiyama showed us the diverse nature of boiling processes by mapping out the pool boiling curve. By then the foundations of latent heat transfer had been securely laid.

1.2 Calorimetry: the measure of heat

Despite the importance of the practical advances stemming from Black's discoveries, his contributions to the quantitative science of heat were no less. This is well illustrated by the development of calorimetry. His work on sensible heat can be described through the simple expression

$$Q = C\Delta T \tag{1.1}$$

where Q is the heat supplied to a body and measured through ΔT, the change in temperature; the quantity C is the *heat capacity*. Black invariably measured the heat capacity relative to the same mass of water. Thus, by dropping warm iron into cold water and measuring their initial and final temperatures, he could conclude that 'the heat of iron ... is to that of water, as one to eight'. It is a short step to note that

$$C = Mc_p \tag{1.2}$$

where M is the body mass and c_p the specific heat.

With a quantitative measure of heat, Black introduced the idea of measuring *any* form of heat, regardless of its source or effect. We saw earlier how he used this equivalence principle in measuring latent heat. It was common to express this in relation to the specific heat, i.e. by calculating the ratio of the latent heat λ to the sensible heat $c_p\Delta T$ required for a temperature change of one degree. The large values thus obtained drew attention to the significance of latent heat during phase change processes. Today, the ratio of sensible heat to latent heat during melting or solidification is known as the Stefan number *Ste*. Thus

$$Ste = \frac{c_p \Delta T}{\lambda_{SL}}. \tag{1.3}$$

The corresponding ratio during evaporation or condensation is known as the Jakob number *Ja*; thus

$$Ja = \frac{c_p \Delta T}{\lambda_{LV}}, \tag{1.4}$$

or

$$Ja = \frac{\Delta h}{\lambda_{LV}} \tag{1.5}$$

which, like the Stefan number, is an enthalpy ratio. It would be fitting if the enthalpy ratio was more generally known as the Black number in honour of the man who quantified both the sensible and latent contributions. It is therefore suggested that the Black number be defined by

$$Bl = \frac{\Delta h}{\lambda}, \tag{1.6}$$

regardless of which pair of phases is being considered.

Since those early days, the science of calorimetry has grown substantially through the use of the transient version of eqn (1.1). If, for simplicity, we take C to be constant, then

$$\dot{Q} = C\frac{dT}{dt}. \tag{1.7}$$

This may be usefully applied to material in a calorimeter where the heat transfer rate can be carefully regulated. For example, if a liquid is cooled to, and then below, its freezing point, with the rate of heat withdrawal being fixed, the temperature–time curve takes the form illustrated schematically in Fig. 1.1. When either liquid or solid is present, the specific heat may be determined from

$$c_p = \frac{\dot{Q}}{M\frac{dT}{dt}} = \frac{\dot{Q}\Delta t}{M\Delta T}. \tag{1.8}$$

For a changing mixture, the latent heat is given by

$$\lambda_{SL} = \frac{1}{M}\int_{t_1}^{t_2} \dot{Q}\, dt = \frac{\dot{Q}(t_2 - t_1)}{M}. \tag{1.9}$$

The cooling curve for the liquid may sometimes overshoot, as illustrated by the dashed curve. This supercooling phenomenon was first observed by Gabriel Fahrenheit in 1724.

Experiments of this type provide valuable information on many forms of transition. In particular, they may be used to study chemical reactions, as illustrated in Fig. 1.2. This is a representation of a chemical reaction taking place between T_1 and T_2 while the specimen is being heated. In the absence of the reaction, the temperature rise would simply reflect the addition of sensible heat along a monotonic base curve AB. The departure above this between t_1 and t_2 thus represents the heat released in an exothermic reaction. The dashed curve represents an endothermic reaction. The enthalpy of reaction $\mathcal{H}$ may thus be measured from

$$\mathcal{H} = \frac{1}{M}\int_{t_1}^{t_2} \dot{Q}\, dt - c_p(T_2 - T_1), \tag{1.10}$$

again assuming the specific heat to be constant. This enables us to establish the enthalpy ratio $\mathcal{H}/\lambda$ which arises during latent heat transfer in the presence of a chemical reaction, e.g. the combustion of liquid hydrocarbons which evaporate before they burn.

1.3 Enthalpy, energy, and structure

Latent heat transfer problems are characterized by various enthalpic changes. The three common changes identified above imply two non-dimensional ratios: $\mathcal{H}/\lambda$ and the Black number $\Delta h/\lambda$; in general, these influence phase change behaviour. In this section we examine the physical and chemical basis of these enthalpic changes.

The breaking or forming of *inter*molecular bonds signifies a phase

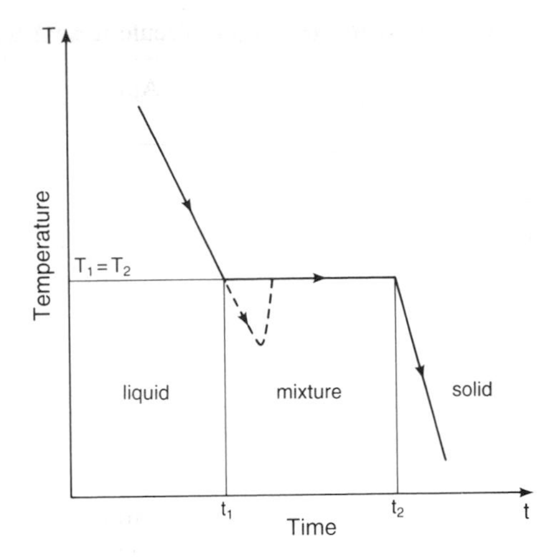

Fig. 1.1 Cooling curve with solidification phase change.

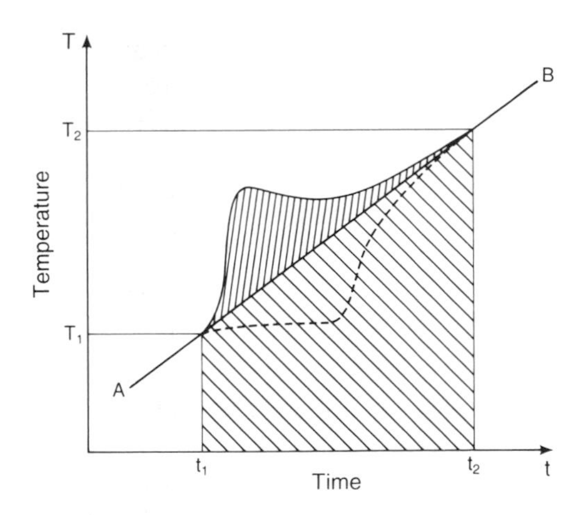

Fig. 1.2 Heating curve with chemical reaction.

change. For example, when the intermolecular bonds in a solid are all completely broken, and a vapour is formed, we speak of sublimation. This is illustrated in Table 1.1 which lists the approximate energy levels (in electron-volts) identified for the H_2O molecule in the vicinity of 273°K. The hydrogen bonds in the ice lattice each require an energy of about 0.29 eV for a molecule to break free of its neighbours, and since there are two bonds per molecule the latent heat of sublimation should be of the order of 0.58 eV; the value measured is 0.49 eV. Vaporization requires 0.39 eV while melting requires only 0.06 eV. However, both of these are much greater than the lattice vibrational energy of 0.54×10^{-2} eV. The

Table 1.1 Energies associated with the H_2O molecule near 273 K

Type of energy	Approximate magnitude (eV) per molecule
Lattice vibration	0.54×10^{-2}
Hydrogen bond breaking (intermolecular)	0.58
Valence bond breaking (intramolecular)	10
Enthalpy of melting	0.06
Enthalpy of vaporization	0.39
Enthalpy of sublimation	0.49
Enthalpy of formation	9.6

comparisons reveal why latent heats are usually much greater than specific heats. They also point to circumstances where latent heat will not dominate sensible heat: namely, where intermolecular bonds are not strong, e.g. near the critical point.

This intermolecular bond model of latent heat, although qualitative, provides us with a useful interpretation of macroscopic events. It implies, for example, that the bond energy and the molecular configuration will characterize the phase. In other words, it suggests that a phase transition is a change from one level or type of intermolecular ordering to another. A phase change may thus be viewed as a re-ordering process which will be reflected in the specific entropy of the material. For evaporation, melting, and sublimation, there is a well-established discontinuity in entropy Δs between the phases. Thus we might expect that

$$\Delta s = \frac{\lambda}{T} = \text{constant} \tag{1.11}$$

during a phase change at temperature T if s and λ refer to a single molecule. This leads to a rough guide. For vaporization, Trouton's rule states that

$$\tilde{s}_{\mathrm{V}} - \tilde{s}_{\mathrm{L}} = \frac{\tilde{\lambda}_{\mathrm{LV}}}{T_{\mathrm{vap}}} \simeq 83.7\ \mathrm{J\,mol^{-1}\,K^{-1}}.$$

For melting, Richards' rule states that

$$\tilde{s}_{\mathrm{L}} - \tilde{s}_{\mathrm{S}} = \frac{\tilde{\lambda}_{\mathrm{SL}}}{T_{\mathrm{melt}}} \simeq 8.37\ \mathrm{J\,mol^{-1}\,K^{-1}}, \tag{1.12}$$

while for sublimation

$$\tilde{s}_{\mathrm{V}} - \tilde{s}_{\mathrm{S}} = \frac{\tilde{\lambda}_{\mathrm{SV}}}{T_{\mathrm{sub}}} \simeq 92.1\ \mathrm{J\,mol^{-1}\,K^{-1}}. \tag{1.13}$$

It is important to keep in mind that these empirical rules do not fit every substance; there are many significant exceptions, as Tables 1.2 and 1.3 reveal. Most notable is H_2O.

In a similar way, *intra*molecular bonds provide a useful way of interpreting events during chemical reactions where molecules may be broken apart and re-formed with the addition or removal of heat. The energy of the bonds required to form the molecule in the first place thus appears as the enthalpy of formation. Again, using the water molecule data in Table 1.1 to illustrate, the two valence bonds each have an energy of 5 eV, whereas the enthalpy of formation is about 9.6 eV per molecule. But it is again worth noting that the model is only qualitative.

In general, the bond energies at the root of latent heats and enthalpies of reaction represent thresholds which vary according to the circumstances. Even so, their relative magnitudes help us distinguish between the physical change associated with latent heat and the chemical change associated with an enthalpy of reaction. As we have seen, the three types of enthalpic

Table 1.2 Comparison with Trouton's Rule. Molecular entropy change during evaporation expressed as a fraction of 83.7 J mole^{-1} K^{-1} (source: Gray 1963)

Substance	Molecular entropy change	Substance	Molecular entropy change
N_2	0.865	CCl_4	1.02
Ne	0.875	HCl	1.02
CH_4	0.875	H_2S	1.05
A	0.890	Na	1.06
K	0.895	N_2O	1.07
O_2	0.905	Hg	1.12
F_2	0.920	CH_2O	1.12
C_2H_6	0.955	SO_2	1.13
CS_2	1.00	NH_3	1.16
Cl_2	1.02	H_2O	1.30

Table 1.3 Comparison with Richards' Rule. Molecular entropy change during melting expressed as a fraction of 8.37 J mole^{-1} K^{-1} (source: Gray 1963)

Substance	Molecular entropy change	Substance	Molecular entropy change
K	0.825	Hg	1.17
Na	0.840	CCl_4	1.18
CF_4	0.935	CH_4	1.24
Pb	0.950	Zn	1.27
Fe	1.01	Al	1.37
Ag	1.09	H_2O	2.63
Cu	1.14	C_2H_6	3.80

change *in a given substance* are usually different. Typically, $c_p \Delta T \ll \lambda \ll \mathcal{H}$, except as noted above. These inequalities reflect the differences between unbroken bond energy, the breakage of intermolecular bonds, and the breakage of intramolecular bonds.

1.4 Interfaces

This leads us naturally to the boundary separating any two phases in thermal equilibrium, i.e. under saturated conditions. It is clear from the discussion above that this boundary marks an abrupt change between two different structural entities. The advance or retreat of the boundary into either phase may thus take place only if energy is supplied or withdrawn; only then can the intermolecular bonding energy be provided or removed. Nature abhors discontinuities and establishes a thin transitional zone between the distinct phases. More will be said about this zone in the next chapter.

On the molecular level, this description implies that a phase change requires the addition or subtraction of one molecular layer at a time, and this is not inconsistent with the way in which many pure solids and liquids grow in saturated conditions. In a strongly ordered, crystalline solid, for example, the interface geometry follows the outline of the crystal structure. In a loosely ordered liquid, on the other hand, the interface geometry is largely controlled by surface tension. In both instances the interface is macroscopically smooth, though not necessarily free of edges and corners.

Supercooling may destroy the macroscopic smoothness of the interface. For example, it creates conditions in which growth normal to a freezing interface is thermally unstable, thus leading to the appearance of dendrites. This does not occur at a liquid–vapour interface which may, however, become unstable for a different reason; namely, the imbalance of interfacial forces. More precisely, when the restraining effect of surface tension no longer dominates the disturbing effect of inertia or buoyancy, interfacial ripples appear and grow.

To place interfacial behaviour in a clearer perspective, it is useful to distinguish between *bipartitioned* and *dispersed* systems. The bipartitioned system, exemplified by condensation on a liquid film, or freezing in a warm water pipe, is characterized by an extensive, continuous interface on which flow conditions may or may not induce a dynamic type of instability while supercooling may or may not induce a thermal instability. Generally, the conditions within both (continuous) phases are clearly defined. For a dispersed system, exemplified by a boiling liquid or a burning spray, conditions in the continuous phase are seldom easy to define; it is also difficult to specify the interaction between the dispersed pockets or particles. On the other hand, the surface area of the dispersed phase, although usually large, is broken into discrete amounts which may be small enough to eliminate interfacial instabilities.

1.5 Origin and significance of latent heat transfer problems

1.5.1 The environmental context

This planet, along with the others in the solar system, was formed through a series of phase changes, and it is therefore not surprising to find remnants of these still occurring today. In recent years, volcanoes in Hawaii, Washington State, Iceland, and Sicily have been particularly active, the flow of molten lava providing spectacular examples of solidification. Figure 1.3 is one illustration. Equally spectacular, though less destructive, are the boiling manifestations of geothermal energy seen in geysers. In harnessing this type of energy, the city of Reykjavik in Iceland is able to offer benign and inexpensive domestic heating to its residents.

Widespread evaporation from large bodies of water is a principal determinant of our weather, changing from day to day. Evaporative changes from century to century have played a major role in shaping the topography and, over time, in defining continents. Water vapour in the atmosphere ultimately forms the clouds in which the process of precipitation begins. This is usually a complex latent heat process in which water or ice, or both, may form.

Fig. 1.3 Molten lava flowing from the fiery cauldron of an Iceland volcano (courtesy of H. R. Bardarson).

Water vapour in the atmosphere may form frost or snow, usually following a sequence of complicated events. Similar complexities arise in the formation of ice covers on rivers, lakes, and oceans. In the ground, ice grows and decays in the pores of soil, taking forms ranging from an interconnected matrix (permafrost) to massive intrusions and extrusions which sculpt the earth's surface. On the largest scales, massive ice is encountered as icebergs, glaciers, and ice sheets. These ice forms reflect our climate and its shifts over millennia. They are also useful measures of current shifts in global climate; in particular, they are sensitive indicators of global warming.

Ice on a large scale both serves and threatens humankind. During the winter in northern nations, for example, terrain which is impassable for much of the year may be engineered into snow roads and ice bridges, thus creating temporary but vital transportation routes. Similarly, the creation of frozen ground using evaporative thermosyphons makes it possible to support engineering structures permanently. Figure 1.4 shows a long stretch of vertical thermosyphon tubes supporting an oil pipeline in Alaska. Defence against ice on the scale of an iceberg or a glacier is often impossible but some limited measures may be taken, for example, during breakup of a river in the vicinity of a bridge. Protection against the atmospheric icing of engineered structures is another example of partial success. This ice may form on fixed structures such as overhead cables, towers, and offshore platforms, but has it most insidious effect on aircraft and ships where it

Fig. 1.4 Evaporative thermosyphons maintaining frozen ground beneath a pipeline in Alaska.

often leads to loss of life. Figure 1.5 illustrates the severity of the problem. Anti-icing and de-icing techniques pose very complex latent heat transfer problems, especially if evaporative thermosyphons are used as a heat supply.

Desalination is an example of the benign production of ice in the creation of drinkable water. The same result may also be reached through evaporation if energy is supplied. Figure 1.6 provides a schematic illustration of a solar still using this evaporative principle. Evaporation on a large scale has also been proposed for ocean thermal energy conversion (OTEC) systems.

1.5.2 The industrial context

On a more human scale, latent heat transfer problems frequently arise from the demands of industry. Casting, for example, is one of many thermal pro-

Fig. 1.5 Superstructure icing on board a ship (courtesy of L. Makkonen).

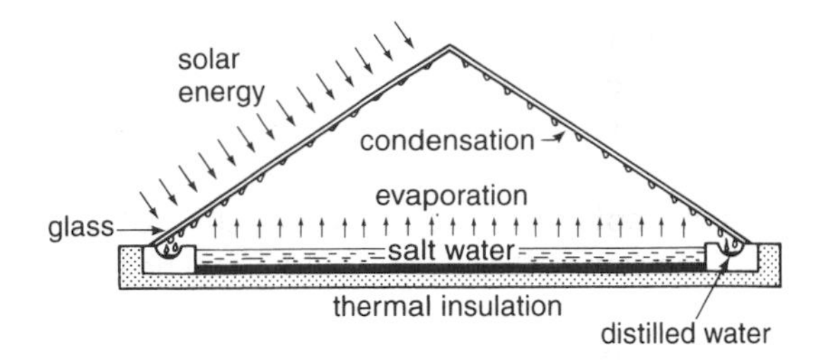

Fig. 1.6 Solar energy being used to distill salt water.

cesses used in the production of components, particularly in metal. Joining of metals is another technique widely used in manufacturing. Welding and soldering both typify the latent heat transfer problem encountered. On the other hand, the protective use of the latent heat of fusion is well illustrated by the spacecraft ablation shield.

Steam generators are perhaps the most familiar example of evaporation. The huge variability in the demand for steam has led to boiler sizes ranging, in terms of the heat supply rate, from less than 1 kW to more than 1 GW. A large boiler is illustrated in Fig. 1.7. For those boilers burning fossil fuels, the combustion process may provide a further example of latent heat transfer in the burning of solid coal or sprayed liquid hydrocarbons; both entail the generation of a vapour phase which subsequently combines with oxygen in the combustor.

Not surprisingly, the need for industrial condensers matches that for boilers. And again, the wide variation in particular needs is reflected in the various sizes and designs of condensers to be found. Figure 1.8 provides a cutaway view of a typical large steam condenser. The steam condenses on the outside of the tubes while cooling water flows inside. Another application is found during the liquefaction of industrial gases such as hydrogen, oxygen, nitrogen, carbon dioxide, and methane. Distillation columns in the process industries provide many examples of a vapour condensing directly on its own liquid. An interesting variation on this theme is provided by the spray condenser shown schematically in Fig. 1.9. Direct contact condensation has a wide range of applications, including its use in large-scale OTEC systems.

In most industrial applications, boiling or condensation takes place in some form of heat exchanger. Many of these resemble the heat exchangers used in single-phase conditions. Shell and tube arrangements are common with boiling or condensation occurring either inside or outside the tubes. Some situations call for the use of the thermosyphon principle, in which boiling and condensation occur at different locations in the same tube. An interesting application of this principle is illustrated in Fig. 1.10, which

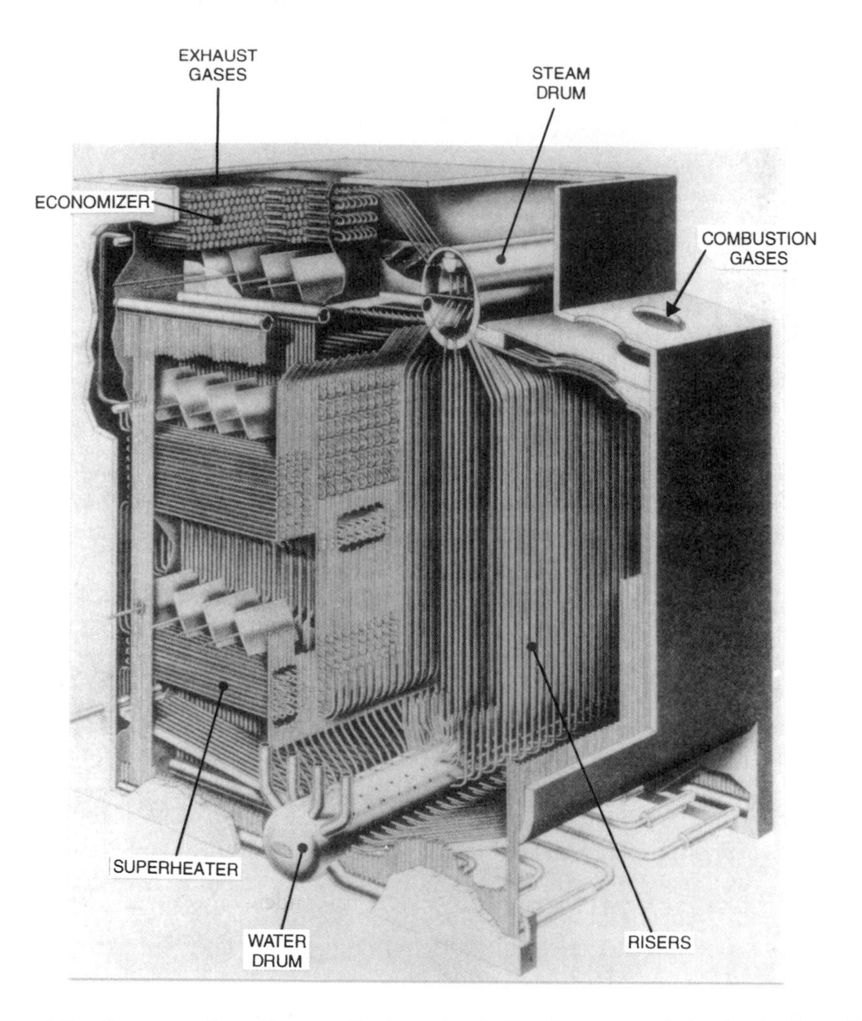

Fig. 1.7 Cutaway view of a top-fired marine boiler (courtesy of the Institution of Mechanical Engineers).

shows boiling, condensation, and melting arranged in series to provide a means of storing thermal energy at a fixed (fusion) temperature.

Evaporative heat exchangers are essential components in vapour compression refrigerators and heat pumps. The air conditioning industry, in particular, makes extensive use of evaporative devices in controlling the temperature and humidity of air spaces. Evaporative cooling is an excellent example of latent heat transfer in the service of human society. It is found on scales ranging from the domestic humidifier to the industrial cooling tower illustrated in Fig. 1.11. Increasing emphasis on energy conservation and water conservation coupled with stricter pollution controls have greatly

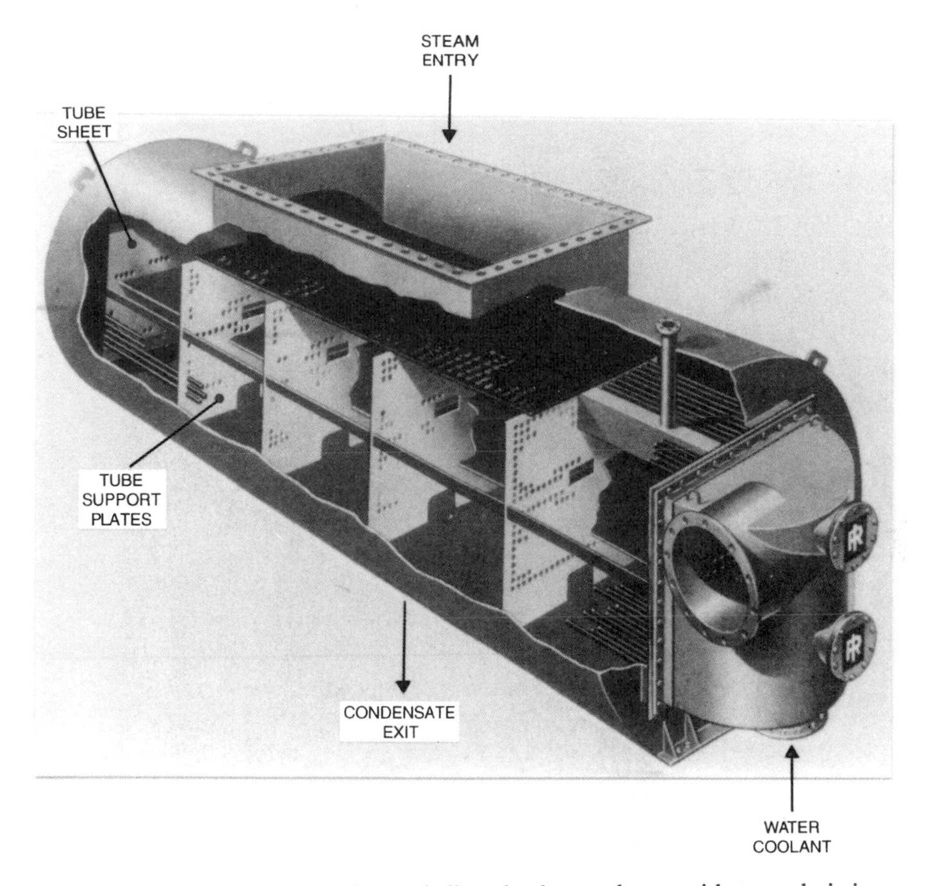

Fig. 1.8 Cutaway view of a large shell-and-tube condenser with top admission.

expanded the role of cooling towers and cooling ponds in recent years. This industrial response brings us full circle, back into the environmental context.

1.6 Normalization and scale analysis

Throughout this book we will encounter a great many different physical situations most often described by partial differential equations. Typically, the problems posed will be too difficult to solve exactly. For the present purpose, which is to introduce the fundamentals of latent heat transfer, exact solutions are not necessary. It is usually sufficient to develop first-order approximations which reveal the essential features of the situation. These enable us to estimate heat transfer and phase change rates. They also identify the relevant variables on which these rates depend, and determine

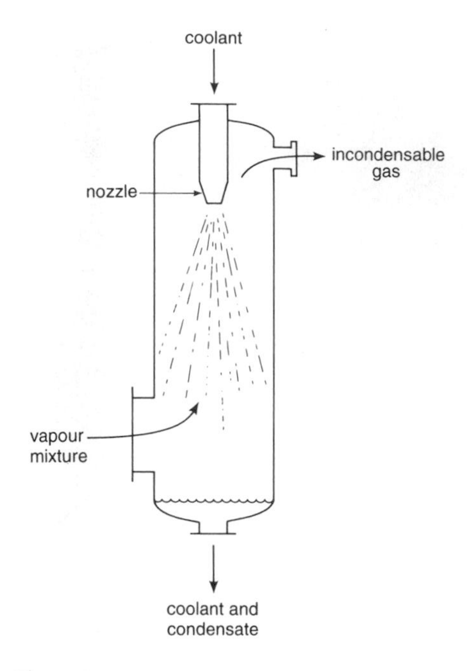

Fig. 1.9 Schematic of a spray condenser.

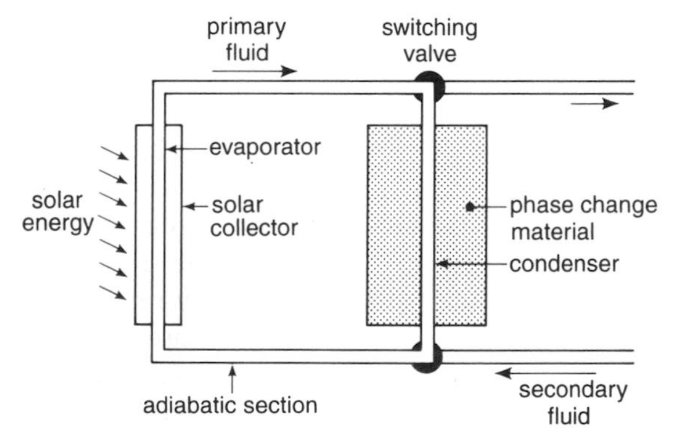

Fig. 1.10 The use of a loop thermosyphon in the storage of solar energy by melting a solid. After switching the valve positions, the latent heat may be removed by a secondary fluid.

the structure of their interrelation. This section is devoted to a technique which achieves these ends.

1.6.1 Normalization

Normalization is the process in which the physical variables in an equation, or set of equations, are transformed into non-dimensional quantities having

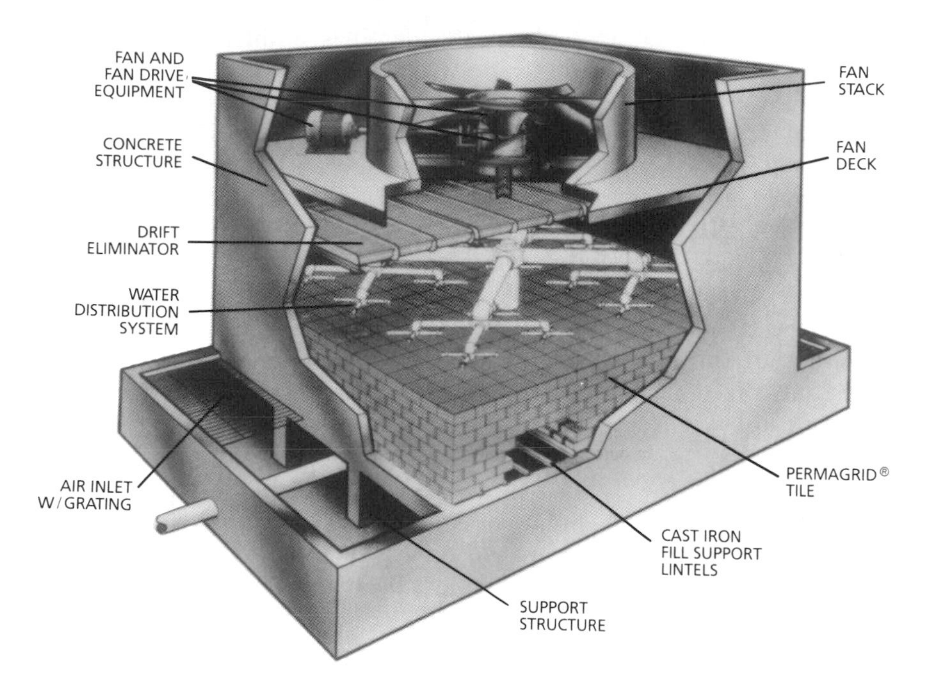

Fig. 1.11 Industrial cooling tower using forced draught and ceramic filling (courtesy of the Ceramic Cooling Tower Company).

a magnitude range of order one, written $O(1)$. This simple process enables us to form reduced, or normalized, variables having *two* characteristics: they are non-dimensional and they are of unit order. The first characteristic is very precise and brings with it the advantage that we may discuss each variable without reference to any particular system of units. The second characteristic is equally definite but is not precise. It results from the fact that each normalized variable is defined by a physical variable divided by the range, or scale, over which it varies; these scales are not always known precisely.

An order of magnitude is a qualitative measure, usually specified on a logarithmic scale, the base being 10 by convention. Thus $O(1)$ means $O(10^0)$, i.e. between $10^{-1/2}$ and $10^{+1/2}$. Similarly, $10^{-3/2} \leqslant (10^{-1}) \leqslant 10^{-1/2}$, and $10^{1/2} \leqslant O(10^1) \leqslant 10^{3/2}$. Orders of magnitude are not generally meant to be quantitative. Frequently, it is sufficient to state an order of magnitude in relation to $O(1)$. Thus, $0.009 = O(10^{-2}) \ll O(1)$, and $163 = O(10^2) \gg O(1)$. Such comparisons enable us to decide the *relative* importance of two terms in an equation. Thus, in relation to the magnitude of the dominant terms, which are $O(1)$, 'small' means $\ll O(1)$ while 'comparable' means $O(1)$.

The range, or scale, of a variable will be designated by a superscript c, for 'characteristic'. Strictly, the scale identifies a difference, as when we measure temperatures in relation to 273 K or pressures relative to atmospheric. However, this is not always necessary, as when we measure the radial displacement R in a pipe of radius R_0; it is superfluous to write $R^c = R_0 - 0$. Some scales are known from the way the physical situation is prescribed, but some are not. Boundary layer convection near a sphere at T_0 surrounded by air at T_∞ implies a temperature (difference) scale of $T_0 - T_\infty$, but the boundary layer thickness scale δ is not known at the outset; despite this, the scale must exist. Unknown scales may be determined through the essential physics. Through a series of examples, we will show how the physics implicitly determines both the variable scales and the orders of magnitude of various terms in the governing equations.

1.6.2 Newtonian cooling

First consider Newtonian cooling of a body, mass M and surface area A. Let the (uniform) body temperature be $T > T_\infty$, the temperature of the fluid bulk surrounding the body. A heat balance requires that

$$Mc_p \frac{dT}{dt} = -hA(T - T_\infty), \tag{1.14}$$

where c_p is the specific heat of the body material, and h is the heat transfer coefficient at its surface. The temperature scale is prescribed by $T_\infty < T < T_i$, where T_i is the initial body temperature. Hence, if we take $\theta = T - T_\infty$ we may define the normalized temperature (difference) by

$$\phi = \frac{\theta}{\theta^c},$$

where $\theta^c = T_i - T_\infty$. The time scale, however, is unknown. It may be estimated from eqn (1.14) by defining the normalized time as

$$\tau = \frac{t}{t^c},$$

and substituting to obtain

$$\left(\frac{Mc_p\theta^c}{t^c}\right)\frac{d\phi}{d\tau} = -(hA\theta^c)\phi$$

in which the bracketed expressions both have the same dimensions. Hence

$$\frac{d\phi}{d\tau} = -\left[\frac{hAt^c}{Mc_p}\right]\phi. \tag{1.15}$$

Since the normalized variables ϕ and τ are $O(1)$, by definition, it is reasonable to expect that

$$\frac{d\phi}{d\tau} \simeq \frac{\Delta\phi}{\Delta\tau} = O(1),$$

and therefore the coefficient of ϕ in eqn (1.15) must determine the relative importance of the right-hand side; the coefficient of the left-hand side is 1.0. But the physical necessity stated in eqn (1.14) requires the retention of both sides; without either side, the equation is reduced to a trivial statement. It therefore follows that

$$\frac{hAt^c}{Mc_p} = O(1)$$

and hence the time scale must be given by

$$t^c = O\left(\frac{Mc_p}{hA}\right), \tag{1.16}$$

often called the response time or the time constant of the system.

In this particular example, eqns (1.14) and (1.15) have simple, exact solutions. It is important to recognize, however, that the time scale estimated from eqn (1.16) did not require a solution. This feature may provide considerable insight when the equation or equations cannot be easily solved.

1.6.3 Transient conduction

In the above situation, heat conduction within the body was neglected. When internal temperature gradients are not small, the energy equation near the body surface must be written

$$\frac{\partial T}{\partial t} = \kappa \frac{\partial^2 T}{\partial X^2} \tag{1.17}$$

where X is the depth beneath the surface and κ is the body thermal diffusivity. If the initial (uniform) temperature of the body is again T_i we may define the temperature scale $\theta^c = T_i - T_0$, where T_0 ($< T_i$) is the temperature imposed at the surface $X = 0$. In this situation, neither the length scale X^c nor the time scale t^c are known unless further conditions are specified. Even so, the scales must exist and we may therefore define the normalized variables

$$\tau = \frac{t}{t^c} \quad \text{and} \quad x = \frac{X}{X^c}$$

which may be substituted into eqn (1.17) to yield

$$\left[\frac{(X^c)^2}{\kappa t^c}\right] \frac{\partial\phi}{\partial\tau} = \frac{\partial^2\phi}{\partial x^2}. \tag{1.18}$$

The non-dimensional coefficient

$$\frac{\kappa t^{c}}{(X^{c})^{2}} = Fo$$

is the Fourier number.

Since x, τ, and ϕ are $O(1)$, the Fourier number dictates the relative importance of the transient term on the left-hand side; the coefficient of the right-hand side is 1.0. Thus, when $Fo \gg 1$, it follows that

$$\frac{\partial^{2}\phi}{\partial x^{2}} = 0 \tag{1.19}$$

which has a simple, linear solution. Such a situation requires either a very long time scale, a very small length scale, or both: for example, when T_0 applied to one side of a thin slab is held indefinitely below T_i applied to the other side.

On the other hand, when $Fo = O(1)$, both terms must be retained in eqn (1.18). In that case,

$$X^{c} = O(\kappa t^{c})^{1/2} \tag{1.20}$$

defines either the length or time scale in terms of the other. Thus, for example, if we are dealing with a slab of width W,

$$t^{c} = O(W^{2}/\kappa)$$

provides an estimate of the time for a thermal disturbance to move across the slab. Alternatively, if T_0 is a periodic function of time. e.g. $\sin \omega t$, with a half period $t^{c} = \pi/\omega$, the penetration depth is given by

$$X^{c} = O\left(\frac{\pi\kappa}{\omega}\right)^{1/2}.$$

The third possibility, $Fo \ll 1$, generates only a trivial solution of eqn (1.18) by suppressing the cause of change, heat conduction.

It is again worth emphasizing that the above insights stem from physical prescriptions and do not require a solution of the governing eqn (1.17). Also worthy of note is the assumption that normalization of the variables in the governing equation is taken to imply normalization of the derivatives. This assumption is not rigorous. However, it is a reasonable expectation in the absence of essential singularities, i.e. when the derivatives are well behaved. Any doubts about this assumption may be removed later through reference to a full solution or an experiment.

1.6.4 Blasius flow

We turn now to a two-dimensional situation, the Blasius problem, in which a fluid with a free stream velocity U_∞ flows parallel to a flat plate of length $X = L$. This is described by the equations

$$\frac{\partial U}{\partial X}+\frac{\partial V}{\partial Y}=0 \tag{1.21}$$

and

$$U\frac{\partial U}{\partial X}+V\frac{\partial U}{\partial Y}=\nu\left(\frac{\partial^2 U}{\partial X^2}+\frac{\partial^2 U}{\partial Y^2}\right). \tag{1.22}$$

Normalizing, we define

$$x=\frac{X}{X^c},\quad y=\frac{Y}{Y^c},\quad u=\frac{U}{U^c},\quad v=\frac{V}{V^c}, \tag{1.23}$$

in which two scales are given: $X^c=L$ and $U^c=U_\infty$. We therefore need two equations, based on physical requirements, to determine the two unknowns Y^c and V^c. Equation (1.21) may now be rewritten

$$\frac{\partial u}{\partial x}=-\left[\frac{V^cX^c}{Y^cU^c}\right]\frac{\partial v}{\partial y}.$$

For a non-trivial solution, both sides of this equation must be retained. Hence

$$\frac{V^cX^c}{Y^cU^c}=O(1). \tag{1.24}$$

Equation (1.22) may thus be written

$$\left[\frac{U^c(Y^c)^2}{\nu X^c}\right]\left(u\frac{\partial u}{\partial x}+v\frac{\partial u}{\partial y}\right)=\frac{\partial^2 u}{\partial y^2}+\left[\frac{Y^c}{X^c}\right]^2\frac{\partial^2 u}{\partial x^2}, \tag{1.25}$$

using eqn (1.24). We now recognize the physical requirement that the driving inertial forces (the left-hand side) are balanced by the viscous forces (the right-hand side), the latter being represented by the term $\partial^2u/\partial y^2$ which we anticipate to be larger. Hence

$$\frac{U^c(Y^c)^2}{\nu X^c}=O(1), \tag{1.26}$$

since the coefficient of $\partial^2u/\partial y^2$ is 1.0.

Equations (1.24) and (1.26) then yield

$$\frac{Y^c}{L}=\frac{\delta}{L}=O\left(\frac{\nu}{U_\infty L}\right)^{1/2}=O\left(\frac{1}{Re_L^{1/2}}\right), \tag{1.27}$$

where δ is defined as the boundary layer thickness, and

$$\frac{V^c}{U_\infty}=O\left(\frac{1}{Re_L^{1/2}}\right). \tag{1.28}$$

From these results we gain a number of insights, none of which require a solution of eqns (1.21) and (1.22). Firstly, from eqn (1.27), it is evident that the boundary layer thickness varies with the square root of the surface length, and that this thickness is much less than the surface length when $Re_L \gg 1$. Secondly, eqn (1.28) shows that the lateral velocity is much less than the longitudinal velocity under the same conditions. Thirdly, the coefficient of a $\partial^2 u/\partial x^2$ in eqn (1.25) is $1/Re_L$, from eqn (1.27). Therefore, when $Re_L \gg 1$, eqn (1.25) reduces to

$$u\frac{\partial u}{\partial x} + v\frac{\partial u}{\partial y} = \frac{\partial^2 u}{\partial y^2}, \tag{1.29}$$

the error in neglecting the longitudinal viscous term having now been determined.

1.6.5 Boundary layer convection

Finally, consider the same boundary layer flow when the surface temperature T_0 is higher than the bulk fluid temperature T_∞. We must now add to eqns (1.21) and (1.22) the corresponding energy equation

$$U\frac{\partial \theta}{\partial X} + V\frac{\partial \theta}{\partial Y} = \kappa\left(\frac{\partial^2 \theta}{\partial X^2} + \frac{\partial^2 \theta}{\partial Y^2}\right). \tag{1.30}$$

Introducing $\phi = \theta/\theta^c$, where $\theta^c = \theta_0$, this may be rewritten

$$\left[\frac{U^c(Y^c)^2}{\kappa X^c}\right]\left(u\frac{\partial \phi}{\partial x} + v\frac{\partial \phi}{\partial y}\right) = \frac{\partial^2 \phi}{\partial y^2} + \left[\frac{Y^c}{X^c}\right]^2\frac{\partial^2 \phi}{\partial x^2}, \tag{1.31}$$

where Y^c is now the *thermal* boundary layer thickness. If the boundary layer is interpreted as a balance between the longitudinal thermal flux and the lateral diffusion of heat (the first term on the right-hand side), then

$$\frac{U^c(Y^c)^2}{\kappa X^c} = O(1)$$

or

$$\frac{Y^c}{L} = O\left(\frac{1}{Pe_L^{1/2}}\right), \tag{1.32}$$

where $Pe_L = U_\infty L/\kappa$ is the Peclet number. By comparison with eqn (1.27) it appears that thermal and momentum boundary layer thicknesses are equal when $Re_L = Pe_L = PrRe_L$, i.e. when $Pr = 1$. This requires more careful examination.

Let δ^M and δ^T be the momentum and thermal boundary layer thicknesses, respectively. Equations (1.27) and (1.32) may then be written

$$\delta^{\mathrm{M}} \simeq \frac{L}{Re_{\mathrm{L}}^{1/2}} \tag{1.33}$$

and

$$\delta^{\mathrm{T}} \simeq \frac{L}{Pe_{\mathrm{L}}^{1/2}}. \tag{1.34}$$

Hence

$$\frac{\delta^{\mathrm{M}}}{\delta^{\mathrm{T}}} \simeq Pr^{1/2} = \left(\frac{\nu}{\kappa}\right)^{1/2} \tag{1.35}$$

which provides us with the insight that the relative thickness of the momentum boundary layer is a simple, monotonically increasing function of the fluid property ratio ν/κ. When $\nu \gg \kappa$, eqn (1.35) may be refined by taking a better approximation for the longitudinal thermal flux in eqn (1.30). Instead of balancing $U^{\mathrm{c}}\theta^{\mathrm{c}}/X^{\mathrm{c}}$ with $\kappa\theta^{\mathrm{c}}/(\delta^{\mathrm{T}})^2$, we take

$$\frac{U^{\mathrm{c}}}{X^{\mathrm{c}}}\frac{\delta^{\mathrm{T}}}{\delta^{\mathrm{M}}}\theta^{\mathrm{c}} \sim \kappa\frac{\theta^{\mathrm{c}}}{(\delta^{\mathrm{T}})^2},$$

in which case

$$\frac{\delta^{\mathrm{T}}}{L} = O\left(\frac{Pr^{1/6}}{Pe_{\mathrm{L}}^{1/2}}\right) \tag{1.36}$$

in place of eqn (1.34), and

$$\frac{\delta^{\mathrm{M}}}{\delta^{\mathrm{T}}} \simeq O\left(\frac{\nu}{\kappa}\right)^{1/3} \tag{1.37}$$

in place of eqn (1.35). The two sets of results differ by a factor of $Pr^{1/6}$.

1.6.6 Summary

Normalization is a useful way of redefining physical variables free of units and dimensions. The range of a normalized variable is, by definition, restricted to $O(1)$; this is especially useful when seeking numerical solutions. Extending this idea to derivatives, the orders of magnitude of various terms in differential equations may be estimated from their non-dimensional coefficients, relative to $O(1)$. Any terms with coefficients of this order are of equal importance to the dominant terms determined from the essential physics; terms with much smaller coefficients may be neglected, the error being estimated from the coefficient itself, relative to $O(1)$. The result is a first-order model of the situation. Typically, it will provide a reduced form of the original problem and may therefore be easier to solve. The assumption that derivatives are well behaved may be checked *a posteriori*.

The insights made possible by scale analysis are considerable. At the very least, unknown scales may be estimated and then used to assess system behaviour; for example, in predicting a response time, a penetration depth, or a boundary layer width. Such results are implicit in the physical statement of the problem and its mathematical counterpart; they flow not only from the equations themselves but from the initial and boundary conditions associated with them. Some scales are derived from the equations, some from the associated conditions.

In the course of establishing the orders of magnitude, various non-dimensional groups appear as coefficients. Above we identified the Fourier number, the Reynolds number, the Peclet number, and the Prandtl number. It was not necessary to define or introduce these separately or to identify them at the outset. Normalization automatically generates the minimum set of appropriate non-dimensional groups. In a familiar situation, the groups are recognizable immediately. In a new situation, new groups will appear.

1.7 Purpose and scope

The main aim of this book is to introduce the reader, student and practitioner alike, to the subject of latent heat transfer. It assumes a basic knowledge of heat transfer, conduction and convection in particular, such as might be acquired in a first course at university. Introductory courses usually treat conduction as a new subject, essentially the application of Fourier's law within the energy equation. They also give a cursory coverage of single phase fluid mechanics before discussing convection. Understandably, that is the main thrust of a first course, but it often leaves little time for buoyant flows, and often less for situations where a change of phase occurs.

Single phase behaviour is a mere prelude to the material in this book. Latent heat transfer problems are significantly different from sensible heat problems; they must accommodate an interface across which properties vary discontinuously and where local behaviour is complex and crucial in determining overall behaviour. Yet the elementary treatment of heat conduction does provide a sound basis for entry into problems of melting and freezing. Likewise, single phase fluid mechanics provides a useful starting point in the understanding of multi-phase flows even though it must be extended considerably to accommodate such disparate systems as bubbles and sprays. The expanded requirements have dictated the layout of this book.

Chapter 2 is a review of thermodynamics applied to changes of phase. In particular, it develops the Clausius equation and introduces the equation governing energy transfer across the interface. Macroscopic and microscopic descriptions are both used to discuss nucleation and surface tension. Chapter 3 is an overview of heterogeneous fluid dynamics divided into the two major categories: bipartitioned and dispersed. For the former, emphasis is placed on the hydrodynamic stability of the interface; for the latter, the treatment covers both the origin and the behaviour of bubbles and drops.

The four subsequent chapters are the main ones. Each builds on the earlier chapters and provides a detailed coverage of only one particular aspect of latent heat transfer: solidification (and fluidification), evaporation, condensation, and direct contact situations. The last of these differs from the others only through the fact that intermediate walls separating the phases are completely absent.

Every chapter uses a blend of analysis and experiment to reveal the essential characteristics of a given situation and thus develop a sense of the main variables controlling it. While particular applications are frequently mentioned, the emphasis is on fundamentals. Mathematics and physics are used to establish and interpret empirical relations and provide a sound basis for extrapolation. More importantly, they help develop strategies which may be employed with new problems. Upon completion of the book, the reader may not expect to understand every conceivable situation in latent heat transfer but, faced with a particular problem, should be able to identify its basic structure and, with confidence, be able to make a significant contribution towards its solution, however complex.

To accomplish this objective, normalization and scale analysis are used repeatedly. This technique is very powerful and generously repays the effort required to understand it. Numerical methods, on the other hand, have been deliberately excluded. They too are powerful but provide no more insight than scaling analysis, if as much, even though they require considerably more effort. It is better to deal with numerical analysis comprehensively in a separate course and then apply it in latent heat transfer after the essential structure has been made clear. We are now ready to explore this essential structure, in its component parts and in its entirety. We begin with the foundations on which it will be built.

Selected bibliography

Bejan, A. (1993). *Heat transfer*. Wiley, New York.

Bergles, A. E. (1981). Two-phase flow and heat transfer, 1756–1981. *Heat Transfer Engineering*, **2**(3–4), 101–14.

Black, J. (1803). General effects of heat. In *Lectures on the elements of chemistry* (ed. J. Robinson), Vol. 1, Part 1. Mundell and Son, Edinburgh.

Desaguliers, J. T. (1729). An attempt to solve the phenomenon of vapours, formation of clouds and descent of rain. *Philosophical Transactions*, **36**, 6–22.

Gray, D. E. (ed.) (1963). *American Institute of Physics handbook*. McGraw-Hill, New York.

Hemminger, W. and Höhne, G. (1984). *Calorimetry: fundamentals and practice*. Verlag Chemie, Weinheim.

Joule, J. P. (1861). On the surface condensation of steam. *Transactional Proceedings of the Royal Society of London*, **151**, 133–60.

Kelvin, Lord (1870). On the equilibrium of vapour at a curved surface of a liquid. *Proceedings of the Royal Society of Edinburgh*, **7**, 63.

Leidenfrost, J. G. (1966). On the fixation of water in diverse fire (transl). *International Journal of Heat Mass Transfer*, **9**, 1154–66.

Lienhard, J. H. (1983). Notes on the origins and evolution of the subject of heat transfer. *Mechanical Engineering*, **105**(6), 20–7.

Lock, G. S. H. (1986). Normalization. *International Journal of Mechanical Engineering Education*, **14**(3), 193–204.

McKie, D. and Heathcote, N. H. deV. (1935). *The discovery of specific and latent heats*. Edward Arnold, London.

Nishikawa, K. (1987). Historical developments in the research of boiling heat transfer. *JSME International Journal*, **30**, 897–905.

Nukiyama, S. (1934). The maximum and minimum values of heat Q transmitted from metal to boiling water under atmospheric pressure. *Journal of the JSME*, **37**, 367–74.

Nusselt, W. (1916). Die Oberflächen Kondensation des Wasserdampfes. *Z. ver. Dtsch. Ing.*, **60**, 541–46, 549.

Rayleigh, Lord (1917). On the pressure developed during the collapse of a spherical cavity. *Philosophical Magazine*, **34**, 94–99.

Stefan, J. (1891). Ueber die theorie der eisbildung, inbesondere uber die eisbildung im polarmeere. *Ann. Phys. Chem. Neue Folge*, **42**, 269–86.

2
THERMODYNAMICS

Latent heat transfer is rooted in thermodynamics, and is therefore better understood in terms of thermodynamic concepts and principles applied specifically to phase change processes. This chapter provides a brief review and summary of those particular thermodynamic ideas and results which are prerequisites for the full discussion of latent heat transfer: surface energy, interface dynamics, stability, nucleation, etc. Much of this fundamental information assumes that individual phases may be treated as continua; the associated macroscopic analysis is widely applicable and very powerful. On occasion, insights are also provided through microscopic analysis. This is especially valuable during the discussion of interfacial phenomena and nucleation.

2.1 The behaviour of continua

2.1.1 The energy equation

Figure 2.1 depicts part of a continuous medium bounded by a surface area $A(t)$ enclosing a volume $\mathcal{V}(t)$. A stress vector $\mathbf{\Pi}$, having the normal and tangential components indicated, acts on the surface element $\mathrm{d}A$. Surface motion with velocity $\mathbf{V}$ thus corresponds to a work transfer through $\mathrm{d}A$. Similarly, the body force per unit mass $\mathfrak{F}$ is also capable of transferring work to or from the volume. The *heat flux density* vector $\mathbf{j}_Q$ represents the rate of heat loss through $\mathrm{d}A$. According to the first law of thermodynamics, heat and work transfer alter the system energy. Thus

$$\frac{\mathrm{d}}{\mathrm{d}t}\int_{\mathcal{V}(t)} \rho(u+\tfrac{1}{2}\mathbf{V}\cdot\mathbf{V})\,\mathrm{d}\mathcal{V} = -\int_{A(t)} \mathbf{j}_Q\cdot\mathrm{d}\mathbf{A} + \int_{A(t)} \mathbf{\Pi}\cdot\mathbf{V}\mathrm{d}A + \int_{\mathcal{V}(t)} \rho\mathfrak{F}\cdot\mathbf{V}\mathrm{d}\mathcal{V} \tag{2.1}$$

where u and $V^2/2$ are the specific values of the internal and kinetic energy, respectively, and ρ is the material density.

This general energy equation may be simplified by neglecting terms which are seldom important during latent heat transfer. Thus, if we neglect changes in potential and kinetic energy, we may reduce eqn (2.1) to

$$\frac{\mathrm{d}}{\mathrm{d}t}\int_{\mathcal{V}(t)} \rho u\,\mathrm{d}\mathcal{V} = -\int_{A(t)} \mathbf{j}_Q\cdot\mathrm{d}\mathbf{A} + \int_{A(t)} \mathbf{\Pi}\cdot\mathbf{V}\,\mathrm{d}A \tag{2.2}$$

or, in point form,

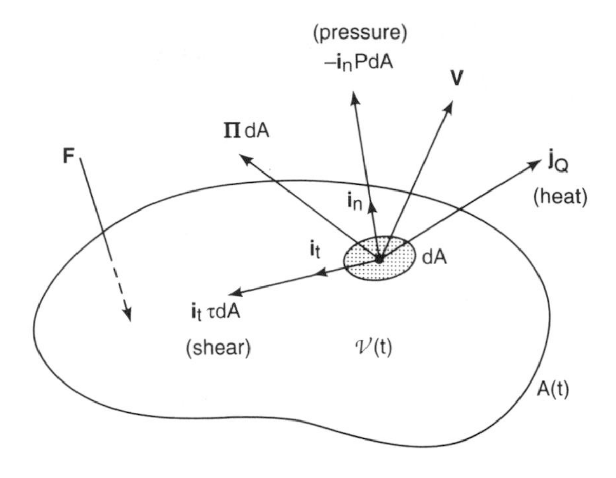

$\Pi dA = -\mathbf{i}_n PdA + \mathbf{i}_t \tau dA$

Fig. 2.1 Surface and interior forces acting on a closed volume from which heat is being lost through the surface $A(t)$ as it moves outwards with a velocity $\underset{\sim}{\mathrm{V}}$.

$$\rho \frac{\mathrm{d}u}{\mathrm{d}t} = -\nabla \cdot \mathbf{j}_Q + \nabla \cdot (\pi \cdot \mathbf{V}) \tag{2.3}$$

in which $\mathbf{\Pi} = \mathbf{i}_n \cdot \boldsymbol{\pi}$, where $\boldsymbol{\pi}$ is the stress tensor. The treatment here will be limited to constant property, Newtonian fluids for which $\boldsymbol{\pi}$ will be specified later. At this point, it is sufficient to note that

$$\nabla \cdot (\pi \cdot \mathbf{V}) = -P\nabla \cdot \mathbf{V} + \Phi \tag{2.4}$$

where Φ, the viscous dissipation function, is a heat source. Equation (2.3) may now be written

$$\rho \frac{\mathrm{d}h}{\mathrm{d}t} = \frac{\mathrm{d}P}{\mathrm{d}t} - \nabla \cdot \mathbf{j}_Q + \Phi \tag{2.5}$$

where $h = u + Pv$ is the specific enthalpy. In many latent heat transfer problems, neither viscous dissipation nor the pressure field exert much influence on energetics. Under such conditions, the energy equation reduces further to

$$\rho \frac{\mathrm{d}h}{\mathrm{d}t} = -\nabla \cdot \mathbf{j}_Q. \tag{2.6}$$

This is sometimes called the *conduction equation*. It represents a thermal energy balance which, for a control volume $\mathcal{V}$, may be expressed as

$$\int_{\mathcal{V}} \frac{\partial}{\partial t}(\rho h)\, d\mathcal{V} + \int_{A} \dot{\mathbf{q}} \cdot \mathrm{d}\mathbf{A} = 0 \tag{2.7}$$

in which

$$\dot{\mathbf{q}} = \mathbf{j}_{\mathrm{Q}} + h\mathbf{G} \tag{2.8}$$

is the thermal energy flux *density* (Wm^{-2}) consisting of two parts: a diffusive (heat) contribution

$$\mathbf{j}_{\mathrm{Q}} = -k\nabla T, \tag{2.9}$$

expressing Fourier's law; and a bulk (convective) contribution $h\mathbf{G}$ in which $\mathbf{G} = \rho\mathbf{V}$ is the total net mass flux density. Substituting these into eqn (2.7) we obtain the point form

$$\frac{\partial}{\partial t}(\rho h) + \nabla \cdot (h\mathbf{G}) = -\nabla \cdot \mathbf{j}_{\mathrm{Q}} \tag{2.10}$$

which is equivalent to eqn (2.6).

2.1.2 The equations of motion, continuity, and diffusion

The first law may also be used to derive the equations of motion and continuity. Assuming eqn (2.1) to be true in any inertial frame, i.e. invariant under a uniform translation of velocity **B**, the first law may be rewritten with **V** replaced by **V** + **B**. Substracting the two statements, we obtain

$$\mathbf{B} \cdot \left[\frac{\mathrm{d}}{\mathrm{d}t}\int_{\mathcal{V}(t)} (\rho\mathbf{V} + \rho\mathbf{B})\, \mathrm{d}\mathcal{V} - \int_{A(t)} \mathbf{\Pi}\mathrm{d}A - \int_{\mathcal{V}(t)} \rho\mathfrak{F}\, \mathrm{d}\mathcal{V}\right] = 0. \tag{2.11}$$

Since **B** is an arbitrary vector, this equation may only be satisfied if

$$\frac{1}{2}\frac{\mathrm{d}}{\mathrm{d}t}\int_{\mathcal{V}(t)} \rho\mathbf{B}\, \mathrm{d}\mathcal{V} = \frac{\mathrm{d}}{\mathrm{d}t}\int_{\mathcal{V}(t)} \rho\, \mathrm{d}\mathcal{V} = 0 \tag{2.12}$$

and

$$\frac{\mathrm{d}}{\mathrm{d}t}\int_{\mathcal{V}(t)} \rho\mathbf{V}\, \mathrm{d}\mathcal{V} = \int_{A(t)} \mathbf{\Pi}\, \mathrm{d}A + \int_{\mathcal{V}(t)} \rho\mathfrak{F}\, \mathrm{d}\mathcal{V}. \tag{2.13}$$

The first of these, expressing the conservation of mass, is the *continuity equation*, while the second is the *equation of rectilinear motion*. Both of these results may be restated for a control volume $\mathcal{V}$ and thus be reduced to point form. Hence

$$\rho\frac{\mathrm{d}\mathbf{V}}{\mathrm{d}t} = \nabla \cdot \boldsymbol{\pi} + \rho\mathfrak{F}, \tag{2.14}$$

for the equation of motion, and

$$\frac{d\rho}{dt} + \rho\nabla \cdot \mathbf{V} = 0 \tag{2.15}$$

for the continuity equation.

Commonly, we must deal with a multicomponent mixture for which it is often more convenient to re-write the continuity equation as

$$\frac{\partial\rho}{\partial t} + \nabla \cdot \mathbf{G} = 0 \tag{2.16}$$

where, in general, $\mathbf{G} = \rho\mathbf{V}$ must be decomposed into

$$\mathbf{G} = \sum_k \dot{\boldsymbol{m}}_k \tag{2.17}$$

in which

$$\dot{\boldsymbol{m}}_k = \mathbf{j}_{Mk} + \mathbf{G}_k \tag{2.18}$$

is the kth *component* mass flux density. The two parts of $\dot{\boldsymbol{m}}_k$ are

$$\mathbf{j}_{\mathrm{Mk}} = -\rho\mathfrak{D}_k \nabla \mathfrak{m}_k, \tag{2.19}$$

a diffusive contribution expressing Fick's law in which $\mathfrak{m}_k = \rho_k/\rho$ is the mass fraction, and

$$\mathbf{G}_k = \mathfrak{m}_k\mathbf{G} = \rho_k\mathbf{V} \tag{2.20}$$

which is a bulk (convective) contribution.

Substituting the above into the continuity equation (2.16) we obtain

$$\sum_k \left(\frac{\partial\rho_k}{\partial t} + \nabla \cdot \dot{\boldsymbol{m}}_k\right) = 0, \tag{2.21}$$

summing over all the components. For a single component which, in general, is not conserved, this takes the form

$$\frac{\partial\rho_k}{\partial t} + \nabla \cdot \mathbf{G}_k = -\nabla \cdot \mathbf{j}_{Mk} + \dot{s}_k, \tag{2.22}$$

where $\dot{s}_k$ is the volumetric rate at which component k is generated. Using the continuity equation (2.16), eqns (2.10) and (2.22) may now be rewritten as

$$\rho\left[\frac{\partial h}{\partial t} + (\mathbf{V} \cdot \nabla)h\right] = -\nabla \cdot \mathbf{j}_{\mathrm{Q}} \tag{2.23}$$

and

$$\rho\left[\frac{\partial m_k}{\partial t} + (\mathbf{V}\cdot\nabla)m_k\right] = -\nabla\cdot\mathbf{j}_{Mk} + \dot{s}_k, \tag{2.24}$$

in which form they describe the *diffusion* of heat and mass, respectively. A heat source $\dot{s}_Q$, e.g. if $\Phi \neq 0$, may be added to eqn 2.23 if necessary.

2.2 Surface energy, surface tension, and wetting

2.2.1 The transitional zone

Nature abhors discontinuities and this is true, in particular, when two phases co-exist in stable, thermodynamic equilibrium. Established between the two phases is a transitional zone in which each phase accommodates the other. On the molecular level, the structure and behaviour of one phase gradually adjusts to take on the structure and behaviour of the other. On the macroscopic level, this is almost impossible to see, thus giving rise to the conceptual model of an interfacial surface of separation. Across this surface, phase properties are assumed to change discontinuously. The surface itself is also assumed to have measurable properties. Such a model of reality has the great merit of simplicity and has found widespread use.

From a mechanical point of view, the forces attributed to the interface are equivalent to the unknown forces acting in the transitional zone. Similarly, the molecular energy in the transitional zone is ascribed to the equivalent interface. This energy differs from that possessed by molecules in the bulk of the phase because the molecular structure of the zone is different. The total internal energy of a homogeneous phase must therefore be taken as the sum of the bulk energy E_b and the transitional zone energy E_I. In macroscopic terms,

$$\begin{aligned} E &= E_b + E_I \\ &= e_v \mathcal{V} + \sigma A \end{aligned}$$

where σ is the interfacial energy density and A is the interfacial area. The transitional zone energy has thus been converted from

$$E_I = e_{vI}\,\mathcal{V}_I,$$

where e_{vI} is the unknown average energy per unit volume of the transitional zone, to

$$E_I = (e_{vI}\,\delta)A \tag{2.25}$$

where δ is the unknown zone thickness. However, $e_{vI}\,\delta = \sigma$ is the measurable energy per unit interfacial area. Between two phases, E_I 'belongs' to neither.

Under adiabatic (or isothermal) conditions, any change in the interfacial energy is attributable to interfacial work. Thus, for example, the work (or energy) required to establish a disc-shaped interface of radius R is

$$E_{\mathrm{I}} = W_{\mathrm{I}} = \pi R^2 \sigma \tag{2.26}$$

and hence for a virtual expansion from R to $R + \delta R$, the incremental transfer of virtual work is given by

$$\delta W_{\mathrm{I}} = (2\pi R\sigma)\delta R.$$

This may be interpreted as a radial force

$$F_{\mathrm{I}} = 2\pi R\sigma$$

acting over a virtual displacement δR. The force, in turn, may be interpreted as a surface tension per unit length σ acting over the length $2\pi R$. This is the most familiar interpretation of the interfacial energy density.

In applying the macroscopic model of surface tension at phase boundaries it is always worthwhile bearing in mind that σ is a surface energy density. It represents not only the molecular energy difference between the bulk phases individually, but the steep energy gradient in the transitional zone. It thus reflects the energetics of molecular accommodation, and is therefore altered by events and effects which alter this accommodation: thermal and chemical effects, in particular. Thus, for example, an increase in temperature will lower the demands of molecular accommodation, and the surface tension will tend to decrease. At the critical temperature, the surface tension vanishes completely. Indeed, the critical condition may be defined as the point where molecular energy in the transitional zone no longer differs from that in the bulk. Accommodation is no longer required.

2.2.2 *Wetting and spreading*

Latent heat transfer problems are not always restricted to two phases, as Fig. 2.2 illustrates. This provides a planar representation of three co-existing fluid phases. The three interfacial energy densities are represented by the three surface tensions shown acting vectorially at the common 'point' 0. For static equilibrium, the sum of these forces must be zero. Hence

$$\boldsymbol{\sigma}_{12} + \boldsymbol{\sigma}_{13} + \boldsymbol{\sigma}_{23} = 0 \tag{2.27}$$

in which $\boldsymbol{\sigma} = \mathbf{i}_{\mathrm{t}}\sigma$, where $\mathbf{i}_{\mathrm{t}}$ is a unit tangent vector. The surface tensions thus influence the shapes of the interfaces near their juncture. The shapes also depend upon body force fields, gravity in particular.

Consider the static equilibrium of the liquid drop shown as a spherical cap in Fig. 2.3. If this lies on a horizontal solid substrate but is otherwise surrounded by a vapour, the surface tensions will act as shown. Horizontally eqn (2.27) reduces to

$$\sigma_{\mathrm{LV}} \cos\theta + \sigma_{\mathrm{SL}} - \sigma_{\mathrm{SV}} = 0,$$

giving Young's equation

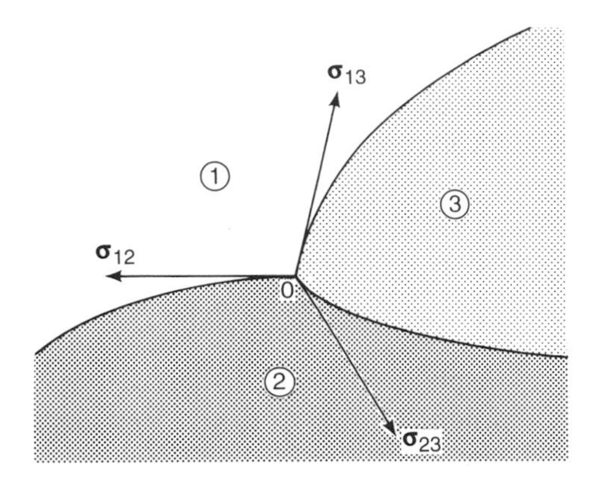

Fig. 2.2 Surface tensions acting at the junction of three fluid phases.

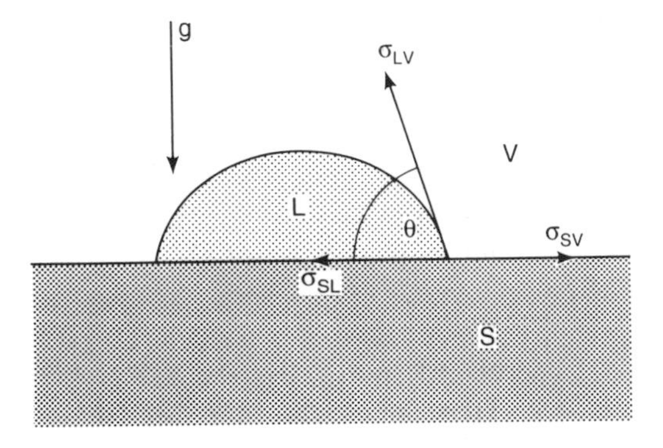

Fig. 2.3 Surface tensions acting at the junction of a vapour, a liquid drop, and a solid substrate.

$$m = \cos\theta = \frac{\sigma_{SV} - \sigma_{SL}}{\sigma_{LV}}, \tag{2.28}$$

where θ is the *contact angle*. The quantity m is known as the degree of wetting or *wetting coefficient*, while $\sigma_{SV} - \sigma_{SL}$ is often described as the *wetting* or *adhesion tension*. The contact angle is readily measured under these steady conditions. It is important to note, however, that when the contact line moves over a substrate the contact angle changes. Referring to Fig. 2.3, when the contact line advances, i.e. the substrate is being wetted further, the contact angle increases. The reverse is true of a receding contact line. Wetting and de-wetting, considered together, thus introduce hysteresis.

The above descriptions emphasize the contributions which *both* adjacent phases make to surface tension as the representation of interfacial energy density. Theoretically, one phase alone could be ascribed a surface energy density or surface tension induced by adhesive forces. Let σ_A and σ_B represent the surface tensions *in vacuo* of phases *A* and *B*, respectively. The surface tension at the interface separating them may be represented by σ_{AB}. The relation between $\sigma_B - \sigma_A$, which is a measure of molecular adhesion, and σ_{AB}, which is a measure of molecular cohesion, is given by

$$S_{AB} = (\sigma_B - \sigma_A) - \sigma_{AB}. \tag{2.29}$$

The quantity S_{AB} is known as the *spreading coefficient*, particularly when applied to two immiscible liquids. When $S_{AB} > 0$, adhesion dominates and liquid *A* will spread as a film on the surface of liquid *B*; however, if $S_{AB} < 0$, cohesion dominates and *B* will not spread over *A* but will instead form drops or lenses.

Finally, consider the role of the substrate in wetting. When the substrate surface is perfectly smooth it affects the wetting process solely through its composition. Different substrate materials have different adhesive abilities in holding vapour or liquid molecules on their surface. As Fig. 2.3 and eqn (2.28) indicate, efficient wetting corresponds to a small contact angle (small surface tension σ_{LV}); conversely, poor wetting corresponds to a large contact angle (large surface tension). Engineered surfaces may be treated with wetting or anti-wetting agents to deliberately alter the contact angle.

When the substrate surface is microrough, as the result of manufacturing, corrosion, etc., the macroscopic description of surface tension and contact angle may need modification. At the very least, surface irregularities will create local variations in the direction of the vector $\boldsymbol{\sigma}_{LV}$; the wetting coefficient must then be regarded as an averaged value. More importantly, the initial wetting process commonly traps pockets of air in surface microcavities. The size and shape of these pockets, which are crucially important during the nucleation events which precede boiling, for example, reflect both the microcavity geometry and the magnitude of the surface tension σ_{LV}.

2.3 Growth at the interface

2.3.1 Phase and interface motion

Consider the result of a phase change on the displacement of the plane interface shown in Fig. 2.4. During the time interval dt, let the growing phase g increase in extent by an amount dY_I where Y_I marks the location of the interface *relative to the growing* phase. Also, let the corresponding shift of an arbitrary point Y_p buried in the parent phase be dY_p. If $\rho_g < \rho_p$, as

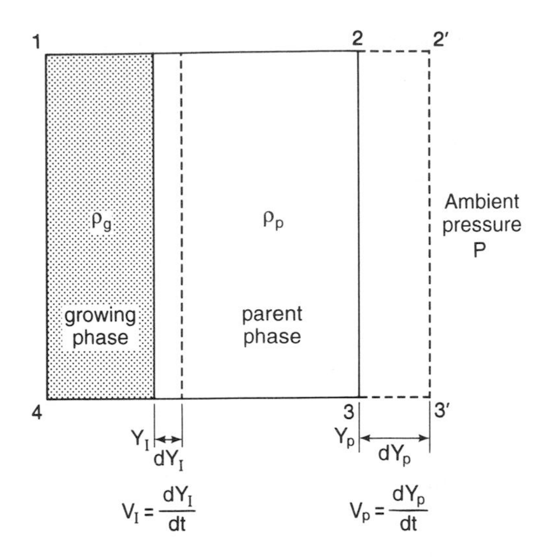

Fig. 2.4 Phase change causing interfacial motion V_{I}, and phase motion V_{p}, both measured with respect to a point buried in the growing phase.

suggested in Fig. 2.4, $\mathrm{d}Y_{\mathrm{p}} > \mathrm{d}Y_{\mathrm{I}}$. The parent phase is then pushed away from the interface unless constraints are placed on its motion, e.g. during freezing in a water pipe where the pressure may become high enough to burst the pipe. The effect of constraint is explored in problems Q2.1, 2.2, and 2.3 in Chapter 8. Here we will assume that the phases are unconstrained and the system pressure remains fixed at P.

Since Y_{p} moves with the parent phase, the mass contained in the rectangle 1234 equals that contained in the rectangle 12′3′4. The material gained by the growing phase, $\rho_{\mathrm{g}}\,\mathrm{d}Y_{\mathrm{I}}$, therefore equals that lost by the parent phase, $\rho_{\mathrm{p}}\mathrm{d}Y_{\mathrm{I}} - \rho_{\mathrm{p}}\,\mathrm{d}Y_{\mathrm{p}}$. Hence

$$V_{\mathrm{p}} = \left(\frac{\rho_{\mathrm{p}} - \rho_{\mathrm{g}}}{\rho_{\mathrm{p}}}\right) V_{\mathrm{I}}, \tag{2.30}$$

where $V_{\mathrm{I}} = \mathrm{d}Y_{\mathrm{I}}/\mathrm{d}t$ is the interface velocity and $V_{\mathrm{p}} = \mathrm{d}Y_{\mathrm{p}}/\mathrm{d}t$ is the parent phase velocity, both measured relative to the growing phase.

2.3.2 *The interface equation*

Turning to the application of the energy equation, it is apparent from eqn (2.30) that compression work is transferred during a phase change. The net compression work flux *density* leaving the interface is

$$\dot{w} = P\left(\frac{\rho_{\mathrm{p}} - \rho_{\mathrm{g}}}{\rho_{\mathrm{p}}}\right) V_{\mathrm{I}}. \tag{2.31}$$

At the same time, the net heat flux *density* leaving the interface is given by

$$(j_Q)_{net} = k_g \left(\frac{\partial T_g}{\partial Y}\right)_I - k_p \left(\frac{\partial T_p}{\partial Y}\right)_I. \tag{2.32}$$

Now consider the closed system bounded between the plane Y_p and a fixed plane immediately to the left of Y_I. The specific energy difference across the moving interface implies a system energy rate of change per unit interfacial area of

$$\frac{de_A}{dt} = V_I \rho_g (u_g - u_p) \tag{2.33}$$

if we neglect potential and kinetic energy changes. Applying the first law of thermodynamics to the same closed system as $Y_p \to Y_I$,

$$\frac{de_A}{dt} = (j_Q)_{net} - \dot{w},$$

in which the right hand is provided by eqns (2.31) and (2.32). It thus follows that

$$(j_Q)_{net} = -V_I \rho_g (u_g - u_p) - P\left(\frac{\rho_p - \rho_g}{\rho_p}\right) V_I = \rho_g V_I [(u_p + Pv_p) - (u_g + Pv_g)].$$

Therefore,

$$k_g \left(\frac{\partial T_g}{\partial Y}\right)_I - k_p \left(\frac{\partial T_p}{\partial Y}\right)_I = \rho_g V_I (h_p - h_g) = \rho_g \lambda V_I, \tag{2.34}$$

where λ is the latent heat per unit mass. This is known as the *interface equation.*

It is important to recognize that the interface equation is not merely a heat balance. It is an energy balance which lies at the heart of every latent heat transfer problem. In the absence of phase change, it simply expresses continuity in the heat flux across the interface but, in general, it intrinsically couples heat transfer in one phase to heat transfer in the other: during evaporation, condensation, melting, freezing, sublimation, and deposition.

2.3.3 Interfacial resistance

The imbalance between the incoming ($\dot{m}_i$) and outgoing ($\dot{m}_o$) mass flux densities on one side of an interface may be interpreted as an interfacial thermal resistance. Between a pure vapour bulk at T_V and P_V, and a *quasi-stationary* interface at T_I and P_I, this resistance may be estimated from the condensation (or deposition) rate $\dot{m}_{cond} = \beta_i \dot{m}_{Vi} - \beta_o \dot{m}_{Vo}$, where β is the fraction of the vapour flux density $\dot{m}_V$ actually captured (i) or escaping (o). Using eqn (2.18) for $\dot{m}_V$, with $G_i = m_i \dot{m}_{cond}$ and $G_o = 0$, we may re-write $\dot{m}_{cond}$ as

$$\dot{m}_{\text{cond}} \simeq \frac{2\beta}{2-\beta}(j_{\text{Mi}} - j_{\text{Mo}}) \simeq \frac{2\beta}{2-\beta}\left[\frac{P_{\text{V}}}{T_{\text{V}}^{1/2}} - \frac{P_{\text{I}}}{T_{\text{I}}^{1/2}}\right] / (2\pi R_{\text{V}})^{1/2}, \quad (2.35)$$

bearing in mind that $\beta_{\text{i}} \simeq \beta_{\text{o}}$ near equilibrium when $m_{\text{o}} \simeq m_{\text{i}} \simeq 0.5$ and $\dot{m}_{\text{cond}} \ll j_{\text{M}} = (P/2\pi RT)^{1/2}$ from kinetic theory.

Under typical conditions, $(T_{\text{V}} - T_{\text{I}}) \ll T_{\text{I}}$ and $(P_{\text{V}} - P_{\text{I}}) \ll P_{\text{I}}$. Hence

$$\frac{P_{\text{V}}}{T_{\text{V}}^{1/2}} - \frac{P_{\text{I}}}{T_{\text{I}}^{1/2}} \simeq \frac{(P_{\text{V}} - P_{\text{I}})}{T_{\text{V}}^{1/2}} - \frac{P_{\text{I}}(T_{\text{V}} - T_{\text{I}})}{2T_{\text{I}}^{1/2}}.$$

Using the Clapeyron equation developed in section 2.5, this may be simplified to

$$\frac{P_{\text{V}}}{T_{\text{V}}^{1/2}} - \frac{P_{\text{I}}}{T_{\text{I}}^{1/2}} \simeq \frac{\lambda P_{\text{V}}(T_{\text{V}} - T_{\text{I}})}{R_{\text{V}} T_{\text{V}}^{5/2}}, \quad (2.36)$$

if $RT_{\text{V}} \ll \lambda$, in which case the interfacial heat transfer coefficient is

$$h = \frac{\lambda \dot{m}_{\text{cond}}}{T_{\text{V}} - T_{\text{I}}} \simeq \frac{2\beta}{2-\beta} \frac{\lambda^2 P_{\text{V}}}{(2\pi R_{\text{V}}^3 T_{\text{V}}^5)^{1/2}}, \quad (2.37)$$

the reciprocal of which is the interfacial resistance. Typically, $\beta \simeq 1$ and h_{I} is very large, as noted in Section 5.4.1 and illustrated in problem Q7.2 of Chapter 8.

2.4 Thermodynamic surfaces

The shape of any thermodynamic surface of stable states is constrained by the laws of thermodynamics. The first and second laws applied to a given body of material dictate that, in general,

$$\mathrm{d}E + đW - T\,\mathrm{d}S = -T\dot{S}_{\text{g}}\,\mathrm{d}t,$$

where $\dot{S}_{\text{g}}$ is the entropy generation rate of the body. Since T, $\dot{S}_{\text{g}}$, and $\mathrm{d}t$ are all taken to be positive, this requires that

$$\mathrm{d}E + đW - T\,\mathrm{d}S \leqslant 0 \quad (2.38)$$

which contains the well-established fact that an isolated system increases its entropy in a natural tendency to seek a stable equilibrium state. The inequality sign thus refers to processes during *unstable* equilibrium conditions for which

$$\mathrm{d}U + đW - T\mathrm{d}S < 0$$

in most phase change situations, when variations in kinetic and potential energy may be ignored.

The above statements may be written in a variety of forms. Using the Gibbs function $\mathcal{G} = H - TS$, for example,

$$\mathrm{d}\mathcal{G} - \mathcal{V}\,\mathrm{d}P + S\,\mathrm{d}T \leqslant 0 \tag{2.39}$$

when $đW = P\,\mathrm{d}\mathcal{V}$, the compression work. More generally, when $đW = P\,\mathrm{d}\mathcal{V} - \Sigma_i F_i\,\mathrm{d}X_i$, where F_i and X_i form other work pairs, e.g. surface tension with area or chemical potential with mole fraction,

$$\mathrm{d}\mathcal{G} - \mathcal{V}\,\mathrm{d}P + S\,\mathrm{d}T - \sum_i F_i\,\mathrm{d}X_i \leqslant 0. \tag{2.40}$$

This expression is particularly useful during a phase change when P and T are often fixed. Under stable conditions, when the equality sign applies, it is known as the Gibbs equation.

The most common thermodynamic surfaces used in the study of latent heat transfer are the stable $P\mathcal{V}T$ and TSP surfaces for a pure substance restricted to $đW = P\,\mathrm{d}\mathcal{V}$. Figure 2.5 illustrates the $P\mathcal{V}T$ surface for a typical substance which contracts on freezing. Latent heat is represented by the ruled saturation surfaces. Figure 2.6 illustrates a typical TSP surface. These surfaces will be used repeatedly in later chapters. Less familiar is the corresponding $\mathcal{G}PT$ surface shown in Fig. 2.7. By contrast with the $P\mathcal{V}T$ and TSP surfaces this does not contain ruled saturation surfaces. As the Gibbs equation (2.40) implies, $\mathcal{G} = \mathcal{G}(T, P)$ for a pure substance with $đW = P\,\mathrm{d}\mathcal{V}$. It therefore follows that saturation conditions with $P_{\mathrm{sat}} = P(T_{\mathrm{sat}})$ will create a single curve for $\mathcal{G}_{\mathrm{sat}}(T)$ or $\mathcal{G}_{\mathrm{sat}}(P)$ separating single phase surfaces on either side. More will be said about these surfaces later.

The unique relation between T_{sat} and P_{sat} on a saturation curve no longer exists if the substance consists of several components. The concentration x of each component then influences the range of temperatures and pressures over which phase change occurs. Even in the simplest situation, with stable states being occupied by only two components, this converts a saturation curve into a saturation surface, following the phase rule. The TxP surfaces of binary solutions are usually presented as Tx curves in isobaric planes. Two representative examples are given below.

Figure 2.8 illustrates the phase diagram of a condensed binary mixture in which a solute of concentration x_2 is dissolved in a solvent of concentration x_1. Suppose the mixture is a liquid, with an initial solute concentration x_2^{a}, and assume it is cooled until freezing begins at the point a on the liquidus curve AE. The first solid to appear will have a solute concentration of x_2^{b} determined by the point b at the same temperature on the solidus curve AC. When the solute is insoluble in the solid phase (e.g. salt in ice) the curve AC coincides with the vertical AB and the solid formed is pure solvent, i.e. $x_1 = 1$. Since freezing depletes the liquid solution of solvent, the solute concentration in the remaining solution increases above x_2^{a}. This has the effect of lowering the solution freezing temperature to some point further down the liquidus. For freezing to continue, the solution temperature must be progressively lowered, a process which may be continued until

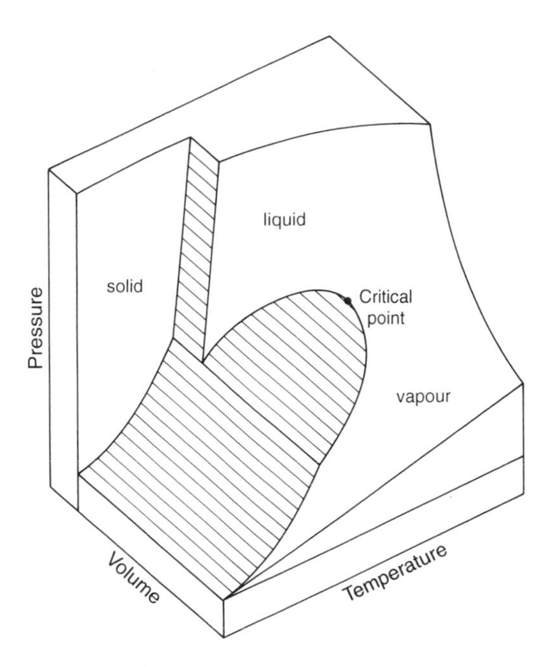

Fig. 2.5 Stable equilibrium *P*V*T* surface for a pure substance showing the ruled saturation surfaces.

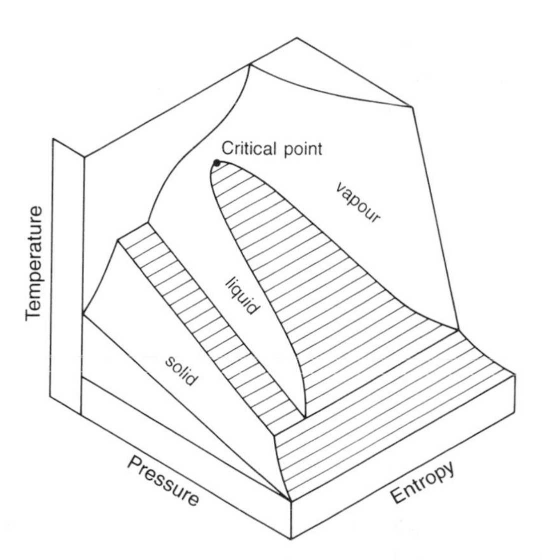

Fig. 2.6 Stable equilibrium *TSP* surface for a pure substance showing the ruled saturation surfaces (after Faires 1970).

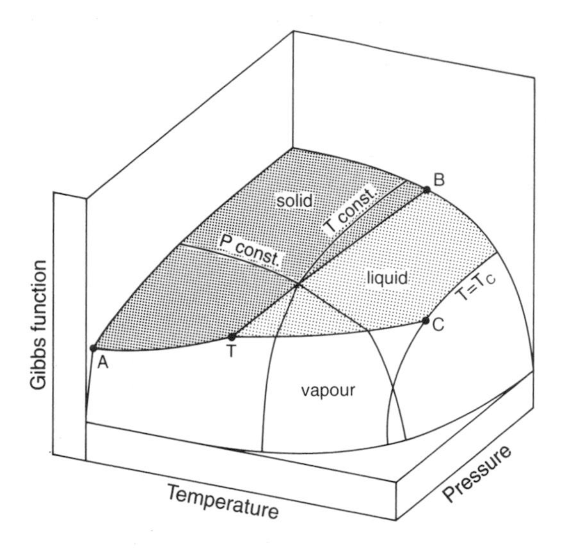

Fig. 2.7 Stable equilibrium $\mathcal{G}TP$ surface for a pure substance showing the saturation curves AT, BT, and CT. Note the absence of ruled surfaces.

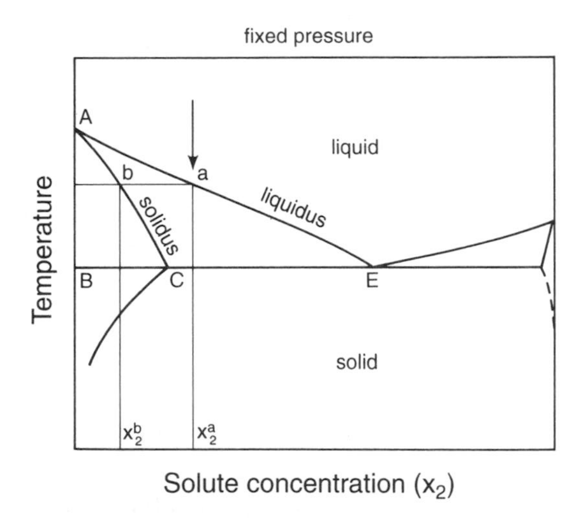

Fig. 2.8 Phase diagram for a binary solution of condensed phases.

the eutectic temperature T_E is reached at E. Below this temperature the mixture is completely solid.

A second example is provided by Fig. 2.9 which illustrates a liquid binary mixture undergoing evaporation. Again assuming the initial solute concentration to be x_2^a, the effect of evaporation at a (on the bubble point curve) is to produce a vapour at b (on the dew point curve) with a solute concentration $x_2^b < x_2^a$. Since this strengthens the solute concentration in the

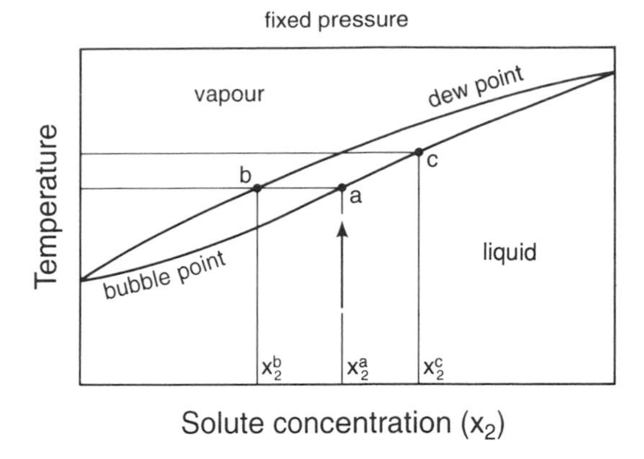

Fig. 2.9 Binary phase diagram for fluid phases.

remaining liquid solution, the temperature must be raised for evaporation to continue. Similar observations apply to condensation from a vapour of concentration x_2^a; the first liquid to condense has a solute concentration of x_2^c. For a non-volatile solute, the vapour is pure solvent. The resulting bubble point curve is explored in problems Q2.4 and 2.5 in Chapter 8.

From both of the above examples, it is evident that phase change in multicomponent systems occurs over a temperature range, even under stable, isobaric conditions. From a heat transfer standpoint this complicates the analysis, particularly near the interface where mass transfer will generally be occurring.

2.5 The Clapeyron equation

The ruled saturation surfaces discussed above are a fundamental feature of latent heat transfer processes because they refer to the co-existence of phases. The extent of these surfaces, and the relationships between the principal thermodynamic variables which govern them, are thus central to our purpose. In this section, the equations of these surfaces will be explored for a pure substance.

To begin with, it is important to recall that the surfaces represent maps of stable, equilibrium states for which, in particular,

$$d\mathcal{G} - \mathcal{V}\,dP + S\,dT = 0.$$

During a phase change, neither the pressure nor the temperature vary, and hence

$$d\mathcal{G} = 0 \tag{2.41}$$

represents stable equilibrium. Consider a mixture of two pure phases 1 and 2 with a total mass M. If x_1 is the mass fraction of phase 1,

$$\mathcal{G} = Mx_1 g_1 + M(1 - x_1)g_2,$$

where g_1 and g_2 are the specific Gibbs functions of phases 1 and 2, respectively. Any virtual change δx_1 in x_1 creates a change in $\mathcal{G}$ given by

$$\mathrm{d}\mathcal{G} = Mg_1\,\delta x_1 - Mg_2\,\delta x_1 = M(g_1 - g_2)\delta x_1 .$$

However, since $\mathrm{d}\mathcal{G} = 0$ under stable equilibrium conditions, this relation may only be satisfied for an arbitrary change δx_1 if

$$g_1 = g_2 . \tag{2.42a}$$

That is, stable equilibrium requires equality of the two specific Gibbs functions. Stated differently, it requires no discontinuity in g at the interface. This feature is present in Fig. 2.7 where $\mathcal{G} = gM$. More generally, for component k in a multicomponent system,

$$\mu_1^k = \mu_2^k, \tag{2.42b}$$

where $\mu^k = (\partial\mathcal{G}/\partial M_k)_{p,T}$ is the chemical potential of component k.

Now if, $\mathcal{G} = \mathcal{G}(T, P)$ then $g = g(T, P)$ in either phase. Therefore

$$\mathrm{d}g = \left(\frac{\partial g}{\partial T}\right)_p \mathrm{d}T + \left(\frac{\partial g}{\partial P}\right)_T \mathrm{d}P,$$

where

$$v = \left(\frac{\partial g}{\partial P}\right)_T \quad \text{and} \quad s = -\left(\frac{\partial g}{\partial T}\right)_p$$

define the specific volume and specific entropy, respectively. These change discontinuously at the interface, as indicated in Fig. 2.7. The situation may thus be described as a *first-order* phase change because discontinuity occurs not in the specific Gibbs function but in its first derivatives. This is the only type of phase change considered in this book. It is the only one exhibiting the entropy change which is characteristic of latent heat and expressed in the simple relation $\lambda = T\Delta s$. Figure 2.10 illustrates typical variations of g in the vicinity of saturation conditions. These follow from Fig. 2.7.

Under saturated conditions, $T = T_{\mathrm{sat}}$, $P = P_{\mathrm{sat}}$, and the specific Gibbs function at any point on the saturation curve is given by $g(T_{\mathrm{sat}}, P_{\mathrm{sat}})$ for both phases; $g_1 = g_2$. Close by on the saturation curve, if $g = g_1 + \mathrm{d}g_1 = g_2 + \mathrm{d}g_2$, it follows that $\mathrm{d}g_1 = \mathrm{d}g_2$ describes an incremental change during which $P_1 = P_2$ and T_1 and T_2, and hence $\mathrm{d}P_1 = \mathrm{d}P_2$ while $\mathrm{d}T_1 = \mathrm{d}T_2$. But

$$\mathrm{d}g = v\,\mathrm{d}P - s\,\mathrm{d}T,$$

and therefore

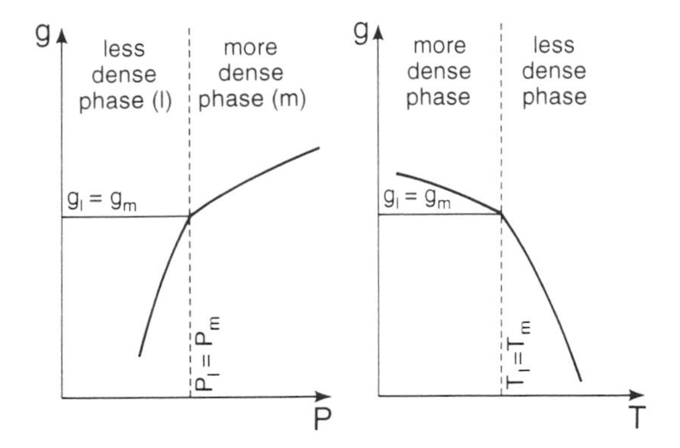

Fig. 2.10 Specific Gibbs function dependence on temperature and pressure.

$$(v_1 - v_2)\,\mathrm{d}P = (s_1 - s_2)\,\mathrm{d}T,$$

or

$$\frac{\mathrm{d}P}{\mathrm{d}T} = \frac{s_1 - s_2}{v_1 - v_2}.$$

Substituting

$$\lambda_{12} = T(s_2 - s_1),$$

we obtain

$$\frac{\mathrm{d}P}{\mathrm{d}T} = \frac{\lambda_{12}}{T(v_2 - v_1)} \tag{2.43}$$

as the equation of the saturation curve for a first-order phase change. It is usually called the *Clapeyron equation*.

The three saturation curves for a pure substance are shown in Fig. 2.11. Typically, if a supply of heat converts phase 1 into phase 2, it is found that $v_2 > v_1$ and $s_2 > s_1$. From eqn (2.43), this implies that $\mathrm{d}P/\mathrm{d}T > 0$ as shown. The ice (1)–water (2) transition, shown as a broken line, is an important exception in which $v_2 < v_1$, even though $s_2 > s_1$.

When phase 2 is a vapour, $v_2 \gg v_1$ and eqn (2.43) may be approximated by

$$\frac{\mathrm{d}P}{\mathrm{d}T} \simeq \frac{\lambda_{12}}{Tv_2} \simeq \frac{P\lambda_{12}}{RT^2} \tag{2.44}$$

if the ideal gas equation is used for v_2. This expression, often called the *Clausius–Clapeyron* equation, provides a convenient and simple representation of the sublimation and evaporation curves but it must be borne in mind

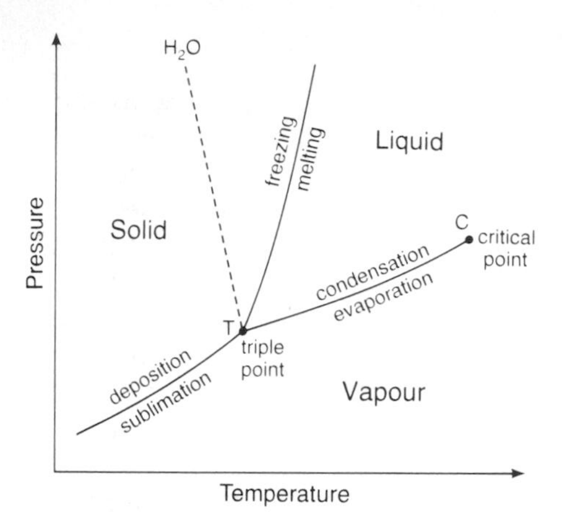

Fig. 2.11 Ruled surfaces of a pure substance as viewed along the volume axis.

that the ideal gas equation is not very accurate except at low pressures. Equation (2.44) is a much better approximation near the triple point than at the critical point.

2.6 Systemic stability

2.6.1 Equilibrium and stability

Equilibrium is a difficult concept to explain but is best approached in terms of balance: specifically, the balance, or lack of balance, between a system and its immediate environment. In *unstable* equilibrium, which follows the dictates of inequality (2.38), a spontaneous and substantial change in the system may occur with no change in the environment. This usually corresponds to the non-uniformity of at least one intensive variable within the system causing alterations in at least one extensive variable, thereby relieving the non-uniformity. Such behaviour is irreversible. For *stable* equilibrium, the intensive variables are either uniform or held unchanged by internal constraints which prevent alterations in the corresponding extensive variable. Under these conditions, systemic changes are caused only by environmental changes and vice versa. Systemic changes then occur in exact and reproducible correspondence to environmental changes. Behaviour is reversible. Processes following stable equilibrium states satisfy the limiting form of relation (2.38) in which

$$\mathrm{d}E + đW - T\,đS = 0. \tag{2.45}$$

We define *metastable* equilibrium as stable equilibrium restricted to small systemic and environmental changes which, if exceeded, lead to instability. The interpretation of ‘small’ is considered below.

2.6.2 *Metastable and unstable equilibrium*

In practice, systemic changes commonly occur as the result of unstable equilibrium. Imbalances in the intensive variables become great enough to overcome constraints. Friction ceases to prevent motion; anticatalysts cease to prevent chemical reactions; stratification ceases to prevent convection; boundary walls cease to prevent expansion. In these circumstances, rates of change may be very high; time is then an important variable. However, there are also many circumstances where events proceed slowly enough that departures from stable equilibrium are small. In the limit, a series of unstable states may closely approximate a stable equilibrium path; time then ceases to be important. The thermodynamic surfaces discussed earlier satisfy these conditions.

Figure 2.12 is an isothermal slice through a ruled surface on the $P\mathcal{V}T$ diagram displayed earlier in Fig. 2.5. In this familiar representation, the more dense phase, starting at point *M*, may be made to expand along the isotherm MA under stable conditions. Also under stable conditions, the less dense phase, starting at point L, may undergo compression on the same isotherm along LC. These single-phase paths are reversible, as is the 'ruled' phase change path AC connecting them at the saturation pressure. However, isothermal compression from L to C may be extrapolated beyond to the point C′: that is, CC′ is a metastable path. Depending upon the circumstances, this 'small' excursion beyond stable equilibrium conditions may be enough to trigger a phase change, causing a rapid alteration along

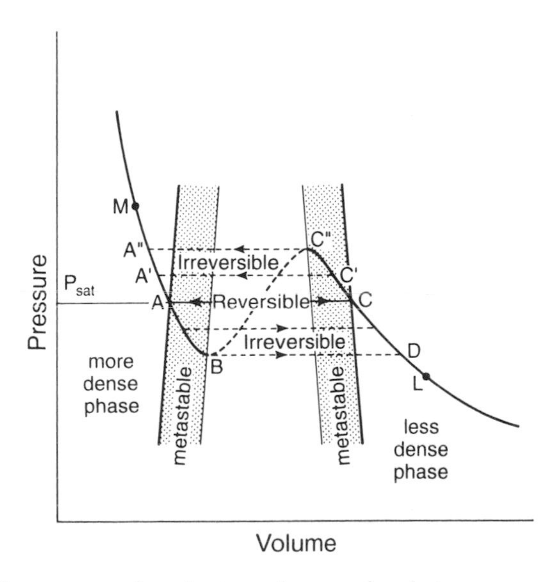

Fig. 2.12 Isotherm traversing the two-phase region between a more dense phase and a less dense phase.

the unstable (and irreversible) path C′A′. Metastable excursions like this exhibit a natural limit at which phase change occurs spontaneously; this limit is indicated by the point C″ in Fig. 2.12.

Similar arguments may be used to explain the metastable extension AB and the possible irreversible paths from the curve AB to curve CD. A different path is shown broken from B and C″. Along this path $(\partial P/\partial \mathcal{V})_T > 0$, and thus in contrast to the stable and metastable paths, where $(\partial P/\partial \mathcal{V})_T < 0$, the path BC″ is unstable. For completeness, the metastable extensions must be added to the $P\mathcal{V}T$ amd TSP stable equilibrium surfaces. Figure 2.13 provides a summary of these extensions in relation to the stable equilibrium saturation surfaces discussed earlier (see Fig. 2.11). Extensions of vapour states beyond the saturation curves are usually described as supersaturation (pressures higher than the saturated value) or supercooling (temperatures lower than the saturated value). Supercooling is the more common term for extensions of liquid states below the fusion curve. Metastable extensions of condensed phases, whether solid or liquid, are invariably described as superheating. It is interesting to note that the ruled surfaces themselves are capable of metastable extensions, as suggested by the dotted curves in Fig. 2.13.

Extensions of the $\mathcal{G}TP$ surface also provide valuable insight, especially in the study of nucleation. For a pure substance, $g = \mu$, and hence the curves presented earlier in Fig. 2.10 may be redrawn in terms of μ. Figure 2.14 provides an example using the same labelling of points as employed in Fig. 2.12. The (stable) curve MA, for instance, extrapolates to the (metastable) curve AB. Similarly, the stable curve LC may be extended along the metastable curve CC″. The points A and C are coincident in the μP plane. The irreversible paths in this plane are all downward. That is, the chemical potential of the metastable phase is always greater than that of the stable phase, thus creating a situation which is *potentially* unstable. This is true for any phase combination. Metastable excursions may therefore be viewed as chemical potential excesses. These excesses may be relaxed artificially through nucleation promoters or, if great enough, they may be relieved spontaneously. Such events are the subject of the next section.

2.7 Nucleation

Any metastable excursion beyond the saturation point, whether in the solid, the liquid, or the vapour phase, can proceed only so far before a natural limit is reached. This limit, indicated by points B and C″ on Figs. 2.12 and 2.14, represents a threshold beyond which it is physically impossible for the substance to remain in a metastable condition. On the threshold, the substance is in unstable equilibrium and will undergo a spontaneous and irreversible change to another phase. The threshold therefore defines the

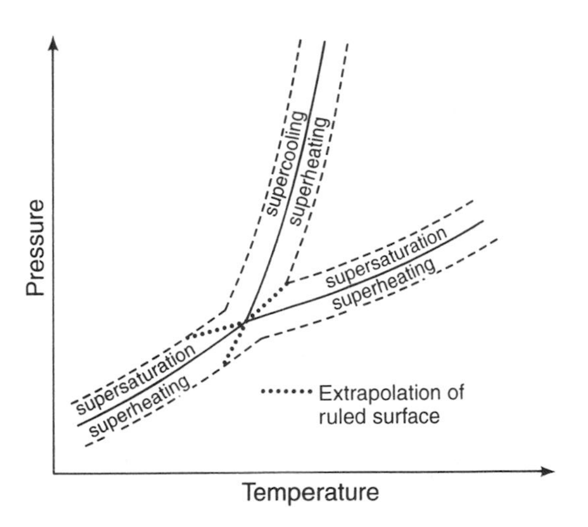

Fig. 2.13 Metastable excursions of the $P\mathcal{V}T$ surfaces for a pure substance.

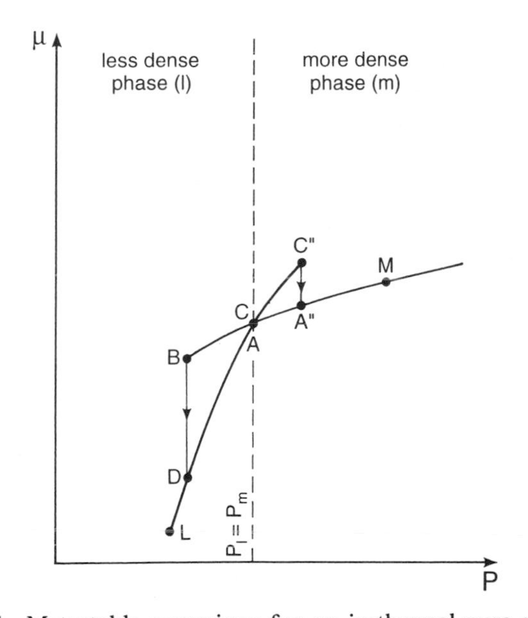

Fig. 2.14 Metastable excursions for an isothermal pure substance.

maximum extent of the metastable departure from saturated conditions ($\mu - \mu_{sat}$) and thus prescribes the meaning of 'small' in the definition of metastable equilibrium. Below this threshold, a phase change may be catalysed or promoted by means which will be discussed later in this section.

A spontaneous change of phase in the bulk of a substance is known as *homogeneous nucleation*. Its molecular origin is best understood by examining the formation of a pure condensed phase. Consider, for example, the transition from vapour to liquid during condensation. Random molecular motion in the vapour phase causes chance encounters between vapour molecules thereby giving rise to two opposing effects:

(1) the tendency for molecules to join or cluster together in the liquid phase;

(2) the tendency of collisions to break up or disintegrate the clusters.

The relative importance of these two effects determines the most probable size and life span of the molecular clusters. As the vapour temperature is lowered, or the pressure increased, the first effect grows while the second reduces; the most probable size then increases, along with the cluster life span. Eventually, the clustering effect dominates and liquid *embryos* grow spontaneously into droplets.

To determine the circumstances under which this spontaneous growth occurs, consider the equilibrium of a liquid cluster surrounded by its metastable vapour. The Gibbs eqn (2.40) for this system requires that while (meta)stable conditions exist

$$d\mathcal{G} = \mathcal{V}dP - SdT + \sum_i F_i dX_i, \tag{2.46}$$

where F_i, X_i are appropriate pairs of work variables, other than P and $\mathcal{V}$ which are already incorporated through compression work. When the vapour pressure and temperature are fixed at given supersaturated values, the chemical potential difference $\Delta g = g_L - g_V < 0$ is also fixed. Variations in the cluster Gibbs function are then attributable to chemical work (vapour molecules joining or leaving the cluster) and surface tension work (changes in the surface area of the cluster). Equation (2.46) may then be written

$$d\mathcal{G} = \Delta g \, dM + \sigma \, dA \tag{2.47}$$

in which the cluster mass and surface area are represented by M and A respectively; σ is the surface tension.

For a cluster of any given size, eqn (2.47) may be integrated to yield

$$\mathcal{G} = n\Delta\mu + \sigma A, \tag{2.48}$$

where μ is the chemical potential (or Gibbs function) per molecule and n is the number of molecules in the cluster. Since $n = 4\pi R^3 n_\mathcal{v}/3$, where R is the cluster radius and $n_\mathcal{v}$ is the number of liquid molecules per unit volume, eqn (2.48) may be rewritten

$$\mathcal{G} = n\Delta\mu + Kn^{2/3}, \tag{2.49}$$

noting that $A = 4\pi R^2$ and $K(n, \sigma)$ is a geometrical coefficient depending on the cluster shape. The molecular chemical potential difference is given by $\Delta\mu = \mu_L - \mu_V < 0$.

Figure 2.15 illustrates the form of eqn (2.49) for various values of μ_V measured relative to saturated conditions where $\mu_L = \mu_{sat}$. When $\mu_V = \mu_L$, i.e. saturated conditions exist, the curve is monotonically increasing. Under supersaturated conditions ($\mu_V > \mu_L$), the curve exhibits a maximum where $d\mathcal{G}/dn = 0$, i.e.

$$\mathcal{G}_{max} = \frac{16\pi\sigma^3}{3n_v^2(\Delta\mu)^2} \tag{2.50}$$

and

$$R_{max} = \left(\frac{3n_{max}}{4\pi n_v}\right)^{1/3} = -\frac{2\sigma}{n_v\Delta\mu}. \tag{2.51}$$

In the vicinity of this maximum, $d\mathcal{G} < 0$ which, according to the general relation (2.40), implies that the cluster has become an unstable embryo: the loss of a few molecules causes spontaneous disintegration while a corresponding gain creates spontaneous growth. When $\mu_V \simeq \mu_{sat}$, and the most probable cluster size is much less than R_{max}, such growth is limited to very few clusters. However, as μ_V rises above μ_{sat}, R_{max} decreases while the most probable cluster size increases until many embryonic clusters become unstable. We thus arrive at the homogeneous nucleation point when μ_V reaches the threshold value $\mu_V{}^*$ and $\mathcal{G}_{max} = \mathcal{G}^*$ while $R_{max} = R^*$.

The basic predictions of cluster theory also apply to melting, freezing, evaporation, and sublimation, although the details may vary. During solidification in particular, intermolecular forces not only bind the cluster

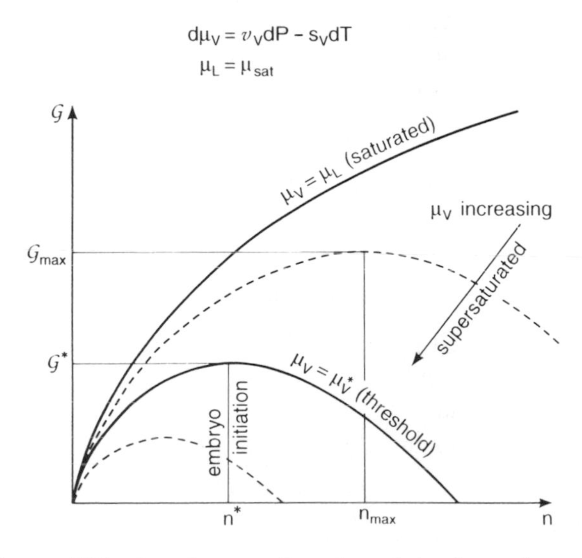

Fig. 2.15 Cluster Gibbs function as a function of size for various value of μ_V per molecule.

together, but act selectively at the cluster surface. Crystal embryos, for example, are assembled in an ordered and definite way determined by the geometry and force fields of the molecules. Since these are unlikely to produce a spherical solid embryo the geometrical coefficient K in eqn (2.49) must be modified accordingly. Intermolecular forces at the surface of a cluster may also establish energy barriers which inhibit growth and thus remove the homogeneous nucleation point further from saturated conditions. Even so, the basic clustering process is unaltered.

Foreign materials also influence the way in which molecules may be accommodated into the cluster. They may, for example, dissolve at the surface, altering the interfacial energy density, i.e. the surface tension. This alters both the critical cluster size R^* and the energy barrier $\mathcal{G}^*$, as suggested by eqns (2.50) and (2.51). The most common effect of foreign materials is to promote nucleation by their presence in the bulk of the metastable phase as dispersed particles or, even more commonly, as bounding surfaces such as tube walls. They then reduce the potential energy barrier mentioned above and thus facilitate the growth of clusters in the presence of small, and often negligible, departures from saturated conditions.

The catalytic process which moves the phase transition from the homogeneous nucleation point towards the equilibrium saturation point defines *heterogeneous nucleation*. The influence of a foreign substrate complicates the details of nucleation but the process may still be described by cluster theory. It has been suggested, for example, that a cluster may then be treated as a cap-shaped island on the surface of the substrate. As with the spherical cluster, the Gibbs function may again be written in terms of the chemical and surface tension work, now referred to the island. An optimum value of $\mathcal{G}$ is once more observed if $\mu_{sat} - \mu_V < 0$ but the free energy threshold now depends upon more than the surface tension σ_{CM} between the cluster and the metastable phase. It is also influenced by the substrate–cluster surface (σ_{SC}) and the surface between the substrate and the metastable phase (σ_{SM}); that is, the degree of wetting given in eqn (2.28) becomes important.

Finally, it is important to distinguish carefully between two very different reasons why nucleation may be delayed or suppressed. The first of these concerns a shift in the *stable, equilibrium saturation condition*: for example, boiling point raising caused by solute addition. The system is always in stable equilibrium. For the second, the equilibrium saturation condition is fixed but an alteration occurs in the *metastable range*: for example, through the introduction of particulates or substrate vibration. Unlike a change in stable saturation conditions, a change in the metastable range may alter the hysteretic characteristics accompanying supersaturation and superheating, as illustrated by the various paths noted in Fig. 2.12.

Selected bibliography

Arpaci, V. S. and Larsen, P. S. (1984). *Convection heat transfer.* Prentice Hall, Englewood Cliffs, New Jersey.

Defay, R. and Prigogine, L. (1966). *Surface tension and adsorption.* (trans. D. H. Everett) Wiley, New York.

Faires, V. M. (1970). *Thermodynamics.* MacMillan, New York.

Fletcher, N. H. (1970). *The chemical physics of ice.* Cambridge University Press, Cambridge.

Green, A. E. and Rivlin, R. S. (1964). On Cauchy's equations of motion. *ZAMP*, **15**, 290–92.

Hsieh, J. S. (1975). *Principles of thermodynamics.* Scripta, Washington, DC.

Lock, G. S. H. (1990). *The growth and decay of ice.* Cambridge University Press, Cambridge.

Malvern, L. E. (1969). *Introduction to continuum mechanics.* Prentice Hall, Englewood Cliffs, New Jersey.

Pippard, A. B. (1961). *Classical thermodynamics.* Cambridge University Press, Cambridge.

Pruppacher, H. R. and Klett, J. D. (1980). *Microphysics of clouds and precipitation.* Reidel, Dordrecht.

Zemansky, M. W. (1957). *Heat and thermodynamics.* McGraw-Hill, New York.

Exercises

1. Consider a virtual displacement δR of the spherical interface surrounding a vapour bubble. Using the concept of an interfacial energy $E_I = 4\pi R^2 \sigma$, show that the pressure difference across the interface is given by $\Delta P = 2\sigma/R$.
2. Write down steady-state expressions for the velocity and mass flux *density* of material moving *relative to the interface*:

   (a) as it enters the interface and leaves the growing phase;

   (b) as it leaves the interface and enters the parent phase.
3. Use the above expressions and the steady flow version of the first law of thermodynamics applied to an interfacial control volume to confirm eqn (2.34). Assume that the interface acceleration does not affect the result.
4. Demonstrate that in the vicinity of the interface formed during condensation of a pure vapour, $\beta_i - \beta_o \to 0$ as $\dot{m}_{cond} \to 0$. Hence confirm eqn (2.35).
5. Construct a μT diagram illustrating isobaric, metastable excursions from saturated conditions using labelling similar to that in Fig. 2.12. Hence demonstrate that the chemical potential of the metastable phase is greater than that of the stable phase.
6. Calculate dP/dT for H_2O at the triple point and when $P_{sat} = 0.844\, P_C$:

   (a) using the Clausius equation;

(b) using the Clausius–Clapeyron equation.

(Ans: 44.44 Pa K^{-1} and 44.37 Pa K^{-1} at the triple point; 2.25×10^5 Pa K^{-1} and 7.25×10^4 Pa K^{-1} at 0.844 P_C.)

7. Show that the threshold value of the specific chemical potential excess $(\mu_V^* - \mu_{sat})$ during isothermal supersaturation of a vapour is given by

$$\mu_V^* - \mu_{sat} = RT \ln(P_V^*/P_{sat})$$

where P_V^* is the homogeneous nucleation pressure.

8. Show that the threshold value of the specific chemical potential excess during isobaric supercooling of a liquid is given by

$$\mu_L^* - \mu_{sat} = \frac{\lambda}{T}(T_{sat} - T_L^*)$$

where T_L^* is the homogeneous nucleation temperature.

3
THE DYNAMICS OF HETEROGENEOUS FLUIDS

This chapter is devoted to the dynamics of heterogeneous fluid media under steady or quasi-steady conditions. It provides an exploration and, to some extent, a review of the various hydrodynamic phenomena which accompany phase change. The organization is based largely on observations commonly made during condensation and evaporation. While the accompanying fluid dynamic phenomena take many forms, they may be subdivided into two general categories: *bipartitioned* and *dispersed*. In the first of these, two continuous and homogeneous phases are separated by an extensive and continuous interface; this is typified by film condensation. In the second, one phase is continuous but the other is dispersed through it in numerous small pockets; a good example is provided by vapor bubbles during boiling. It is reasonable to expect that bipartitioned and dispersed systems will behave very differently. Sections 3.1, 3.2, and 3.3 deal with the former while Sections 3.4 and 3.5 deal with the latter. Section 3.6 explores systems which combine the characteristics of both.

In any given phase, treated as a continuum with constant properties, the continuity eqn (2.15) may be re-written as

$$\nabla \cdot \mathbf{V} = 0. \tag{3.1}$$

In the same medium, eqn (2.14), the equation of motion, is

$$\rho \frac{\partial \mathbf{V}}{\partial t} + \rho(\mathbf{V} \cdot \nabla)\mathbf{V} = \nabla \cdot \pi + \rho \mathfrak{F}. \tag{3.2}$$

Decomposing the fluid stress π into isotropic (pressure) and deviatoric (shear) parts,

$$\nabla \cdot \pi = -\nabla P + \nabla \cdot \tau$$

in which the shear term simplifies to $\nabla \cdot \tau = \mu \nabla^2 \mathbf{V}$ for a Newtonian fluid satisfying the incompressible form of the continuity eqn (3.1). Thus, under steady conditions,

$$\rho(\mathbf{V} \cdot \nabla)\mathbf{V} = -\nabla P + \rho \mathfrak{F} + \mu \nabla^2 \mathbf{V}. \tag{3.3}$$

Equations (3.1) and (3.3) are sufficiently general to apply to a very wide range of latent heat transfer situations. They will be used repeatedly throughout this book to uncover the basic characteristics of latent heat transfer by analysing behaviour on both sides of the interface.

3.1 Smooth laminar films

3.1.1 The planar falling film

Consider the steady flow of a liquid film slowly running down the plane vertical surface (height L) shown in Fig. 3.1. This is a two-dimensional system for which the longitudinal component of eqn (3.3) applied to the liquid reduces to

$$\rho\left(U\frac{\partial U}{\partial X}+V\frac{\partial U}{\partial Y}\right)=-\frac{\partial P}{\partial X}+\rho g+\mu\left(\frac{\partial^2 U}{\partial Y^2}+\frac{\partial^2 U}{\partial X^2}\right) \tag{3.4}$$

where ρ is the liquid film density. The lateral equation of motion may be neglected if the film is thin. Outside of the film, ρ_∞, U_∞, and P_∞ are the density, velocity, and pressure, respectively, of ambient vapour, here taken to be an incondensable and insoluble gas. When the gas is quiescent, the application of eqn (3.3) yields $\mathrm{d}P_\infty/\mathrm{d}X=\rho_\infty g$, which describes the hydrostatic gas pressure then imposed on the liquid at the interface. Since the liquid is allowed to fall freely, $\partial P/\partial X\simeq \mathrm{d}P_\infty/\mathrm{d}X$, and hence eqn (3.4) may be re-written

$$\rho\left(U\frac{\partial U}{\partial X}+V\frac{\partial U}{\partial Y}\right)=g\Delta\rho+\mu\left(\frac{\partial^2 U}{\partial Y^2}+\frac{\partial^2 U}{\partial X^2}\right) \tag{3.5}$$

in which $g\Delta\rho=g(\rho-\rho_\infty)$ is the *Archimedean buoyancy* driving the liquid downwards. Typically, $\rho\gg\rho_\infty$ but it is important to recognize that the buoyancy vanishes whenever $\rho=\rho_\infty$: for example, at the critical point of a liquid and its saturated vapour.

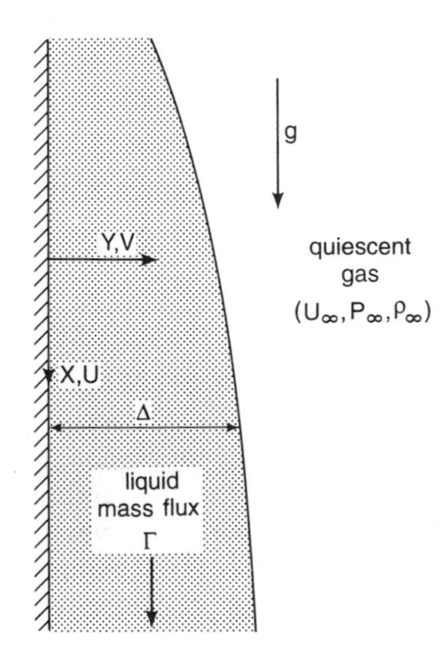

Fig. 3.1 Thin liquid film of width Δ falling over a plane, vertical surface of height L.

Without solving eqn (3.5), we may use the scale analysis developed in Chapter 1 to determine the film characteristics by introducing the natural length and velocity scales X^c, Y^c, U^c, and V^c. Only the first of these is known: that is $X^c = L$, the specified height of the plane. The lateral length scale Y^c is the film thickness Δ which is connected to the longitudinal velocity scale through the expression

$$\Gamma = O(\rho Y^c U^c) \tag{3.6}$$

where Γ is the liquid mass flux per unit surface width (normal to the page). Each of the primitive variables may now be normalized by taking

$$x = \frac{X}{X^c}, \quad y = \frac{Y}{Y^c}, \quad u = \frac{U}{U^c}, \quad v = \frac{V}{V^c},$$

thus enabling eqn (3.5) to be re-written

$$\left[\rho \frac{U^c (Y^c)^2}{\mu X^c}\right]\left(u\frac{\partial u}{\partial x} + v\frac{\partial u}{\partial y}\right) = \left[\frac{(Y^c)^2 g\Delta\rho}{\mu U^c}\right] + [1]\frac{\partial^2 u}{\partial y^2} + \left[\frac{Y^c}{X^c}\right]^2 \frac{\partial^2 u}{\partial x^2} \tag{3.7}$$

after V^c has been eliminated using the continuity eqn (3.1). The square-bracketed coefficients are all non-dimensional. Now since buoyancy is the main cause of motion, while viscous shear is the main impediment, the first coefficient on the right-hand-side of eqn (3.7) must have the same order of magnitude as the second: that is, since the variables are normalized, we may expect

$$\frac{(Y^c)^2 g\Delta\rho}{\mu U^c} = O(1). \tag{3.8}$$

Equations (3.6) and (3.8) now provide enough information to determine the unknown scales Y^c and U^c. We find that

$$\Delta = \left(\frac{\mu\Gamma}{\rho g\Delta\rho}\right)^{1/3} = L(Re_f/Ar)^{1/3} \tag{3.9}$$

and

$$U^c = \left(\frac{g\Gamma^2\Delta\rho}{\rho^3\nu}\right)^{1/3} = \frac{\nu}{L}(Re_f^2 Ar)^{1/3} \tag{3.10}$$

in which $Re_f = \Gamma/\mu$ is the film Reynolds number, while $Ar = L^3 g\Delta\rho/\rho\nu^2$ is the Archimedes number. The thickness and velocity scales are both seen to be independent of position on the surface length; both increase with the mass flux, though not in exactly the same way. It is also evident that whereas an increase in buoyancy increases the liquid velocity, it decreases the film thickness. If required, the lateral velocity scale V^c may be found from the continuity equation.

These scales now make it easier to examine solutions to the equation of motion. For example, with a thin film, i.e. $\Delta \ll L$, or $Re_f \ll Ar$, the second (longitudinal) viscous term is effectively suppressed. Likewise, if the mass flow rate Γ is small, i.e. $Re_f \ll Ar^{1/4}$, the effect of inertia is also suppressed. Under these conditions, eqn (3.7) reduces to

$$0 = 1 + \frac{\partial^2 u}{\partial y^2} \tag{3.11}$$

which has a simple parabolic solution. For a 0.25 mm thick condensate film, for example, we note that $Re_f = 16.7$ while $Ar^{1/4} = 13.4 \times 10^2$ for a 1.0 m high plate.

3.1.2 The annular film

The flow of gas over a liquid film is capable of altering its behaviour. This situation is frequently found in a pipe where the gas may form a moving core surrounded by an annular liquid film. When the pipe is vertical, for example, the principal driving terms in eqn (3.4) may be re-written

$$-\frac{\partial P}{\partial X} + \rho g = -\frac{\partial P_S}{\partial X} + g\Delta\rho = \beta \tag{3.12}$$

where P_S is the gas static pressure in excess of its hydrostatic pressure. This enables us to substitute β for $g\Delta\rho$ in eqns (3.5) to (3.10). Following the same procedure we obtain the more general expressions

$$U^c = \left(\frac{\Gamma^2\beta}{\rho^3\nu}\right)^{1/3} \tag{3.13}$$

and

$$\Delta = \left(\frac{\nu\Gamma}{\beta}\right)^{1/3}. \tag{3.14}$$

From these scale equations it appears that the characteristics of the film are the same whether it is driven by Archimedean buoyancy ($\beta = g\Delta\rho$) or pressure gradient ($\beta = -\partial P_S/\partial X$). On closer examination, however, it is evident that the sign of these driving forces may also be important. For gas flowing upward in a vertical pipe, the pressure gradient opposes buoyancy. Increasing the gas flow rate then decreases the net downward force on the liquid film until, eventually, it is reduced to zero, i.e. $\beta = 0$. Equations (3.13) and (3.14) signal that the film would then be infinitely thick and would cease moving. This simple, qualitative prediction heralds a flow reversal which will be discussed in more detail later.

Once the pressure gradient dominates Archimedean buoyancy, the orientation of the pipe is unimportant. This situation is described in Fig. 3.2. Again assuming a balance between the driving force and the retarding viscous forces, the integral form of the momentum eqn (2.13) may be applied to the annular film under steady conditions. When the film is thin, i.e. $\bar{\Delta} \ll D$, this yields

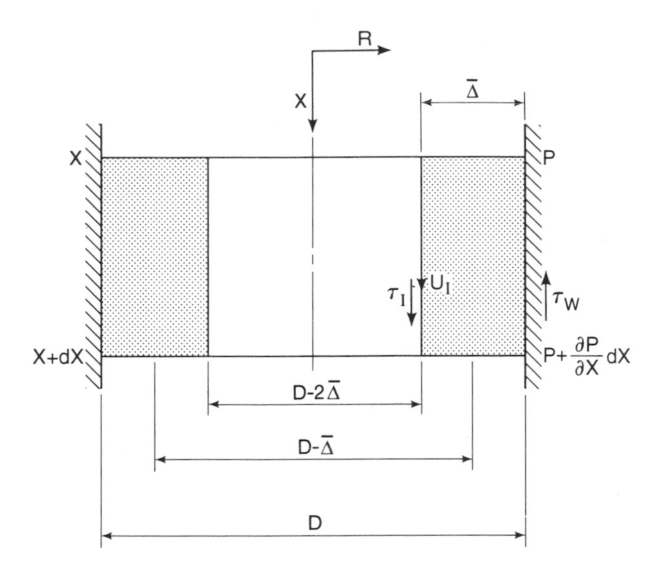

Fig. 3.2 Forces acting on an elementary liquid annulus of thickness $\bar{\Delta}$. Gravity is neglected.

$$\tau_w = \tau_I - \bar{\Delta}\frac{\partial P_g}{\partial X}, \tag{3.15}$$

where $\bar{\Delta}$ is the mean film thickness over the pipe length and τ_w, τ_I are the shear stresses acting at the pipe wall and interface, respectively: $P = P_g$ the gas pressure. The wall shear stress is thus attributable to the interfacial shear stress (the gas pressure gradient acting indirectly) and the pressure gradient acting directly.

Under these conditions, $\partial P_g/\partial X \leqslant 0$ and hence $\tau_w \geqslant \tau_I$. From a gas force balance, $\tau_I \simeq -(D/4)\partial P_g/\partial X$, which enables us to re-write eqn (3.15) as

$$\tau_w \simeq -\frac{\partial P_g}{\partial X}\left(\frac{D}{4} + \bar{\Delta}\right). \tag{3.16}$$

This emphasizes the dominance of interfacial shear over the pressure gradient in a thin film. In such a film, $\tau_w \simeq \mu U_I/\bar{\Delta}$ or, since $\Gamma = \rho U_I \bar{\Delta}/2$,

$$\tau_w \simeq \frac{2\nu\Gamma}{(\bar{\Delta})^2}. \tag{3.17}$$

Combining this with eqn (3.16), we obtain

$$\bar{\Delta} \simeq \left(\frac{-8\nu\Gamma}{D\partial P_g/\partial X}\right)^{1/2} \tag{3.18}$$

providing the film is thin. We thus establish a simple triangular relation between the gas pressure gradient (which controls the gas flow rate), the film flow rate and the film thickness. However, as the film thickens the direct effect of pressure gradient increases, the film velocity profile ceases to be linear and $\tau_w \simeq \tau_I$ no longer.

3.2 Smooth turbulent films

3.2.1 The planar turbulent film

Turbulent flow is generally more difficult to treat than laminar flow but may be modelled satisfactorily after first writing each dependent variable as the sum of a mean and fluctuating part. Thus, for the longitudinal velocity

$$U = \bar{U} + U', \tag{3.19}$$

where $\bar{U}$ is the mean velocity averaged over time and U' is the fluctuation about the mean. Substituting into the longitudinal equation of motion, and averaging over time, we obtain

$$\rho\left(\bar{U}\frac{\partial \bar{U}}{\partial X} + \bar{V}\frac{\partial \bar{U}}{\partial Y}\right) = -\frac{\partial \bar{P}}{\partial X} + \rho g + \mu\left(\frac{\partial^2 \bar{U}}{\partial Y^2}\right) + \frac{\partial}{\partial Y}\rho\,\overline{U'V'} \tag{3.20}$$

for a steady, two-dimensional film flowing over a vertical wall when lateral shear effects dominate. This has the same general form as the laminar eqn (3.4) except for the addition of the Reynolds stress on the right-hand side. It is evident that viscous shear may be neglected whenever

$$\overline{U'V'} \gg \nu\frac{\partial \bar{U}}{\partial Y}. \tag{3.21}$$

This condition applies in the outer reaches of the film but not near the wall, which tends to damp out turbulent fluctuations in a thin region often called the viscous sublayer. In this region, where $U = \bar{U}$, $\tau_w \simeq \mu\partial\bar{U}/\partial Y$ and hence

$$\bar{U} \simeq \frac{\tau_w}{\mu}Y, \tag{3.22}$$

where Y is measured from the wall. The velocity distribution in the viscous sublayer may be written in normalized form using the scales

$$\left.\begin{aligned} U^c &= \left(\frac{\tau_w}{\rho}\right)^{1/2} \\ \text{and} \qquad Y^c &= \frac{\mu}{(\rho\tau_w)^{1/2}} \end{aligned}\right\} \tag{3.23}$$

for velocity and thickness, respectively. Hence from eqn (3.22),

$$\frac{\bar{U}}{U^{c}} = u^{+} = y^{+} = \frac{Y}{Y^{c}} \tag{3.24}$$

within the sublayer.

Beyond the viscous sublayer, the velocity field may also be described using the variables u^{+} and y^{+} if two assumptions are made:

(1) the film is thin enough to be represented as a parallel flow in which the shear stress is constant and given by

$$\tau = \tau_{I} = \tau_{w} = \rho\, \overline{U'V'}; \tag{3.25}$$

(2) the fluctuating velocity components correlate according to the relation

$$U' \sim V' \sim l \frac{d\bar{U}}{dY},$$

where l is a mixing length analogous to the molecular mean free path in the microscopic explanation of viscosity.

With these assumptions, the Reynolds stress may be written

$$\tau = \rho l^{2} \left| \frac{d\bar{U}}{dY} \right| \frac{d\bar{U}}{dY} = \rho \varepsilon \frac{d\bar{U}}{dY}, \tag{3.26}$$

where ε is defined as the eddy diffusivity by analogy with the momentum diffusivity ν. The simplest hypothesis for mixing length assumes that it increases linearly with distance Y from the wall, and ignores damping near the interface. Eqn (3.26) then yields

$$Y \frac{d\bar{U}}{dY} = \text{constant}$$

from which it follows that the form of the velocity profile is given by

$$u^{+} = a \ln y^{+} + b. \tag{3.27}$$

Fitting this form and eqn (3.24) to empirical data obtained from turbulent pipe flow yields the following results for forced flow in turbulent films:

$$\left.\begin{aligned} y^{+} < 5&: \; u^{+} = y^{+} \\ 5 < y^{+} < 30&: \; u^{+} = -3.05 + 5 \ln y^{+} \\ 30 < y^{+}&: \; u^{+} = 5.5 + 2.5 \ln y^{+} \end{aligned}\right\}. \tag{3.28}$$

These monotonic forms are well suited to cocurrent flows driven by the pressure gradient. They are not generally applicable to countercurrent flows. When an upward flowing vapour opposes a downward flowing condensate film, for example, the vapour pressure gradient tends to slow the film. The pressure gradient may reverse both the wall shear stress and the condensate flow. Equations (3.28), and the conditions on which they depend, no longer apply.

3.2.2 The annular film

A turbulent annular film is governed by the same basic considerations as a laminar annular film. Using Fig. 3.2, a film force balance once again leads to eqn (3.15) for the wall shear stress if the film is thin in relation to the pipe diameter. Hence the wall shear stress may again be written in the form of eqn (3.16). However, the relation between wall shear stress and film flow rate is no longer given by eqn (3.17) because of the composite velocity profile provided by eqns (3.28).

The film flow rate may be determined by integrating the velocity profile over the range $0 < y^+ < \delta^+$, where $\delta^+ = (\rho\tau_w)^{1/2}\bar{\Delta}/\mu$. Problem Q3.4 in Chapter 8 provides the result which yields the general relation

$$\tau_w = \tau_w(\Gamma, \bar{\Delta}). \tag{3.29}$$

Alternatively, if the film is thin enough for us to assume that $\tau_w = \tau_I$, we may use the gas-based expression

$$\tau_I = (\rho_g U_g^2/2)f, \tag{3.30}$$

where f is a friction factor, to write

$$\tau_w \simeq \tau_I = \frac{8f\dot{M}_g^2}{\pi^2 D^4 \rho_g} \tag{3.31}$$

in which $\dot{M}_g$ is the gas flow rate. Since f is usually known as a function of the gas Reynolds number (or mass flow rate), eqns (3.29) and (3.31) combine to complete the triangular relation under common circumstances. Problems Q3.2–3.4 in Chapter 8 illustrate the use of the triangular relation for a thick, turbulent film while problems Q5.12 and 5.13 demonstrate its use with a thin laminar film.

3.3 Interfacial stability

When vapour flows over a liquid film, either, neither, or both may be turbulent. Moreover the conditions leading to turbulence in either of the adjacent fluid streams may be independent of conditions in the other. Typically, the origin of turbulence may be traced to the appearance of Tollmien–Schlichting waves which amplify and eventually break up to create the fluctuating components U' and V' mentioned above. To some extent, Tollmien–Schlichting waves may affect the liquid–vapour interface but they do not originate there. Interfacial waves, on the other hand, owe their origin to a changing balance in the disturbing and constraining surface forces at the interface itself. Surface tension forces have a constraining effect, while shear and inertial forces have a disturbing effect. Buoyancy forces may stabilize or destabilize, depending on the circumstances. As we shall see in later chapters, the shifting force balance at the interface exerts a great influence on heat and mass transfer rates, and can lead to the disintegration of the interface itself.

3.3.1 Travelling waves on a horizontal interface

Figure 3.3 shows a horizontal, 'gravity' wave of amplitude A and wavelength Λ driven by the relative motion of a vapour over a liquid. More precisely, the wave is maintained by the centrifugally induced pressure variations in the vapour as it flows over the troughs and peaks. In each trough, for example, the pressure gradient normal to the interface may be estimated from

$$\frac{\partial P}{\partial n} = 0\left(\frac{\rho_V U_V^2}{R}\right), \tag{3.32}$$

where R is the radius of curvature in the trough. For a plane wave with the profile $Y = A \sin(2\pi X/\Lambda)$,

$$\frac{1}{R} = -\left(\frac{d^2Y}{dX^2}\right)_{X=\frac{3\Lambda}{4}} = A\left(\frac{2\pi}{\Lambda}\right)^2.$$

Therefore the centrifugally induced excess pressure, measured from mid-height, is given by

$$\Delta P = 0\left[\rho_V A^2 U_V^2 \left(\frac{2\pi}{\Lambda}\right)^2\right]. \tag{3.33}$$

There is an equal but opposite pressure variation at the wave peak.

Under quasi-steady conditions, the vapour pressure excess in the trough is balanced by a restoring liquid pressure which, neglecting surface tension, is the effective hydrostatic head $\Delta P = Ag\Delta\rho$, where $\Delta\rho = \rho_L - \rho_V$. Mechanical equilibrium thus reveals

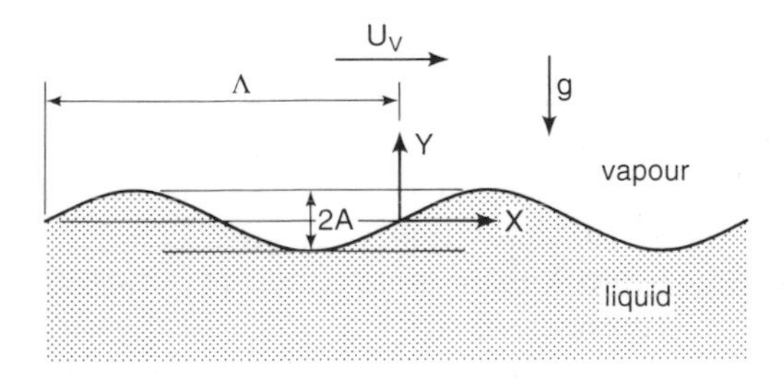

Fig. 3.3 Gravity wave on a liquid surface driven by vapour flowing horizontally with relative velocity U_V.

$$\frac{4\pi^2 \rho_V A U_V^2}{\Lambda^2 g \Delta\rho} = O(1). \tag{3.34}$$

This may also be expressed in terms of the Froude number Fr_Λ based on wavelength. Thus,

$$Fr_\Lambda = \frac{2\pi\rho_V U_V^2}{\Lambda g \Delta\rho} = O\left(\frac{1}{a}\right), \tag{3.35}$$

where the amplitude–wavelength ratio $a = 2\pi A/\Lambda$. This result suggests that smaller wavelengths are less stable than larger wavelengths, thus signalling the importance of short, steep waves: specifically, with $a > 1$.

The stability criterion represented by eqn (3.35) indicates that 'gravity' waves are more likely to appear when the restoring force is reduced, i.e. $g\Delta\rho$ becomes smaller. The interface will therefore become increasingly unstable as the critical point is approached and $\Delta\rho \to 0$. Tilting of the nominally horizontal interface shown in Fig. 3.3 reduces the component of the gravitational acceleration lying normal to the mean interfacial plane. The interface on a vertical liquid film would thus be unconditionally unstable in the presence of a moving vapour were it not for the existence of surface tension.

3.3.2 *Travelling waves on a vertical interface*

This leads us naturally to the discussion of capillary waves for which the disturbing influence of the relative vapour velocity U_V is balanced by the restraining influence of interfacial tension σ_{LV}. Figure 3.4 depicts the

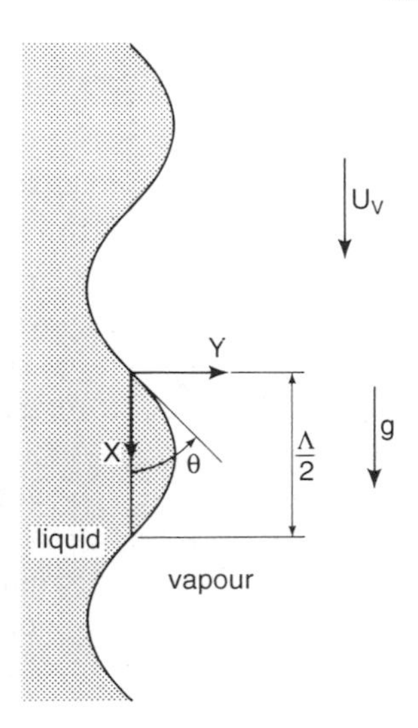

Fig. 3.4 Capillary wave driven by vapour flowing vertically downward with relative velocity U_V over a liquid film.

situation for a nominally vertical interface on a falling liquid film. The centrifugally induced pressure disturbance in each trough of a sinusoidal wave is again provided by eqn (3.33), but for mechanical equilibrium this must now be equated to the capillary induced restoring pressure given by $\Delta P = (4\sigma_{LV}/\Lambda)\sin\theta$ where θ is the inclination of the tangent to the interface as shown. Since $\tan\theta = (\mathrm{d}Y/\mathrm{d}X)_0 = 2\pi A/\Lambda = a$,

$$\Delta P = \frac{4\sigma_{LV}}{\Lambda}\frac{a}{(1+a^2)^{1/2}} \tag{3.36}$$

Hence, for mechanical equilibrium,

$$We_\Lambda = \frac{\rho_V U_V^2 \Lambda}{2\pi\sigma_{LV}} = \frac{2}{\pi a(1+a^2)^{1/2}} \tag{3.37}$$

where We_Λ is the Weber number based on wavelength. As with gravity waves, capillary waves are more likely to occur in the presence of a weak restoring force: specifically, as $\sigma_{LV} \to 0$. It is also evident that short, steep waves, i.e. $a > 1$, are again more likely to predominate.

The above results demonstrate that the interface on a liquid film is capable of developing waves travelling longitudinally when the relative velocity of the adjacent vapour reaches a critical value. To a first approximation, this velocity is independent of viscosity and corresponds to a balance between the disturbing *inertial* force (I) and a restoring force attributable to buoyancy (B) or surface tension (T). The criteria for the development of characteristically short and steep waves are seen to be limiting values of the Froude number (I/B) for 'gravity' waves or the Weber number (I/T) for capillary waves. These limits are particularly helpful to our understanding of annular film behaviour in pipes.

3.3.3 Stationary waves on a horizontal interface

During condensation on a ceiling, or when film boiling takes place above a horizontal substrate, liquid lies *above* vapour. In these circumstances, buoyancy tends to destabilize the interface while surface tension acts restoratively. Figure 3.5 provides a plan view of the resultant wave pattern which eventually develops into drops. The interface resembles the cardboard surface often used in egg packages; the circles shown are meant to suggest the hanging bulges of liquid. These bulges grow into drops which leave the interface periodically. If the wavelength of these bulges is Λ, their diameter is given by $D = 0(\Lambda/2\sqrt{2})$, bearing in mind that the interface rises upward between them. For a spherical cap of volume $\mathcal{V}$ hanging pendant under the restraint of surface tension, $\mathcal{V}g\Delta\rho = 0(\pi D\sigma_{LV}\sin\theta)$. Hence for incipient drop formation when $\theta \simeq \pi/2$ and $\mathcal{V} = \pi D^3/12$, we find that $gD^2\Delta\rho/\sigma_{LV} = O(12)$. After substituting for the drop diameter we may define the stability criterion by the Bond number $Bo_\Lambda = We_\Lambda/Fr_\Lambda$. Hence

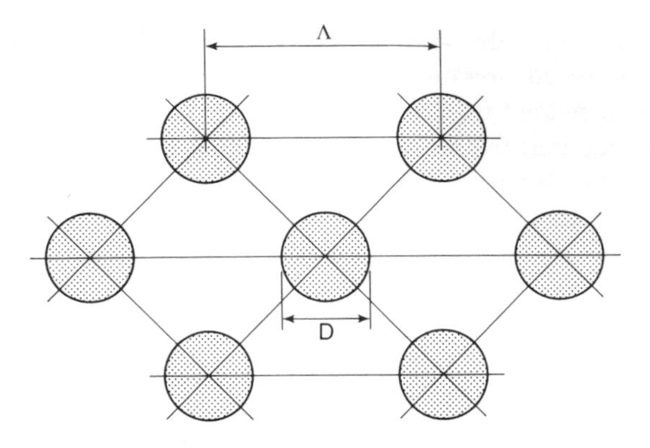

Fig. 3.5 Plan view of incipient drops on a horizontal interface formed by a liquid above a vapour.

$$Bo_{\Lambda} = B/T = \left(\frac{\Lambda}{2\pi}\right)^2 \frac{g\Delta\rho}{\sigma_{LV}} = O(2.4). \tag{3.38}$$

This does not contain a vapour velocity. It is crucial to our understanding of the minimum heat flux in pool boiling.

3.3.4 Rivulet formation

Capillarity may also influence lateral, or spanwise, variations in a flowing film. While these are not usually associated with interfacial waves, they do consist of periodicities controlled by surface tension which becomes increasingly important, for example, when evaporation from a film causes a substantial reduction in its thickness. The net result may be a film which breaks into rivulets separated by stretches of unwetted surface, as illustrated in Fig. 3.6. In transforming into the more stable rivulet configuration, the film creates a number of stagnation points A about which the flow divides. These stagnation points mark the upstream extremity of the resulting dry patches.

Near any stagnation point, the net surface tension force acting on the film is equal to $\sigma_{LV}(1-\cos\theta)$, where θ is the contact angle shown in Fig. 3.6. When this is balanced by the liquid stagnation pressure $\rho_L \bar{U}_f^2/2$ acting on a film of thickness Δ, it follows that the *film* Weber number We_f is given by

$$We_f = \frac{\rho_L \bar{U}_f^2 \Delta}{\sigma_{LV}} = 2(1-\cos\theta) = O(2). \tag{3.39}$$

This provides a useful stability criterion (Hartley and Murgatroyd 1964). If also reveals why the magnitude of the critical Weber number depends

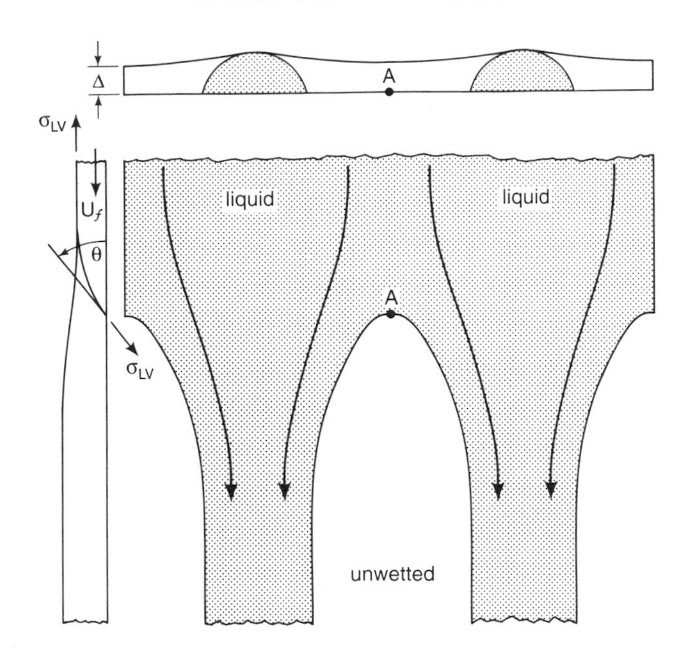

Fig. 3.6 Formation of rivulets in a buoyancy-driven liquid film of thickness Δ. Conditions in a vertical plane through the stagnation point at A are shown on the left.

upon whether the rivulets are under formation (retreating contact angle) or dissolution (advancing contact angle). The rivulet transformation is therefore hysteretic.

We have now introduced all the common forms of interfacial instability which, although they have been treated on a nominally planar interface, are largely independent of overall interfacial geometry. In particular, they may be observed on the curved interfaces of jets, drops, and bubbles. It is evident that they fall into three principal categories which will henceforth be designated by the non-dimensional group that defines the instability. Thus, even though inertially driven instabilities constrained by buoyancy are associated with the work of Helmholtz or Kelvin, they will be described as *Froudian*. Instabilities driven by inertia and constrained by surface tension, and associated with Kelvin or Rayleigh, will be termed *Weberian*. Finally, the buoyancy-driven instabilities constrained by surface tension, and associated with Rayleigh or Taylor, will be termed *Bondian*. With the greatest of respect to the brilliant pioneers, this proposed classification offers the student a simple reminder of the essential physics so valuable in the understanding of evaporation and condensation problems.

3.4 The formation and motion of drops and bubbles

3.4.1 The origin of drops

It will be recalled from Chapter 2 that droplets may grow from embryos formed from the vapour phase under saturated or supersaturated conditions. More will be said about this mechanism during the discussion of condensation in Chapter 5. In this subsection, we will be concerned with drops and droplets formed through mechanical action at an interface. Three mechanisms are of special interest in latent heat transfer. In the first, inertial forces cause a jet or sheet of liquid to disintegrate into globules. In the second, inertial forces act on the crests of waves to generate a spray of drops. In essence, both of these result from Weberian instabilities. The third mechanism is a set of processes originating in the bursting of vapour bubbles.

Consider the behaviour of liquid issuing from a circular orifice into a denser liquid above. This situation is depicted in Fig. 3.7. When the discharge rate is very low, 'bubbles' form and detach periodically, as indicated in Fig. 3.7(a). Increasing the discharge rate increases the 'bubble' frequency until a continuous chain forms into a stable cylindrical jet (Fig. 3.7(b)) with surface undulations attributable to capillarity. If the discharge rate increases further, the surface eventually becomes unstable, the result being the growth of axisymmetric perturbations, as suggested in Fig. 3.7(c). When these have amplitudes comparable with the jet radius, they create drops with diameters of the same order as the jet diameter. The drop formation process has thus been transferred from the orifice to the jet tip.

Provided that the terminal velocity of the (buoyant) drops is greater than

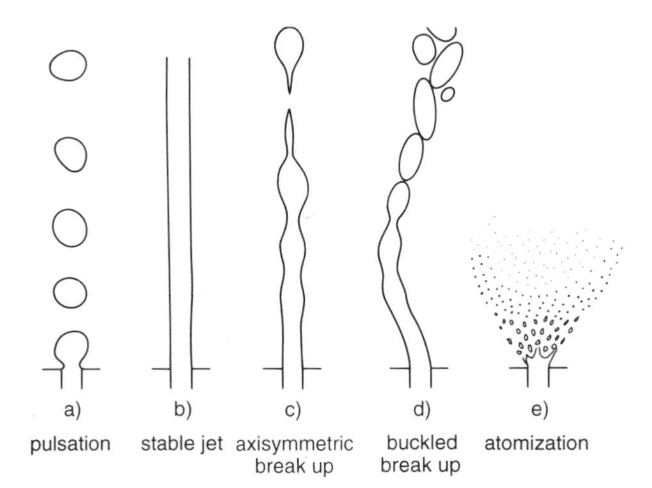

Fig. 3.7 Flow régimes for a liquid discharging from a circular orifice into a denser liquid.

the jet velocity, they remain ahead of the jet tip. Higher jet velocities produce greater inertial and shear forces which oppose forward motion and eventually cause the jet to buckle. Figure 3.7(d) attempts to show how this event leads to a pile up and break up. Drops are thus created over an area beyond the cross sectional area of the orifice. Since the buckling instability has no preferential plane of oscillation, it may produce a wandering motion about the original axis and create a cone-shaped cloud of drops. If the velocity is increased further, as suggested in Fig. 3.7(e), the jet disintegrates and is replaced by a turbulent, cone-shaped region in which larger liquid drops formed at or near the orifice rapidly disintegrate into smaller droplets. The result is an atomized spray.

Sprays may also be produced by gas or vapour blown parallel to a liquid free surface, as suggested in Fig. 3.8. If the surface already exhibits the wave instabilities discussed earlier, separation and recirculation immediately downstream of the crests may give rise to a tip cascade effect in which both drops and bursting bubbles are created. On a lake or river, the resultant spray is associated with familiar whitecaps. In an engineering context, and especially inside tubes where latent heat is being transferred, the waves are much smaller and their behaviour is further complicated by the confining effect of the walls. For example, relative motion of stratified phases inside a horizontal condenser tube may create interfacial waves which, if the liquid occupies more volume than the vapour, may reach the top of the tube. In general, however, the existence of steep waves indicates that $Fr_\Lambda = O(1)$. At the same time, disintegration of the lip into droplets suggests that $We_\Lambda = O(1)$. In combination, these criteria signify that $Bo_\Lambda = We_\Lambda/Fr_\Lambda = O(1)$. Spray generation thus implies that surface tension, buoyancy, and vapour inertial forces are roughly in balance.

Similar comments apply to a film running down inside the wall of a vertical condenser tube with vapour flowing in the centre. Waves are then

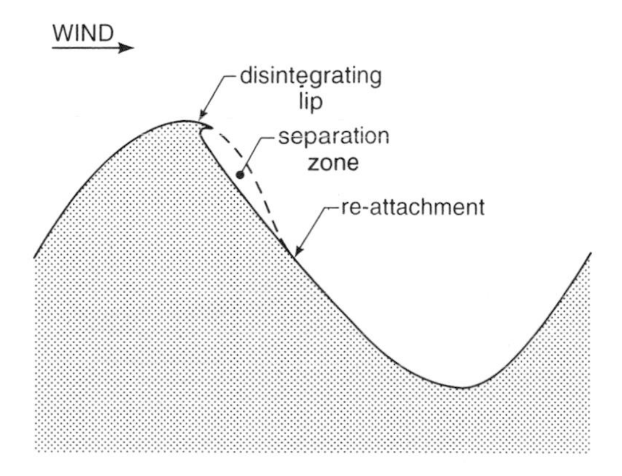

Fig. 3.8 Schematic representation of an incipient whitecap.

initiated through the Weber number criterion and grow until they create a physical blockage. Under counterflow conditions, the back pressure gradient slows the film and may reverse it. The buoyancy force on the 'hanging' waves then produced rises to balance the inertial force created by the constricting wave volume itself: that is, the Froude number criterion is reached. Drops again issue from the wave crests. Once more, the spray-generating mechanism is in full force when the Bond, Weber, and Froude numbers are all of unit order.

In contrast to the above mechanisms, the bursting of vapour bubbles is characterized by strong surface tension forces. Floating between the liquid and the vapour, a bubble may shatter as the result of local thinning. The ruptured film generates a number of very small droplets. However, it is the rush of liquid into the bubble crater which produces the main effect by creating a central jet of liquid which subsequently breaks into one or more larger droplets; these may be ejected a considerable distance. Large and small droplets produced in this manner are either carried away by the vapour or splash on a neighbouring wall. They may also splash back into the liquid and create bubbles as described below.

3.4.2 The origin of bubbles

Bubbles, like droplets, may grow from embryos, as outlined in Chapter 2. This mechanism will be discussed fully in Chapter 6. Here we limit ourselves to mechanical origins beginning with the above-mentioned splashing process during which vapour is entrained with the impacting drop. This may create a plentiful secondary source of evaporation, particularly during boiling in thin liquid films. Figure 3.9 provides a representative map of air bubble entrainment caused by the entry of water drops. The most pervasive form, Mesler entrainment, is not yet fully understood (Pumphrey and

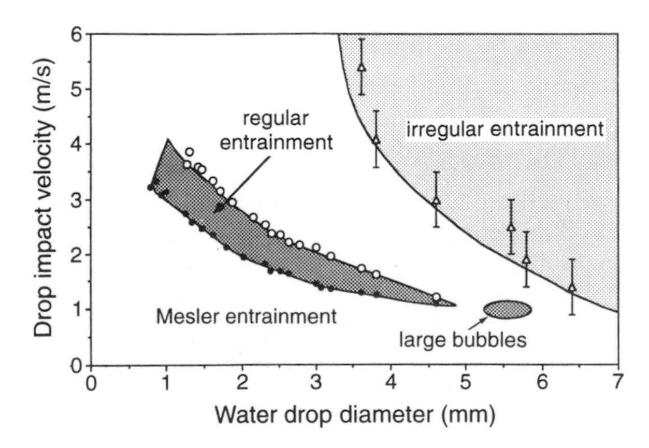

Fig. 3.9 Splash-induced entrainment of air in water. Note the effect of drop size and impact velocity on the type of entrainment (after Pumphrey and Elmore 1990).

Elmore, 1990). It appears to result from separate, though simultaneous, capillary waves induced on impact between the entering drop and the liquid surface. These waves evidently radiate out from the initial point of contact, thus causing toroidal vapour rings between successive crests to be pinched out and carried downward below the surface. Each such torus would instantly break into a ring of discrete bubbles much like those observed.

Bubbles may also evolve from vapour discharging at an orifice (or a nucleation site). Figure 3.7 is thus a useful guide for bubbles also. When the vapour flow rate at the orifice is very low, discrete bubbles grow and detach periodically as shown. As the discharge rate (or nucleation rate) is increased the bubble frequency increases until the chain becomes a continuous jet, assuming no resistance to coalescence. However, in a vapour jet the Reynolds numbers are usually higher than in a liquid jet thus introducing turbulence and accelerating instability. Reversion to a bubbling discharge may therefore be more rapid; the bubble diameters (and numbers) are strongly influenced by coalescence.

3.4.3 The motion of single bubbles or drops

The bubbles or drops which accompany latent heat transfer frequently occur in large numbers. In Section 3.5 we will study their collective, dynamic behaviour but before doing so it is important to understand how they behave individually. This behaviour is well illustrated in Fig. 3.10 which shows various bubble flow régimes plotted in the Reynolds number–Bond number domain. Since there are four independent forces to consider, the information is conventionally presented in terms of three ratios: the Reynolds number (inertial/viscous) defined by $Re = U_{\mathrm{T}}\mathrm{d}/\nu$ in which U_{T} is the terminal velocity of a bubble of equivalent diameter d; the Bond number (buoyancy/surface tension) defined by $Bo = \mathrm{d}^2 g\Delta\rho/\sigma$; and the Morton number defined by

$$Mo = \frac{Bo}{Re^2}\left(\mu\frac{U_{\mathrm{T}}}{\sigma}\right)^2 = \frac{\mu^4 g\Delta\rho}{\rho^2\sigma^3} \tag{3.40}$$

in which $\mu U_{\mathrm{T}}/\sigma$ is the ratio of viscous to surface tension forces. The Morton number contains only fluid properties and the field acceleration. The Bond number reflects the driving force on the bubble while the Reynolds number indicates the relative importance of the two restraining forces.

When the surface tension forces are comparatively large, i.e. $Bo < O(1)$, a bubble behaves as a small, slowly moving sphere. If the Bond number is greater, e.g. $Bo = O(10)$, the effect of increasing the Reynolds number (inertial forces) then converts the bubble shape to an ellipsoid. This shape reflects the altered hydrodynamic conditions, eventually leading to spiralling and wobbling behaviour. Higher Bond numbers set the scene for even greater changes in bubble shape as the Reynolds number increases. The figure indicates that conditions in the wake behind the bubble then cause an indentation which increases with Reynolds number, creating a long skirt.

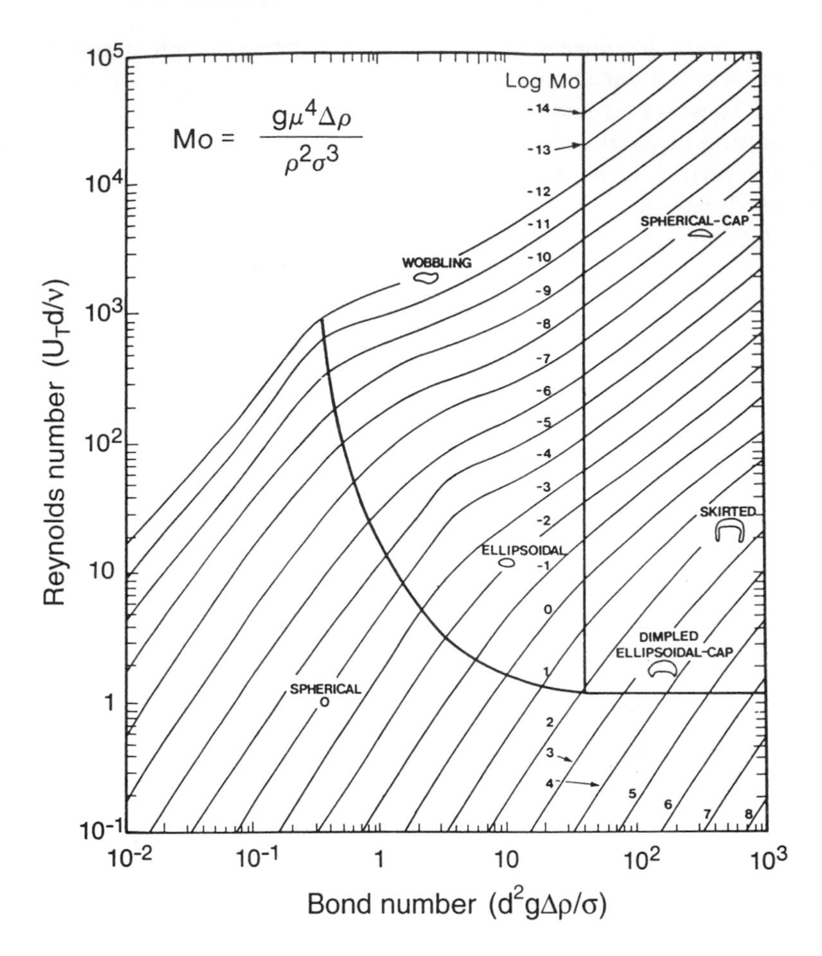

Fig. 3.10 Behavioural régimes for a bubble of diameter d moving at its terminal velocity U_T (after Clift *et al.* 1978).

However, above $Re \simeq 10^2$ the closed wake of the skirted bubble becomes increasingly open as the shape changes to a spherical cap.

The principal hydrodynamic differences between solid particles and fluid pockets in isolation are attributable to two effects. Firstly, surface shear stresses cause circulation within fluid pockets, thereby lowering shear drag and increasing terminal velocity. Secondly, the surface pressure distribution causes fluid pockets to contract in the direction of motion, thus creating the flattened shapes mentioned above. The first effect may be swamped by the opposing effect of surfactants which, even in small quantities, tend to collect near the rear stagnation point where they decrease the local surface tension and reduce circulation. The second effect is also influenced by surface tension, mainly through the role of the Bond number.

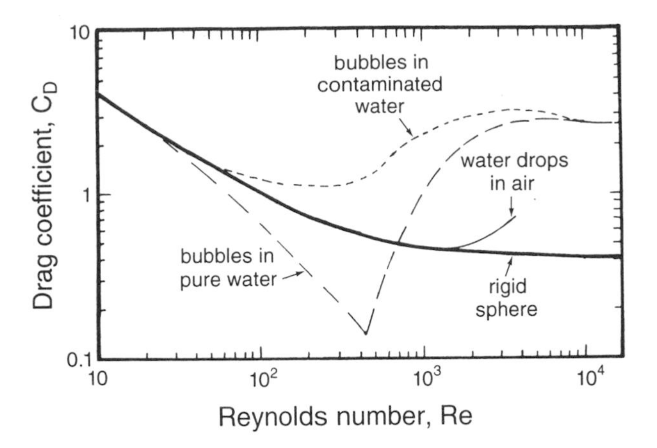

Fig. 3.11 Drag coefficient as a function of Reynolds number for water drops in air, air bubbles in water, and rigid spheres (after Clift *et al.* 1978).

Figure 3.11 compares the influence of the two effects on drag coefficient. Air bubbles in contaminated water, for example, exhibit little internal circulation; their ellipsoidal shape increases drag above that for a rigid sphere. In pure water, on the other hand, the shape effect is overshadowed by the effect of internal circulation which reduces drag, at least until $Re \simeq 450$. Water drops, having a high viscosity ratio when falling in air, exhibit negligible internal circulation and remain almost spherical until $Re > 10^3$.

During its brief lifetime, a drop may shrink through evaporation and vanish; a bubble may grow through evaporation or reduction in hydrostatic pressure. These dimensional changes alter the Reynolds and Bond numbers. Figure 3.11 indicates that an increase in Reynolds number reduces the drag coefficient, at least for $Re \leqslant 200$. For this Reynolds number range, larger drops and bubbles accelerate more rapidly than smaller ones. For higher Reynolds numbers, it is evident that behaviour is more complex but in general terms the effect of size is reversed when $450 < Re \leqslant 3 \times 10^3$, and becomes negligible beyond this range. As we shall see, these relative accelerations alter the size distribution in a collection of drops or bubbles.

The size of a bubble or drop is not unlimited. Falling drops provide a useful illustration of natural break up. When they become large, pressure-induced concavities near the forward stagnation point may grow enough to convert the drop into a 'bag' which then disintegrates (Clift *et al.* 1978). This occurs when

$$Bo = \frac{d_c^2 g \Delta\rho}{\sigma} \simeq 16, \tag{3.41}$$

where d_c is the critical drop diameter. More generally, the relationship between the drop Reynolds number and drag coefficient enables this criterion to be re-stated as a critical Weber number. That is, the relative velocity establishes the limiting drop size, e.g. in water sprays and fuel sprays.

3.5 Collective behaviour of bubbles and drops

3.5.1 Swarms and sprays

Bubbles in proximity to each other are often influenced by Bernouilli and wake effects. Acting on a loose cloud, these effects may create a coherent swarm with a collective identity. Jostling of bubbles, on the other hand, is a different form of interaction which is equally difficult to prescribe. In pure water, for example, bubble contacts invariably lead to coalescence with the resultant bubbles growing by accretion up to the size permitted by the stability limit mentioned immediately above. However, if more than 1 per cent (by weight) of NaCl is added to the water, contacting bubbles do not coalesce, evidently because of concentration of the impurity at the interfaces (Soo 1967). When vapour bubbles originating at a wall become numerous enough, their coalescence may create large pockets which eventually transform into a continuous blanket or film, thus converting a dispersed system into a bipartitioned system.

Similar phenomena may alter the size distribution of a spray during its brief history. It is evident from Section 3.4.1 that sprays created either naturally or artificially vary enormously. Nonetheless, they may be characterized by their size distribution: that is, an instantaneous plot or histogram of number density (drops per unit volume) versus drop size. Many particular forms of size distribution have been suggested, one of the most versatile being

$$n_d = a d^{\alpha} \exp(-bd) \tag{3.42}$$

which determines the number density n_d for drops of diameter d through the three empirical parameters α, a, and b. Integrating over the full size range yields the total number density as

$$n = \frac{a\Gamma(\alpha + 1)}{\mathrm{b}^{\alpha+1}}. \tag{3.43}$$

The mode and median drop diameter may also be found from eqn (3.42), as illustrated in problems Q3.5–3.7 in Chapter 8. For some purposes, such as the calculation of liquid mass fluxes, an averaged drop description may be adequate; for others, such as impact behaviour, it is not.

3.5.2 Accretion on drops and surfaces

In certain circumstances, the contact coalescence mechanism may dramatically alter the size distribution. For example, the free fall of droplets in a cloud typically favours the larger droplets. Following the low Reynolds number behaviour evident in Fig. 3.11, these fall faster as they accrete smaller droplets in their path. Much the same is true for drops of condensate on a vertical surface; any drops large enough to begin a run usually do so with increasing velocity.

The problem of accretion on large free-falling drops is very similar to that for a solid body moving through a cloud of dispersed particles. Figure 3.12 illustrates the situation for a circular cylinder, diameter D, lying in a vapour cross flow containing distributed droplets (diameter d) moving with the same velocity U_∞ as the vapour. Well upstream of the cylinder, the droplet trajectories and streamlines coincide as parallel lines. Within a distance of the order of one diameter D from the cylinder, however, the trajectories diverge from the streamlines as they veer outward with an acceleration of the order (U_∞^2/D) and a velocity of the order of U_∞. This creates a lateral viscous force F_v on each droplet. For a Stokes (viscous) flow around the droplet,

$$F_v = O(3\mu_V \pi U_\infty d),$$

which may be compared with the lateral inertial force

$$F_i = O\left(\frac{\pi d^3 \rho_L}{6} \cdot \frac{U_\infty^2}{D}\right).$$

The ratio is thus given by

$$\frac{F_v}{F_i} = O\left(\frac{18\rho_V D^2}{Re_D \rho_L d^2}\right), \tag{3.44}$$

where Re_D is the cylinder Reynolds number. This ratio is much less than 1.0 for those droplets that collide with the cylinder and are thus confined

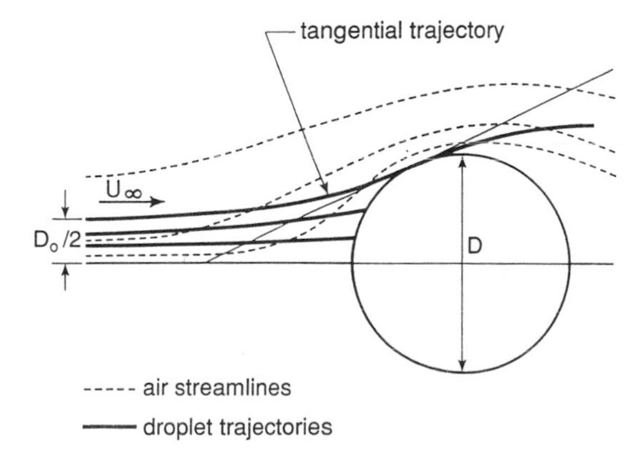

Fig. 3.12 Droplet trajectories near a cylinder of diameter D in a cross flow moving with relative velocity U_∞.

within the tangential trajectory shown in Fig. 3.12. Such droplets come from the region D_0 wide upstream. The ratio D_0/D is usually called the collision efficiency η. The use of these ideas is illustrated in problems Q3.8 and 3.9, and Q4.10–4.13, in Chapter 8.

As eqn (3.44) indicates, smaller droplets veer more than larger droplets, thus varying the droplet size distribution around the cylinder. The behaviour of droplets which do collide depends upon the angle of approach, the impact velocity, and the droplet diameter. For those following a trajectory normal to the surface, i.e. along the stagnation streamline, impact characteristics may be prescribed by the droplet Weber number $We_d = \rho U_\infty^2 d/\sigma$. When We_d is small, surface tension forces may permit the droplet to bounce and immediately leave the surface. As We_d increases, the droplets tend to remain on the surface and spread. If the Weber number becomes great enough the spreading disc will finally disintegrate into small beads. In general, the Reynolds number and Weber number both exert a strong influence on the hydrodynamic processes which become important during spray cooling, for example.

The longer-term behaviour of droplets colliding with solid surfaces depends principally on their number density and the neighbouring vapour velocity. In sufficient numbers, the droplets, following one after another, may produce a liquid film which is sheared by the vapour flowing over it. Should the vapour velocity be high enough, or the film thin enough, rivulets may form in the manner discussed in Subsection 3.3.4. In extreme circumstances, the film will break into new droplets, some being carried away in the vapour while the remainder are dragged over the surface. The details of such behaviour become crucially important in pipe flows, for example, whenever the loss of the film implies the loss of evaporative cooling ability.

3.6 The effects of confining walls

Heat transfer vessels require many surfaces for guiding fluids internally: deflectors, baffles, inlets, bends, adapters, outlets, etc. Most important are the tubular components within or between which most of the heat transfer takes place. The following outline therefore deals only with the confining effects of tubular walls, concentrating on the flow characteristics within circular tubes under steady conditions. These are typified by flow boiling and condensation. Flow between tubes forming a bundle is discussed later in Section 6.4.6.

Figure 3.13 shows a map of the various flow régimes which occur when water and water vapour flow cocurrently upward in a vertical tube. The coordinates are the superficial phase velocities $U^s = G/\rho$, rather than the more conventional momentum flux densities G^2/ρ. This choice enables us to examine the effect of systematically changing the vapour mass fraction $x = 1(1 + \rho_L U_L^s/\rho_V U_V^s)$ by varying the superficial velocities. We begin with

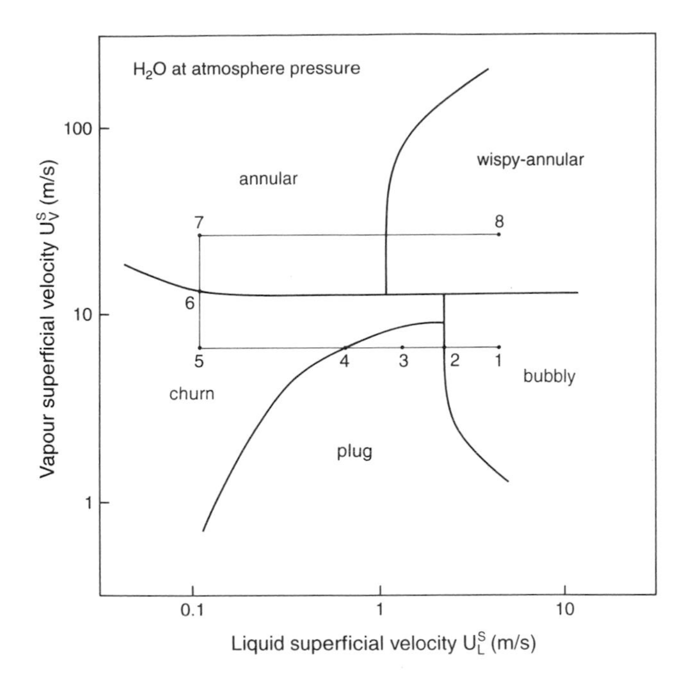

Fig. 3.13 Régime map for the upward flow of saturated water and water vapour in small diameter, vertical tubes (from Hewitt and Roberts 1969).

U_V^s fixed at a low value, e.g. $7\,\mathrm{m\,s^{-1}}$. When x is small (point 1 on Fig. 3.13), a bubbly flow is produced. This is illustrated in Fig. 3.14(a). As x increases, isolated bubbles group into swarms which eventually coalesce (2) into the bullet-like bubbles seen in Fig. 3.14(b). This is plug flow (3).

With further increases in the vapour fraction (lower U_L^s), the interface on the descending liquid annulus around the plug undergoes a Froudian instability, at which point (4) waves appear on the annular interface and mark the onset of an unstable oscillatory régime known as churn flow. This is suggested in Fig. 3.14(c). Both vapour and liquid then behave in a chaotic manner (5). If the *vapour* flow is now increased, the main bubbles tend to occupy more and more of the central region of the tube, displacing the liquid until it is essentially confined to the region immediately adjacent to the tube wall, as indicated in Fig. 3.14(d). This creates the annular régime (7) in which the vapour core is seldom droplet free. Recalling the discussion in Section 3.4.1, $Fr \simeq We \gtrsim O(1)$ under these conditions. Defining the Kutateladze number Ku by

$$Ku^4 = Fr\,We \tag{3.45}$$

we are able to provide a succinct criterion for entry (6) to the annular régime. A value of $Ku \simeq 3$ has been suggested (Hetsroni 1982). Since Ku

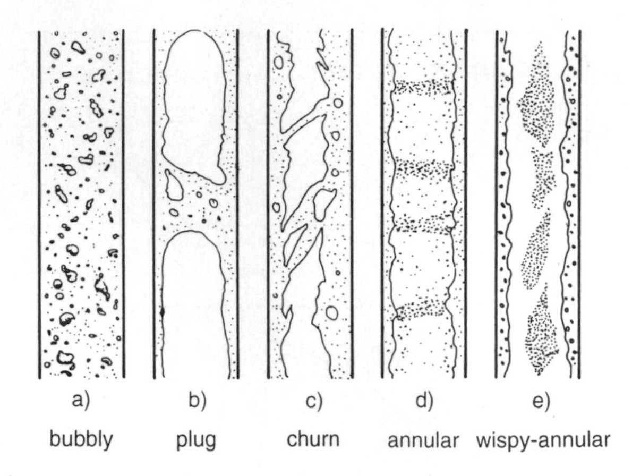

Fig. 3.14 Two-phase flow patterns in a vertical tube (after Collier 1972).

does not contain a length scale it is sometimes a more convenient parameter. Problem Q3.1 in Chapter 8 provides a numerical illustration of its use.

With a high vapour velocity (7–8), the churn flow régime may be bypassed entirely, as evident in Fig. 3.13. The core region of the flow may then retain a significant water content in the form of long threads, thus giving rise to the designation wispy-annular flow (8). For both annular régimes, the liquid–vapour interface exhibits instabilities in a variety of wave forms; large amounts of spray are usually present.

Similar flow behaviour is found when the vapour and liquid flow cocurrently downward, except that the annular régime then divides into three subcategories: a bipartitioned falling film flow; a bubbly falling film flow; and a falling flow with droplets dispersed in the core. These subrégimes follow successively as the flow rate is increased (Bergles *et al.* 1981). On the other hand, a falling film in the presence of an upward vapour flow retains its basic form until flooding produces a churn flow after which a cocurrent upward flow is produced.

The flow régime map for a horizontal tube is illustrated in Fig. 3.15 using an air–water mixture. Certain similarities with the vertical tube régimes are evident; so are several differences. For very low vapour fractions (1), bubbly flow (Fig. 3.16(a)) is once more exhibited but the natural buoyancy of the bubbles now causes them to congregate in the upper half of the tube. As the liquid flux decreases, this tendency creates the intermittent slug (2a) or plug (2b) flow régimes indicated in Figs 3.16(b) and 3.16(e), respectively. The plug bubbles no longer fill most of the tube section, and may be generated at much lower vapour velocities (e.g. $0.1\ \mathrm{m\,s^{-1}}$) when the nominally horizontal interface in the stratified régime (3) develops waves. If the stratified liquid height Δ rises towards the tube diameter d, the waves grow

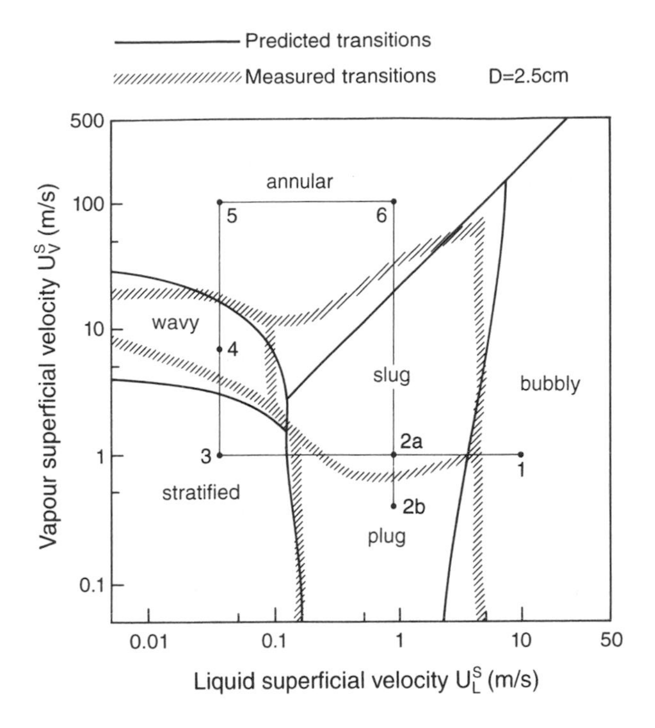

Fig. 3.15 Régime map for air–water flow in horizontal tubes (from Hetsroni 1982).

with increasing liquid velocity to create slug or plug flows depending upon the vapour flux.

Froudian instabilities also account for the transitions to wavy flow and annular flow. When $\Delta \ll d$, i.e. in the wavy régime (4), the effect of increasing the vapour velocity is to amplify ripples until the wave-covered liquid is eventually driven up around the tube wall to create the annular flow régime (5). Whenever the superficial vapour velocity is high enough, e.g. $100\,\mathrm{m\,s^{-1}}$ an annular flow is produced whether the tube is vertical or horizontal, as Figs 3.13 and 3.15 confirm. For the horizontal tube, the annular régime succeeds the stratified and wavy régimes as the vapour velocity increases provided the vapour fraction is high, i.e. $\Delta \ll d$. With moderate vapour fractions (beyond $\Delta \simeq d/2$) the annular régime is preceded by the intermittent slug or plug régimes. The transition from bubbly to annular flow is restricted to high liquid velocities, as in vigorous flow boiling during which the liquid initially fills the tube. Examples of annular flow are given in problems Q6.12 and 6.13 in Chapter 8.

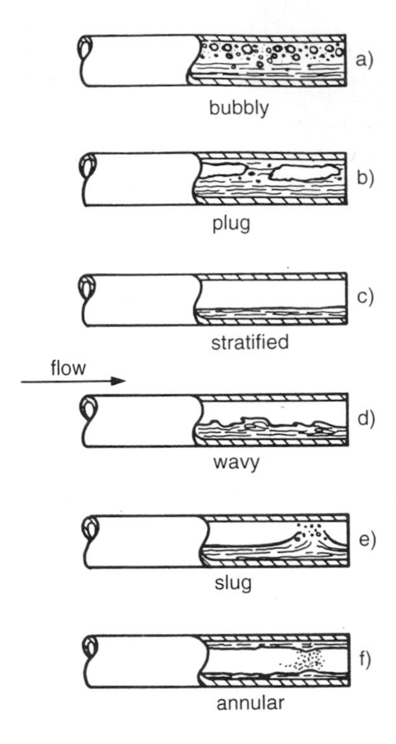

Fig. 3.16 Two-phase flow patterns in a horizontal tube (after Collier 1972).

Selected bibliography

Arpaci, V. S. and Larsen, P. S. (1984). *Convection heat transfer*. Prentice Hall, Englewood Cliffs, New Jersey.

Bejan, A. (1984). *Convective heat transfer*. Wiley, New York.

Bergles, A. E., Collier, J. G., Delhaye, J. M., Hewitt, G. F., and Mayinger, P. (eds.) (1981). *Two-phase flow and heat transfer in the power and process industries*. Hemisphere, Washington.

Clift, R., Grace, J. R., and Weber, M. E. (1978). *Bubbles, drops and particles*. Academic Press, New York.

Collier, J. G. (1972). *Convective boiling and condensation*. McGraw-Hill, New York.

Govier, G. W. and Aziz, K. (1972). *The flow of complex mixtures in pipes*. Van Nostrand Reinhold, New York.

Hartley, D. E. and Murgatroyd, W. (1964). Criteria for the break-up of thin liquid layers flowing isothermally over solid surfaces. *International Journal of Heat Mass Transfer*, **7**, 1003–15.

Hetsroni, G. (ed.) (1982). *Handbook of multiphase systems*. McGraw-Hill, New York.

Hewitt, G. F. and Hall-Taylor, N. S. (1971). *Annular two phase flow*. Pergamon, Oxford.

Lamb, H. (1932). *Hydrodynamics*, 6th edn. Dover, New York.
Langmuir, I. and Blodgett, K. B. (1946). A mathematical investigation of water droplet trajectories. *US Army Air Forces Tech, Rep.* 5418.
Mesler, R. and Mailen, G. (1977). Nucleate boiling in thin films. *AIChE Journal*, **23**, 954–57.
Mikielewicz, J. and Moszynski, J. R. (1976). Minimum thickness of a liquid film flowing vertically down a solid surface. *International Journal of Heat Mass Transfer*, **19**, 771–76
Pumphrey, H. C. and Elmore, P. A. (1990). The entrainment of bubbles by drop impacts. *Journal of Fluid Mechanics*, **220**, 539–67.
Rayleigh, Lord (1916). On the instability of cylindrical fluid surfaces. *Philosophical Magzine*, **5** xxxiv, 177.
Soo, S. L. (1967). *Fluid dynamics of multiphase systems*. Blaisdell, Waltham.

Exercises

1. Show that the square bracketed terms in eqn (3.7) are given by:
   (a) $Re_f \Delta/L$, 1 and $(\Delta/L)^2$;
   (b) $(Re_f^4/Ar)^{1/3}$, 1 and $(Re_f/Ar)^{2/3}$.
2. Find the velocity profile $U(Y)$ for a liquid film flowing down a vertical surface if $\tau_I = 0$. Hence show that:
   (a) $\tau_I = -g(\rho_L - \rho_V)\Delta/2$ when the film surface velocity $U(\Delta)$ is zero,
   (b) $\tau_I = -2g(\rho_L - \rho_V)\Delta/3$ when the net film mass flux is zero.
3. Use Fig. 3.2 with the addition of gravity to derive a more general force balance in which inertia is ignored. Hence:
   (a) show that when $\bar{\Delta} \ll D$,
   $$\tau_w = \tau_I + \beta\bar{\Delta}$$
   where $\beta = g(\rho_L - \rho_V) - \partial P_g/\partial X$,
   (b) find τ_w when $\partial P_g/\partial X = 0$.
   (c) estimate $\partial P_g/\partial X$ for flow reversal of a saturated water film at atmospheric pressure.
   (Ans: (b) $\bar{\Delta}g(\rho_L - \rho_V)$; (c) 9.4 kPa m^{-1}.)
4. Compare the structure of the Archimedes number with that of the Grashof and Rayleigh numbers. Identify:
   (a) the length scale,
   (b) Archimedean and thermal buoyancy (body force per unit mass),
   (c) momentum and thermal diffusivities.
5. If the inertial, viscous, buoyancy, and surface tension forces are represented by $I = \rho L^2 U^2$, $V = \mu UL$, $B = L^3 g\Delta\rho$, and $T = \sigma L$, respectively, with L being a length scale, demonstrate that
   (a) $Re = I/V$
   (b) $Bo = B/T$

(c) $Ar = BI/V^2$

(d) $Mo = \frac{Bo}{Re^2}\left(\frac{V}{T}\right)^2$

(e) $Ku^4 = I^2/BT$

6. From the falling film velocity profile determined in the second question, show that the average film velocity is given by

$$\bar{U}_f = \Delta^2 g(\rho_L - \rho_V)/3\mu$$

when $\tau_I = 0$. Hence estimate the critical film thickness for rivulet formation in a saturated water film at atmospheric pressure when the contact angle $\theta = 30°$.

(Ans: 172 μm.)

7. Air at 20°C flows through a long, 4 cm diameter pipe at a rate of 6×10^{-3} kg s^{-1}. Water is sprayed into the pipe entry at a rate of 50×10^{-3} kg s^{-1}. If the fluids subsequently flow cocurrently, calculate the superficial phase velocities and identify the flow régime:

(a) if the pipe is vertical,

(b) if the pipe is horizontal,

Is the flow stable:

(c) if the pipe is vertical?

(d) if the pipe is horizontal?

(Ans: $U_L^s = 3.99$ cm s^{-1}, $U_V^s = 3.98$ m s^{-1}.)

(a) churn for upflow; annular? for downflow,

(b) stratified,

(c) the churn flow is unstable, and the annular? flow exhibits a Weberian instability,

(d) no Froudian instability.

8. What diameter drops may fall from condensate on a chilled water line overhead?

(Ans: 8.4 mm.)

9. Cloud droplets grow by accretion into large raindrops. Estimate:

(a) their maximum (critical) diameter,

(b) the corresponding terminal velocity assuming a spherical shape.

(Ans: 9.8 mm, 15.5 m s^{-1}.)

10. A water spray with the parameters $\alpha = 6$, $b = 0.6$ μm^{-1} and $a = 10^5\, m^{-3}$ (μm)$^{-6}$ satisfies eqn (3.42) with d in μm. Calculate the mean volume drop diameter if it is 33 per cent greater than the mode diameter (greatest number of drops). Hence find the water flux density if the drops move without slip in an air stream at 20 m s^{-1}.

(Ans: 13.3 μm, 63.3 g m^{-2} s^{-1}.)

4
SOLIDIFICATION AND FLUIDIFICATION

Solidification is defined as the process in which a fluid is transformed into a solid; fluidification is the reverse process. Since fluids take two basic forms, there are four principal phase change processes: solid → liquid, known as melting, and solid → vapour, known as sublimation, are both examples of fluidification; liquid → solid, known as freezing, and vapour → solid, known as deposition, are the two solidification processes. The solid → solid transformation completes the list of commonly encountered latent heat transfer processes involving solid phases.

These processes have many things in common but they also display significant differences. Accordingly, this chapter is organized in such a way as to develop a common thread, physically and mathematically, while at the same time paying attention to unique and particular behaviour. The treatment is thus divided into two parts. In Sections 4.1–4.3, the essential nature of the heat transfer problem is introduced and explored. Following this are three sections dealing with industrial and environmental situations such as the freezing of aqueous solutions, the casting of metals, and atmospheric icing.

Given the presence of a solid phase, hydrodynamic considerations are restricted to one side of the interface only. Frequently, this permits us to incorporate convection by making use of the heat transfer coefficient h at the interface. That is, the fluid heat flux density at the interface is written in the form

$$k(\nabla T)_I = h(T_\infty - T_\mathrm{I}),$$

where T_∞ is the bulk fluid temperature. This procedure greatly simplifies many solidification and fluidification problems. Even so, it must always be borne in mind that h is not a constant; in particular, it is a function of the prevailing velocity and temperature fields.

4.1 Scale analysis and the Stefan problem

4.1.1 Scales

In latent heat transfer, as in sensible heat transfer, it is always useful and frequently insightful to recognize the natural scales of the physical variables; specifically, the scales of temperature, velocity, time and distance. Some of these are usually known from the way the problem is described. Other scales may not be known at the outset but they still exist. To determine them it is first necessary to describe the problem physically and

mathematically; specifically, to write down the differential equations which describe events fully. The general technique was developed in Section 1.6.

The essence of a solidification problem is contained in two equations: the heat conduction equation and the interface equation. To these must be added any boundary or initial conditions; and, if necessary, the details of a phase diagram. For simplicity, here and elsewhere, we will assume that the thermophysical properties of the solid and the liquid are constant, though different; only the densities will be taken equal, or almost equal, so as to eliminate any freeze-induced motion or pressure change. Thus, for one-dimensional freezing of a quiescent, pure liquid, as described in Fig. 4.1, the situation may be summed up by the following:

$$\frac{\partial T_S}{\partial t} = \kappa_S \frac{\partial^2 T_S}{\partial Y_S^2} \tag{4.1}$$

in the solid,

$$\frac{\partial T_L}{\partial t} = \kappa_L \frac{\partial^2 T_L}{\partial Y_L^2} \tag{4.2}$$

in the liquid, and

$$\lambda \rho_S \frac{\mathrm{d}Y_I}{\mathrm{d}t} = k_S \left(\frac{\partial T_S}{\partial Y_S}\right)_I - k_L \left(\frac{\partial T_L}{\partial Y_L}\right)_I \tag{4.3}$$

at the interface where $Y = Y_I$ and $T = T_I$. Let us assume that liquid at temperature $T_\infty > T_I$, initially filled the space $Y \geqslant 0$, and that the process

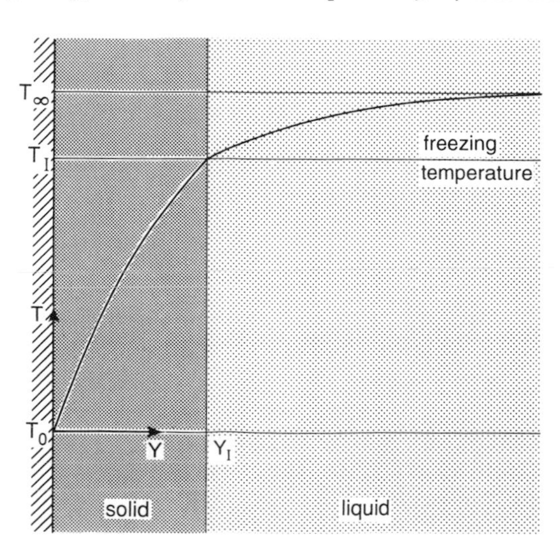

Fig. 4.1 Temperature profiles during one-dimensional freezing of a liquid initially at T_∞.

began at $t=0$ when the surface temperature was suddenly lowered to the fixed value $T_0 < T_I$.

It is evident from Fig. 4.1 that some scales are implied: in the liquid, $T_I < T_L < T_\infty$; in the solid, $T_0 < T_S < T_I$, and $0 < Y_S < Y_I$. Designating the scale of a variable by a superscript c, we take

$$\theta_L^c = T_\infty - T_I, \quad \theta_S^c = T_I - T_0, \quad \text{and} \quad Y_S^c = Y_I \tag{4.4}$$

as prescribed scales; strictly, *difference* scales. The remaining scales are unknown. These are: t^c, Y_L^c, and Y_I^c, the last of which is particularly important because, as the thickness scale of the solid formed, it provides the information most often sought; it also determines the third scale in eqns (4.4).

To begin with, consider the situation where $T_\infty = T_I$ and conduction in the liquid may therefore be ignored. This eliminates eqn (4.2) and simplifies the interface equation which may be re-written after introducing the normalized variables

$$\phi_S = \frac{T - T_0}{\theta_S^c}, \quad y_S = \frac{Y_S}{Y_S^c}, \quad \tau = \frac{t}{t^c}, \quad \text{and} \quad \xi = \frac{Y_I}{Y_I^c}. \tag{4.5}$$

Substituting these variables into the reduced interface equation, we obtain

$$\frac{d\xi}{d\tau} = \left[\frac{k_S \theta_S^c t^c}{\lambda \rho_S (Y_I^c)^2}\right] \left(\frac{\partial \phi_S}{\partial y_S}\right)_I, \tag{4.6}$$

since $Y_S^c = Y_I^c$, as implied in eqns (4.4). The coefficient in square brackets is non-dimensional and its magnitude depends upon the unknown scales t^c and Y_I^c. On closer examination, this magnitude is not as arbitrary as it may appear because we anticipate that the derivatives in the equations, being normalized, are of order one. Subject to this approximation, the only way eqn (4.6) may be satisfied in general is if

$$\frac{k_S \theta_S^c t^c}{\lambda \rho_S (Y_I^c)^2} = O(1), \tag{4.7}$$

which provides a *qualitative* relation between the unknown scales t^c and Y_I^c. This may be converted into a *quantitative* relation by simply exercising our freedom of choice and taking the right-hand side to be 1.0 exactly. The equation then becomes a definition based on the physics of the interface.

To illustrate the usefulness of this result, let us estimate the thickness of the ice cover formed on a lake after four winter months if the air temperature near the ice averages $-10°C$. Re-arranging eqn (4.7),

$$Y_I^c \simeq \left(\frac{k_S \theta_S^c t^c}{\lambda \rho_S}\right)^{1/2}.$$

With the required thermophysical properties given by $k_S = 2.17\ \mathrm{W\,m^{-1}\,K^{-1}}$, $\rho_S = 0.92 \times 10^3\ \mathrm{kg\,m^{-3}}$ and $\lambda = 334\ \mathrm{kJ\,kg^{-1}}$, we find that $X_I^c \simeq 0.86$ m, thick enough to walk on at a temperature of $-10°$C.

4.1.2 *The Stefan number*

Turning to the heat conduction eqn (4.1), we may now substitute the normalized variables of eqn (4.5) to find

$$\left[\frac{(Y_I^c)^2}{\kappa_S t^c}\right]\frac{\partial \phi_S}{\partial \tau} = \frac{\partial^2 \phi_S}{\partial y_S^2}. \tag{4.8}$$

The coefficient on the left-hand side, again non-dimensional, is reminiscent of a reciprocal Fourier number *Fo* but has little meaning unless the length and time scales are known. Equation (4.7) provides these and may be rearranged to yield

$$\frac{(Y_I^c)^2}{\kappa_S t^c} = \frac{c_{pS}\theta_S^c}{\lambda} = Ste_S,$$

the *Stefan number* of the solid layer; it is the ratio of sensible and latent heat. Hence

$$Ste_S \frac{\partial \phi_S}{\partial \tau} = \frac{\partial^2 \phi_S}{\partial y_S^2} \tag{4.9}$$

describes conduction in the solid.

It is evident that the magnitude of the Stefan number determines the relative importance of the left-hand side of eqn (4.9). When $Ste \ll 1$, the transient term may be neglected; conduction is then essentially the same as under steady conditions. That is, whenever the sensible heat in the solid layer is very small compared with the latent heat absorbed at the interface, the temperature distribution in the planar solid will be almost linear. A 10 K temperature difference across an ice layer corresponds to $Ste_S \simeq 0.06$. The neglect of transience always implies the steady-state distribution of temperature in the frozen material, regardless of its geometry. The effect of time is then felt only through the slow movement of the interface, and the problem takes a quasi-steady form.

The structure of the planar freezing problem must be modified slightly if sensible heat transfer from the liquid is to be incorporated. Normalized variables may again be introduced into the interface eqn (4.3) to yield

$$\frac{d\xi}{d\tau} = \left(\frac{\partial \phi_S}{\partial y_S}\right)_I - \left(\frac{Y_S^c}{Y_L^c}\right)\left(\frac{k_L}{k_S}\right)\left(\frac{\theta_L^c}{\theta_S^c}\right)\left(\frac{\partial \phi_L}{\partial y_L}\right)_I \tag{4.10}$$

in which the characteristic scales obtained from eqn (4.7) have been retained. It is now evident that conduction in the liquid may be ignored if the coefficient $Y_I^c k_L \theta_L^c / Y_L^c k_S \theta_S^c \ll 1$. This would require, for example, that the

superheat ratio $\theta_L^c/\theta_S^c = (T_\infty - T_I)/(T_I - T_0)$ and/or the conductivity ratio k_L/k_S be very small. Alternatively, the length ratio Y_I^c/Y_L^c may be very small, as when an ice cover forms over deep water. If the coefficient is not small, the full eqn (4.10) must be retained and the solution of the heat conduction eqn (4.2) in the liquid must be incorporated. Using normalized variables, the latter becomes

$$Ste_S \frac{\kappa_S}{\kappa_L}\left(\frac{Y_L^c}{Y_I^c}\right)^2 \frac{\partial \phi_L}{\partial \tau} = \frac{\partial^2 \phi_L}{\partial y_L^2} \tag{4.11}$$

which reduces to quasi-steady form only if the coefficient on the left-hand side is very small; otherwise, the temperature distribution in the liquid is intrinsically transient. An example is given in Section 4.3.

Freezing and melting problems such as the above have the same mathematical characteristics as deposition and sublimation problems; only the latent heat changes. Collectively, they are usually known as Stefan problems. They are quite similar to the Nusselt problems discussed in the next chapter.

4.2 Simple Stefan problems

4.2.1 Simple planar problems

Figure 4.2 illustrates the Stefan problem in its most elementary planar form. A solid, initially at the uniform freezing temperature T_I, is melted by the application of a fixed temperature $T_0 > T_I$ at its free surface where $Y = 0$. When the Stefan number $Ste_L = c_{pL}(T_0 - T_I)/\lambda$ is small, the melting takes

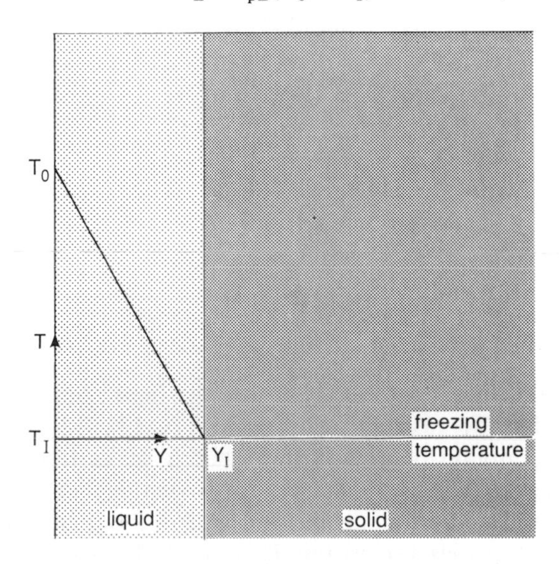

Fig. 4.2 The simple, planar Stefan problem during melting.

place very slowly, as noted above. The heat conduction equation for the liquid, here assumed to be quiescent, then reduces to

$$\frac{\mathrm{d}^2 T_\mathrm{L}}{\mathrm{d}Y^2} = 0$$

which, with the boundary conditions $T_\mathrm{L}(0) = T_0$ and $T_\mathrm{L}(Y_\mathrm{I}) = T_\mathrm{I}$, has the solution

$$T_\mathrm{L}(Y) = \left(\frac{T_\mathrm{I} - T_0}{Y_\mathrm{I}}\right) Y + T_0, \tag{4.12}$$

where Y_I is the melt thickness.

Since the solid is not subcooled, all the heat flowing through the liquid is absorbed at the interface. The interface equation thus reduces to

$$\lambda \rho_\mathrm{L} \frac{\mathrm{d}Y_\mathrm{I}}{\mathrm{d}t} = -k_\mathrm{L} \left(\frac{\mathrm{d}T_\mathrm{L}}{\mathrm{d}Y}\right)_\mathrm{I} \tag{4.13}$$

in which it is worth noting that the temperature gradient is uniform throughout the liquid and may be found from eqn (4.12) to be

$$\frac{\mathrm{d}T_\mathrm{L}}{\mathrm{d}Y} = \frac{T_\mathrm{I} - T_0}{Y_\mathrm{I}}. \tag{4.14}$$

Substituting this into eqn (4.13), we obtain

$$Y_\mathrm{I} \frac{\mathrm{d}Y_\mathrm{I}}{\mathrm{d}t} = \frac{1}{2} \frac{\mathrm{d}Y_\mathrm{I}^2}{\mathrm{d}t} = \frac{k_\mathrm{L}(T_0 - T_\mathrm{I})}{\lambda \rho_\mathrm{L}}$$

which integrates to give

$$Y_\mathrm{I}(t) = \left[\frac{2k_\mathrm{L}(T_0 - T_\mathrm{I})t}{\lambda \rho_\mathrm{L}}\right]^{1/2} \tag{4.15}$$

if no liquid is present initially, i.e. $Y_\mathrm{I}(0) = 0$. This is sometimes called the *Stefan solution*, and was anticipated in the scale relation, eqn (4.7). Should a liquid layer of thickness Y_i be present initially,

$$Y_\mathrm{I}(t) = \left[\frac{2k_\mathrm{L}(T_0 - T_\mathrm{I})t}{\lambda \rho_\mathrm{L}} + (Y_\mathrm{i})^2\right]^{1/2},$$

provided that the energy required to raise the liquid temperature is negligible, as implied by taking $Ste_\mathrm{L} \simeq 0$, i.e. $c_{p\mathrm{L}}(T_0 - T_\mathrm{I}) \ll \lambda$. A similar solution may be obtained for a liquid freezing at its free surface.

The above expressions for the melted (or frozen) layer thickness are often useful in making rapid estimates but they suffer from two major practical limitations: the surface temperature T_0 may not be fixed, or even known; and the solid (or liquid) may not be at the equilibrium freezing temperature initially. When T_0 varies with time in a known way, the interface equation may accommodate the variation by taking

$$\frac{1}{2}\frac{\mathrm{d}}{\mathrm{d}t}(Y_\mathrm{I})^2 = \frac{k_\mathrm{L}}{\lambda\rho_\mathrm{L}}[T_0(t) - T_\mathrm{I}],$$

during melting conditions, for example. Integrating,

$$Y_\mathrm{I}(t) = \left[\frac{2k_\mathrm{L}}{\lambda\rho_\mathrm{L}}\int_0^t [T_0(t) - T_\mathrm{I}]\,\mathrm{d}t\right]^{1/2} \tag{4.16}$$

if no melt layer exists initially. The integral is often called the thaw (or freeze) index, and is used in the determination of the mean effective temperature above (or below) the freezing point.

When the free boundary heat transfer occurs convectively in a gas with a bulk temperature T_A, the situation is described by Fig. 4.3 which also accommodates the general condition of a solid subcooled to $T_{\mathrm{S}\infty}$ initially. Under these conditions, $T_0(t)$ is an unknown function of time and hence eqn (4.16) is no longer useful. It is necessary instead to re-solve the problem using the convective boundary condition

$$h_\mathrm{A}(T_\mathrm{A} - T_0) = -k_\mathrm{L}\left(\frac{\mathrm{d}T_\mathrm{L}}{\mathrm{d}Y}\right)_0$$

combined with eqn (4.14) to obtain $T_0(t)$ as a function of $Y_\mathrm{I}(t)$. Hence

$$T_0(t) = \left[\left(\frac{h_\mathrm{A}Y_\mathrm{I}}{k_\mathrm{L}}\right)T_\mathrm{A} + T_\mathrm{I}\right] \bigg/ \left(\frac{h_\mathrm{A}Y_\mathrm{I}}{k_\mathrm{L}} + 1\right). \tag{4.17}$$

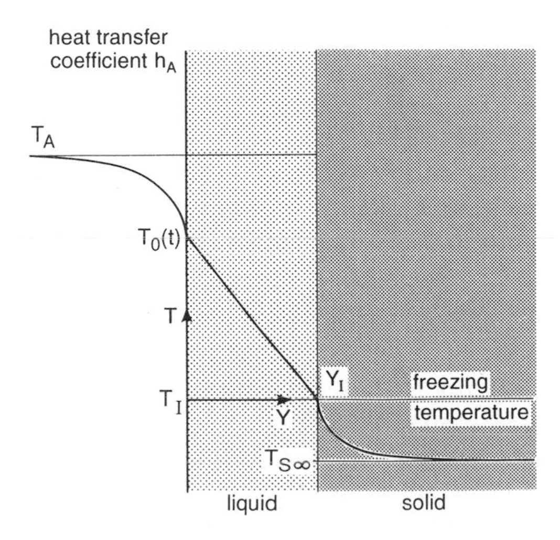

Fig. 4.3 The simple, planar Stefan problem with a convection boundary condition and subcooling.

Substituting this into the interface equation we obtain

$$\left(Y_{\mathrm{I}}+\frac{k_{\mathrm{L}}}{h_{\mathrm{A}}}\right)\frac{\mathrm{d}Y_{\mathrm{I}}}{\mathrm{d}t}=\frac{k_{\mathrm{L}}(T_{\mathrm{A}}-T_{\mathrm{I}})}{\lambda\rho_{\mathrm{L}}}$$

if the solid temperature is T_{I}, throughout. This integrates to give

$$t=\frac{\lambda\rho_{\mathrm{L}}}{2k_{\mathrm{L}}(T_{\mathrm{A}}-T_{\mathrm{I}})}\left[Y_{\mathrm{I}}^{2}+\frac{2k_{\mathrm{L}}Y_{\mathrm{I}}}{h_{\mathrm{A}}}\right]. \tag{4.18}$$

When $h_{\mathrm{A}}\rightarrow\infty$, $T_0\rightarrow T_{\mathrm{A}}$ and the Stefan solution, eqn (4.15), is recovered. On the other hand, when h_{A} is very small the solution takes a different form in which Y_{I} is directly proportional to time. This condition may occur, for example, when freeze-shrinkage creates a high resistance gap between a casting and the mould wall.

If the solid is subcooled, the complete interface equation must be used together with the heat conduction equation for the solid. On the other hand, if the problem is one of freezing at the surface of a superheated liquid in which a heat transfer coefficient h_{L} exists, the complete interface equation may be retained in the form

$$\lambda\rho_{\mathrm{S}}\frac{\mathrm{d}Y_{\mathrm{I}}}{\mathrm{d}t}=k_{\mathrm{S}}\frac{(T_{\mathrm{I}}-T_0)}{Y_{\mathrm{I}}}-h_{\mathrm{L}}(T_{\mathrm{L}\infty}-T_{\mathrm{I}}). \tag{4.19}$$

This is readily integrated after substituting

$$\tilde{Y}_{\mathrm{I}}=\frac{k_{\mathrm{S}}(T_{\mathrm{I}}-T_0)}{\lambda\rho_{\mathrm{S}}}-\frac{h_{\mathrm{L}}(T_{\mathrm{L}\infty}-T_{\mathrm{I}})}{\lambda\rho_{\mathrm{S}}}Y_{\mathrm{I}}.$$

Solutions to some simple planar Stefan problems are given in Q4.1–Q4.3 of Chapter 8.

4.2.2 Simple problems in other geometries

Simple Stefan problems also occur in other common geometries, most notably in spherical and cylindrical coordinate systems. A typical example is given by the freezing of water in a pipe. Figure 4.4 illustrates the situation in which convective cooling around the pipe establishes the heat transfer coefficient h_{A} such that

$$h_{\mathrm{A}}(T_0-T_{\mathrm{A}})=-k_{\mathrm{S}}\left(\frac{\mathrm{d}T_{\mathrm{S}}}{\mathrm{d}R}\right)_{R_0} \tag{4.20}$$

describes the rate of heat loss through the pipe wall. Since the steady-state temperature distribution in the ice is given by

$$\frac{T-T_0}{T_{\mathrm{I}}-T_0}=\frac{\ln R/R_0}{\ln R_{\mathrm{I}}/R_0},$$

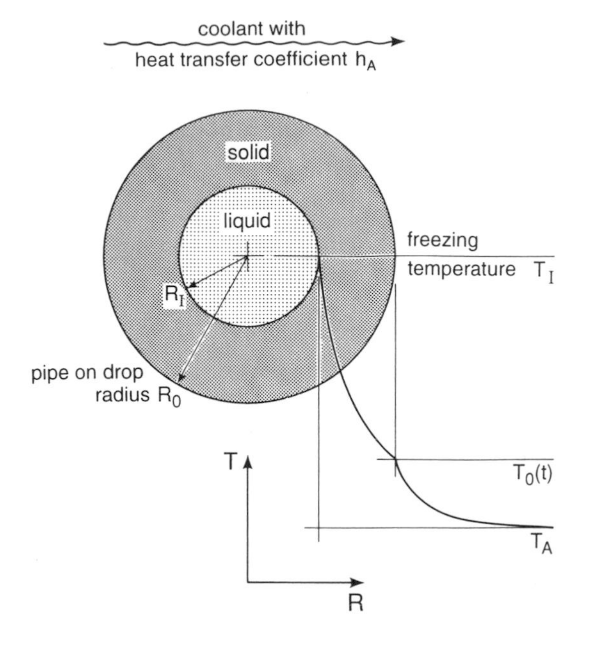

Fig. 4.4 The simple Stefan freezing problem in cylindrical (or spherical) coordinates.

eqn (4.20) may be used to obtain

$$T_0(t) = \frac{T_I - T_A Bi \ln R_I(t)/R_0}{1 - Bi \ln R_I(t)/R_0} \tag{4.21}$$

as the varying pipe wall temperature: $Bi = h_A R_0/k_S$ is a Biot number. Now if the water is initially at T_I, the interface equation becomes

$$\lambda\rho_S \frac{dR_I}{dt} = k_S \left(\frac{\partial T_S}{\partial R}\right)_I = \frac{k_S(T_I - T_0)}{R_I \ln R_I/R_0}$$

which, after substituting $T_0(t)$ from eqn (4.21), may be integrated to give

$$\tau = \frac{1}{2}\left[r_I^2 \ln r_I + \left(\frac{1}{Bi} + \frac{1}{2}\right)(1 - r_I^2)\right] \tag{4.22}$$

if the pipe is initially ice free, $r_I = R_I/R_0$ and $\tau = k_S(T_I - T_A)t/\lambda\rho_S R_0^2$.

When Fig. 4.4 applies to a spherical water drop falling in a subzero atmosphere, a similar analysis leads to

$$t = \frac{\lambda\rho_S R_0^2}{2k_S(T_I - T_A)}\left[\frac{2}{3Bi}(1 - r_I^3) + \frac{1}{3}(1 + 2r_I^3) - r_I^2\right], \tag{4.23}$$

where R_0 now refers to the drop radius. The time required to completely freeze the drop or shut off the pipe is found by putting $r_I = 0$ in the above expressions. For example, if the heat transfer coefficient around a 10 cm diameter, water-filled, copper pipe situated in a cross flow of air at −20°C is 220 W m^{-2}K^{-1}, then $Bi = 5$. Using eqn (4.22), the closure time is found from

$$\tau_f = \frac{k_S(T_I - T_A)t_f}{\lambda\rho_S R_0^2} = \frac{1}{2}\left(\frac{1}{2} + \frac{1}{Bi}\right) = 0.35.$$

Hence $t_f = 103$ minutes.

A comparison of theoretical and experimental results for ice growth in a convectively cooled pipe is shown in Fig. 4.5. The experimental data were obtained with two different interface locators: a mechanical device which touched the interface and an ultrasonic device which measured ice thickness directly. The data are in excellent agreement with the accurate theoretical solution. The prediction of eqn (4.22) overestimates the ice thickness but underestimates the closure time. However, it does demonstrate the usefulness of quasi-steady analyses when $Ste \ll 1$ in the growing phase, especially during the early stages of growth.

4.3 The Neumann method

4.3.1 *The classical solution and its limiting form*

The single most important analytic solution in freezing and melting is the Neumann solution. Unlike the Stefan solution it is not restricted to small values of *Ste*; nor is it confined to heat conduction in the forming phase

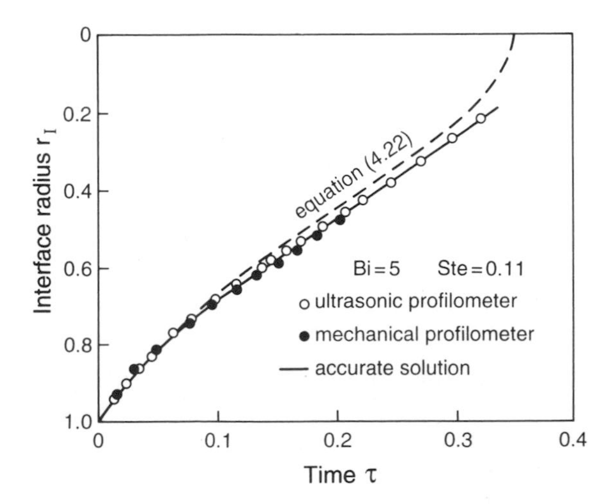

Fig. 4.5 Ice growth in a convectively-cooled pipe (after Lock, 1974).

only. It is, however, limited to a simple temperature boundary condition applied at the free surface of a semi-infinite domain. This condition requires that the surface temperature, once it is altered to initiate freezing (or melting), must remain fixed. The situation was described earlier by Fig. 4.1 together with eqns (4.1), (4.2), and (4.3) which pose a one-dimensional, transient problem: a quiescent liquid, possesses an initial, and uniform, temperature $T_\infty > T_I$. At time $t = 0$, the temperature at the surface $Y = 0$ is suddenly lowered to T_0 where it remains. Equations (4.1) and (4.2) possess a general solution of the form

$$T(X, t) = A\,\mathrm{erf}[Y/2(\kappa t)^{1/2}] + B$$

in which the arbitrary constants A and B are determined from suitable boundary conditions. It is convenient here to replace $T(Y, t)$ by $\theta(Y, t) = T - T_0$.

The temperature in the solid may thus be described by

$$\theta_S(Y, t) = A\,\mathrm{erf}[Y_S/2(\kappa_S t)^{1/2}]. \qquad (4.24)$$

This satisfies the boundary condition at $Y = 0$ and applies in the range $0 < Y_S < Y_I$. The corresponding solution in the liquid $Y_I < Y_L < \infty$, is given by

$$\theta_L(Y, t) = \theta_\infty + B\,\mathrm{erfc}[Y_L/2(\kappa_L t)^{1/2}] \qquad (4.25)$$

in which $\theta_\infty = \theta_L(\infty, t) = T_\infty - T_0$, thus satisfying the boundary condition at $Y_L = \infty$ where $T = T_\infty$. At the interface, where $Y = Y_I$, the above solutions must also satisfy the condition $\theta_S = \theta_L = \theta_I = T_I - T_0$. Thus,

$$A\,\mathrm{erf}[Y_I^2/2(\kappa_S t)^{1/2}] = \theta_I = \theta_\infty + B\,\mathrm{erfc}[Y_I^2/2(\kappa_L t)^{1/2}], \qquad (4.26)$$

from which it follows that the arbitrary constants A and B are given by

$$A = \theta_I/\mathrm{erf}\beta, \quad B = (\theta_I - \theta_\infty)/\mathrm{erfc}[\beta(\kappa_S/\kappa_L)^{1/2}], \qquad (2.27)$$

where $\beta = Y_I/2(\kappa_S t)^{1/2}$ is also a constant, as required by eqn (4.26) in which the arguments of the error functions must be constants. Thus

$$Y_I = 2\beta(\kappa_S t)^{1/2}, \qquad (4.28)$$

the form seen earlier in the Stefan solution.

The temperature distributions in the solid and liquid phases are given by

$$\theta_S(Y_S, t) = \theta_I\,\mathrm{erf}[Y_S/2(\kappa_S t)^{1/2}]/\mathrm{erf}\beta \qquad (4.29)$$

and

$$\theta_L(Y_L, t) = \theta_\infty\,\{1 - (1 - \theta_I/\theta_\infty)\,\mathrm{erfc}[Y_L/2(\kappa_L t)^{1/2}]/\mathrm{erfc}[\beta(\kappa_S/\kappa_L)^{1/2}]\}\,. \qquad (4.30)$$

As indicated in Fig. 4.1, these temperature profiles exhibit a curvature attributable to the non-linear shape of the error functions. It is interesting to compare this feature with the linear profiles found in the Stefan solution.

When β is very small, it follows from eqn (4.28) that since $Y_S \leqslant Y_I$, $Y_S \ll 2(\kappa_S t)^{1/2}$, in which case

$$\mathrm{erf}[Y_S/2(\kappa_S t)]^{1/2} \simeq Y_S/(\pi\kappa_S t)^{1/2}.$$

Equation (4.29) then reduces to

$$\theta_S(Y_S, t) \simeq \theta_I Y_S/(\pi\kappa_S t)^{1/2}\,\mathrm{erf}\beta \tag{4.31}$$

which has the linear temperature profile of the Stefan solution.

In general, β must be determined from the interface eqn (4.3). Substituting the temperature profiles presented above in eqns (4.29) and (4.30) we obtain

$$\frac{\beta\pi^{1/2}}{Ste_S} = \frac{\exp(-\beta^2)}{\mathrm{erf}\beta} - \frac{k_L}{k_S}\left(\frac{\kappa_S}{\kappa_L}\right)^{1/2}\left(\frac{\theta_\infty - \theta_I}{\theta_I}\right)\frac{\exp(-\beta^2\kappa_S/\kappa_L)}{\mathrm{erfc}[\beta(\kappa_S/\kappa_L)^{1/2}]}, \tag{4.32}$$

where $Ste_S = c_{pS}\theta_I/\lambda$. This transcendental equation reveals that β has two basic dependencies: on material properties in both phases, and on the temperature scales in both phases. For the special situation in which conduction may be ignored in the parent phase, i.e. $T_\infty = T_I$, and hence $\theta_\infty = \theta_I$, the equation reduces to

$$Ste_S = \pi^{1/2}\beta\exp\beta^2\,\mathrm{erf}\beta, \tag{4.33}$$

thus providing β in terms of the Stefan number alone. When β is small, this simplifies even further. Expanding the exponential and error functions in series forms, we obtain

$$Ste_S = 2\beta^2\left(1 + \frac{2}{3}\beta^2 \ldots\right)$$

and hence as $\beta \to 0$, $Ste_S \to 2\beta^2$. Equation (4.28) then yields

$$Y_I = (2k_S\theta_I t/\lambda\rho_S)^{1/2},$$

the Stefan solution, which is thus seen to be a limiting form of the Neumann solution when $Ste_S \to 0$. Several applications of the Neumann solution are provided in Q4.4–Q4.7 of Chapter 8.

4.3.2 The Neumann method applied to a supercooled liquid

Another interesting application of Neumann's method occurs when the liquid has an initial temperature T_∞ less than T_I; it is therefore supercooled. In this condition, the latent heat released at the interface will, in general, be conducted away into both the liquid and the solid. However, if the solid is not cooled artificially, as when it is entirely surrounded by the liquid, its temperature remains uniformly at the equilibrium freezing temperature T_I required by successive layers of new solid. This is suggested in Fig. 4.6 using the mirror image formed by taking $Y = 0$ as a plane

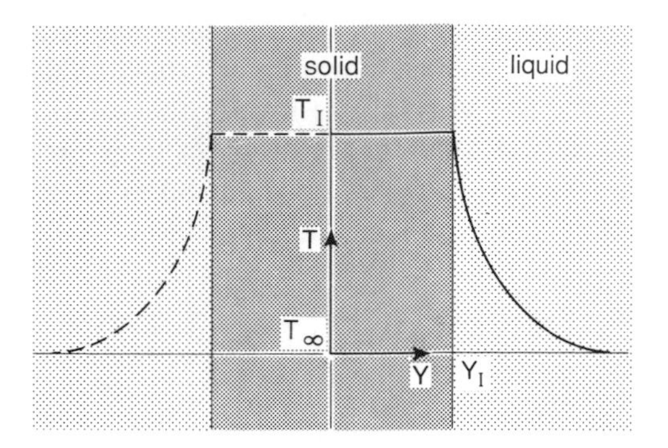

Fig. 4.6 Planar solidification in a supercooled liquid: $T_\infty < T_I$.

of symmetry in a solidifying slab. Conduction in the solid may then be ignored, but eqns (4.2) and (4.3) continue to apply.

Following the above treatment, heat conduction in the liquid leads to the result

$$\Theta_L(Y, t) = \Theta_I \operatorname{erfc}[Y_L/2(\kappa_L t)^{1/2}]/\operatorname{erfc}\beta, \tag{4.34}$$

where $\Theta = T - T_\infty$ and β is now defined by

$$Y_I = 2\beta(\kappa_L t)^{1/2}. \tag{4.35}$$

Substituting these results into the interface equation then yields a transcendental equation for β. Thus

$$\rho_S Ste_L = \rho_L \pi^{1/2} \beta \exp\beta^2 \operatorname{erfc}\beta, \tag{4.36}$$

where $Ste_L = c_{pL}\Theta_I/\lambda$ refers to the liquid.

This problem is intrinsically transient in the fixed coordinate system chosen. Even so, the temperature distribution in the liquid does tend towards a steady-state solution in a coordinate system moving with the interface. Writing this coordinate as $Y' = Y_L - V_I t$, where $V_I = dY_I/dt$ is the *steady* interface velocity, the liquid heat conduction equation,

$$\frac{\partial \Theta_L}{\partial t} = \kappa_L \frac{\partial^2 \Theta_L}{\partial Y_L^2},$$

may be transformed into

$$\kappa_L \frac{\partial^2 \Theta}{\partial Y'^2} + V_I \frac{\partial \Theta}{\partial Y'} = 0$$

which has the solution

$$\Theta(Y') = \Theta_I \exp\left(-\frac{V_I Y'}{\kappa_L}\right), \tag{4.37}$$

satisfying the conditions $\Theta(0) = \Theta_I$, and $\Theta(\infty) = 0$.

The above treatments of freezing in a supercooled liquid leave a very important question unanswered. How long will the initially planar interface remain planar? While the interface is stable, the solutions remain valid; and they always provide a basis for the study of incipient instability. Once stability is lost, however, the freezing process takes a very different form. This is the subject of the next section.

4.4 Supercooling and its effects

4.4.1 Pure supercooling

The interface at Y_I in Fig. 4.6 is potentially unstable because any protuberance on it will penetrate into slightly cooler liquid and thus be in a better position to dissipate latent heat. In such a position, the protuberance will grow faster than the interfacial area immediately surrounding it and will therefore propagate even further ahead. In a pure supercooled liquid, this process is capable of producing crystalline spears, or dendrites, at many points on the interface simultaneously. The crystals are oriented with their fastest growth axes pointing down the temperature gradient.

Dendritic growth occurs when the degree of supercooling is sufficient to bring about nucleation. Most commonly, this takes the form of heterogeneous nucleation either on a confining substrate or on suitable microscopic sites distributed throughout the liquid bulk. Freezing in a domestic water pipe provides an instructive example which is illustrated in Fig. 4.7. As the pipe wall cools beneath 0°C some of the water eventually passes through 4°C, at which point the density reaches a maximum. Water which subsequently reaches temperatures lower than this is lighter and therefore begins to accumulate near the top of the pipe with the result that supercooling becomes greatest in this region. As Fig. 4.7(b) shows, nucleation and dendritic growth begins there. The continual withdrawal of heat from the pipe causes the dendrites to grow rapidly downwards, contacting many points around the pipe circumference in the process; this is indicated in Fig. 4.7(d) and (e). Near the pipe wall, the coldest region in the system, crystal growth is also rapid, creating a continuous polycrystalline coating which eventually forms the annulus seen in Fig. 4.7(e) and (f). Since this annulus prevents further supercooling of the water the existing dendrites grow more and more slowly, their deposition of latent heat gradually reducing the supercooling of the water surrounding them. Finally, growth is limited to the annulus alone. Curiously, the solid annulus may not alter the hydraulic characteristics of the pipe very much whereas the labyrinth of dendritic sheets, some of which extend from wall to wall, frequently make it very difficult to re-start the flow of water.

Crystalline growth may also occur around particles or ice nuclei dispersed throughout a supercooled liquid. Ice and water again provide a natural example, this time in a turbulent river during winter. It has often been

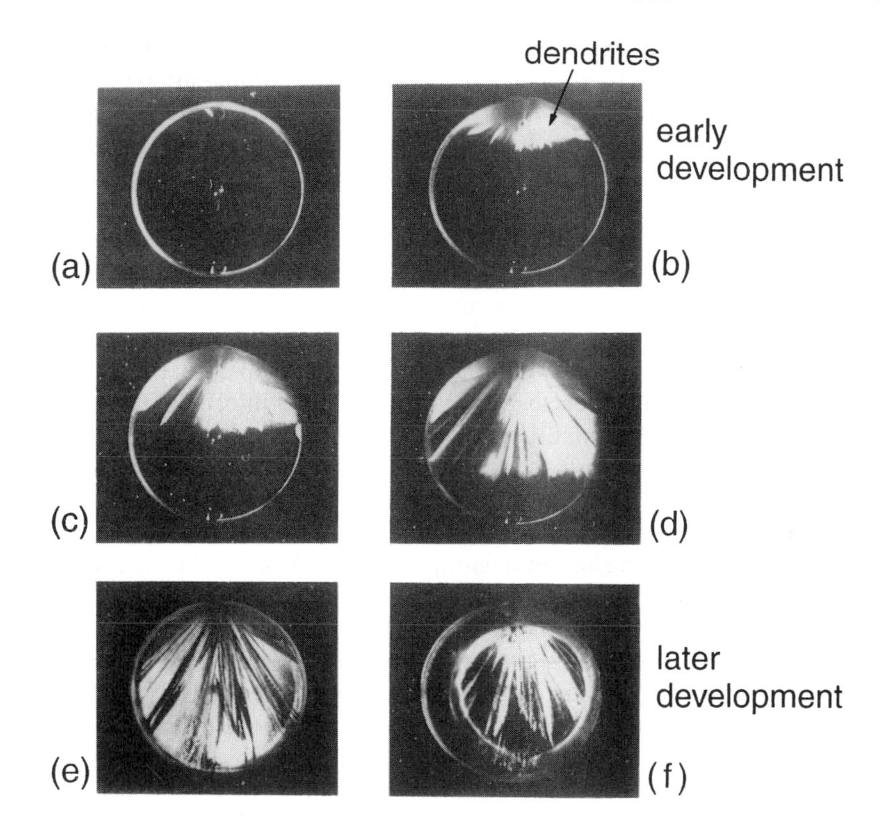

Fig. 4.7 Ice formation in a horizontal, domestic water pipe (after Gilpin 1978).

observed that the turbulent water in rapids can prevent an ice cover from forming, even though the air temperature may be substantially below 0°C. The water, being well mixed, may have a mean temperature only slightly below the equilibrium freezing point, e.g. −0.01°C to 0.1°C, but this is sufficient to allow distributed ice nuclei to grow. The result is a suspension of tiny, disc-like crystals, typically 0.5–5 mm in diameter and about one tenth as thick; these are known as *frazil*. They may appear suddenly in very large numbers and have been known to completely block hydroelectric intakes within a period of one hour. They may also 'settle' on the underside of an ice cover downstream where they are capable of accumulating in such numbers as to create a hanging dam, seriously altering river hydraulics.

4.4.2 Constitutional supercooling

Freezing behaviour undergoes a substantial change when impurities are dissolved in the liquid. It will be recalled from Chapter 2 that the equilibrium freezing temperature varies with the amount of material dissolved.

Figure 4.8 is a typical phase diagram for a solute which is soluble in the liquid but not in the solid phase which therefore consists of pure frozen solvent. Increasing the amount of solute depresses the freezing point from A, the pure liquid point, to E, the eutectic point below which liquid no longer exists.

If the temperature of a substrate immersed in such a solution is lowered below the initial freezing point at b, for example, pure solid solvent begins to form on the substrate; this has the result that the solute concentration in the liquid nearby is increased above the initial value x_I^b to a slightly higher value x_I^a. The equilibrium freezing temperature at the solid–liquid interface is now lower than the initial freezing temperature T_I^b. Figure 4.9(a) suggests the concentration profiles near the interface immediately before and shortly after solidification begins; the interfacial concentration, in changing from x_I^b to x_I^a, then causes diffusion of the solute. Figure 4.9(b) suggests the corresponding temperature profiles T_L^b and T_L^a. For freezing to continue in the presence of solute diffusion, the interface temperature must drop, as indicated, from T_I^b to T_I^a. This requirement stems from the phase diagram.

Now the rate of heat diffusion through the liquid is determined by the thermal diffusivity κ while the mass diffusivity $\mathfrak{D}$ plays the same role in solute diffusion. In the previous section we noted that the thermal wave moving ahead of a slow and steady planar interface could be represented by

$$\Theta(X') = \Theta_I \exp\left(\frac{-V_I Y'}{\kappa}\right),$$

where V_I is the steady interface velocity and Y' is the distance from the interface. The thermal penetration distance (i.e. length scale) is thus measured by $Y_T^c = O(\kappa/V_I)$. The same arguments applied to the diffusion of solute ahead of the interface reveal that the corresponding solute penetration distance is $Y_M^c = O(\mathfrak{D}/V_I)$. It thus follows that the penetration of heat relative to solute is given by $\kappa/\mathfrak{D} = Le$, the Lewis number. When $Le \gg 1$, which is common in liquids, the temperature profile propagates more rapidly, and hence further, than the concentration profile; this is suggested in Fig. 4.9.

The concentration profile in Fig. 4.9(a) may be used to determine a corresponding profile for the equilibrium freezing temperature T_I using Fig. 4.8. This has the same penetration distance as the concentration profile and is shown dashed in Fig. 4.9(b). Adjacent to the interface, the local solution temperature T_L^a seen to be below the local value of T_I; the liquid is locally supercooled. This phenomenon, which reflects both the shape of the phase diagram and the magnitude of the Lewis number, is known as *constitutional supercooling*. Like supercooling in a pure liquid, it creates an unstable interface on which dendritic crystals grow out into the solution.

Constitutional supercooling may be initiated at any point on the liquidus

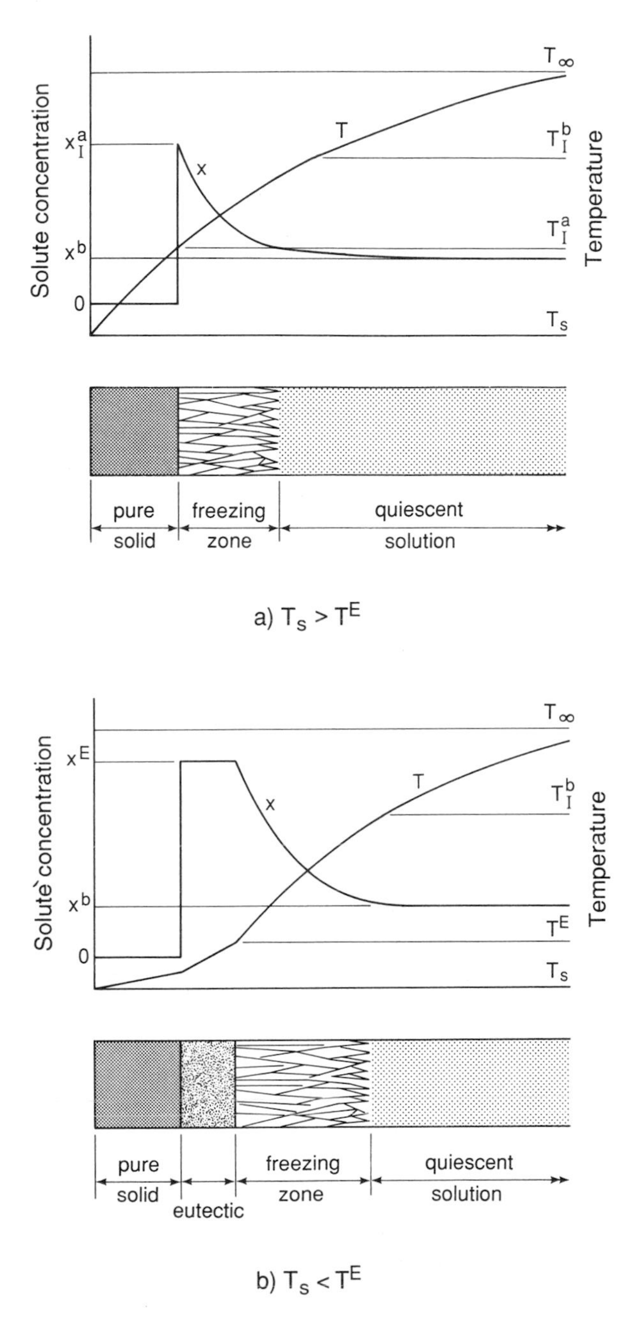

Fig. 4.10 Dendrite growth in a binary solution.

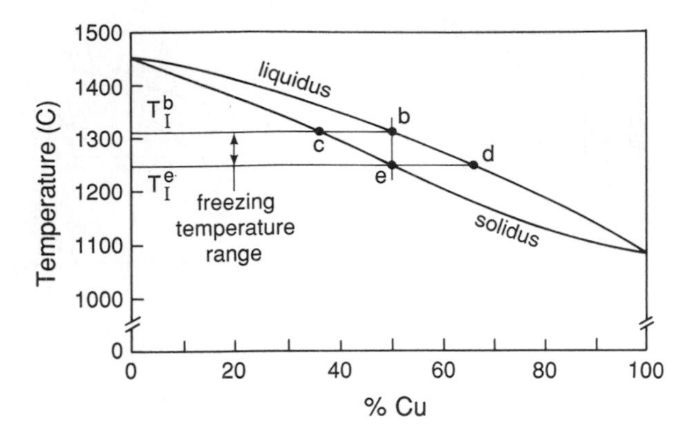

Fig. 4.11 Phase diagram for copper–nickel alloy.

they are often retained in the resulting solid because diffusion rates are usually low.

The heat transferred in the infinitesimal temperature range $T \rightarrow T + \mathrm{d}T$ now consists of two parts: sensible heat $c_\mathrm{p}\mathrm{d}T$ and latent heat. The latter stems from the gradual change in the solid and liquid mass fractions, m_S and m_L, respectively. These are sketched against temperature in Fig. 4.12 which expresses the requirements:

$$m_\mathrm{S} + m_\mathrm{L} = 1 \quad \text{and hence} \quad \frac{\partial m_\mathrm{L}}{\partial T} = -\frac{\partial m_\mathrm{S}}{\partial T} > 0.$$

The heat transferred per unit mass of alloy may thus be written

$$\mathrm{d}q = c_\mathrm{p}\,\mathrm{d}T - \lambda_\mathrm{SL}\frac{\partial m_\mathrm{S}}{\partial T}\,\mathrm{d}T, \tag{4.38}$$

where the latent heat $\lambda_\mathrm{SL}(T)$ is a function of temperature in the freezing range, and $c_\mathrm{p}(T) = c_\mathrm{pS}m_\mathrm{S} + c_\mathrm{pL}m_\mathrm{L}$. This enables us to define the effective specific heat by

$$c_\mathrm{p}' = c_\mathrm{p} - \lambda_\mathrm{SL}\frac{\partial m_\mathrm{S}}{\partial T}. \tag{4.39}$$

A useful approximation to this general description may be obtained when the phase change proceeds uniformly between T_I^b and T_I^e. This is suggested in Fig. 4.12 by the dashed lines, and yields the simple expression

$$c_\mathrm{p}' = c_\mathrm{p} + \frac{\lambda_\mathrm{SL}}{(T_\mathrm{I}^\mathrm{b} - T_\mathrm{I}^\mathrm{e})} = c_\mathrm{p}\left[1 + \frac{\lambda_\mathrm{SL}}{c_\mathrm{p}(T_\mathrm{I}^\mathrm{b} - T_\mathrm{I}^\mathrm{e})}\right], \tag{4.40}$$

where λ_SL is now the average value between T_I^b and T_I^e. For a 50 per cent Cu-Ni alloy, the inverse Stefan number $\lambda_\mathrm{SL}/c_\mathrm{p}(T_\mathrm{I}^\mathrm{b} - T_\mathrm{I}^\mathrm{e}) \simeq 6$, thus indi-

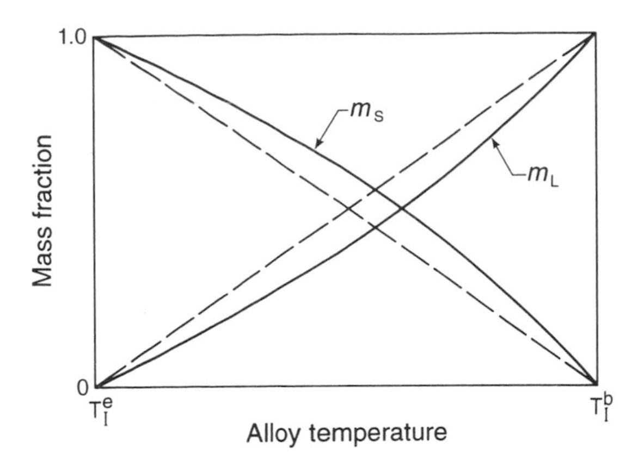

Fig. 4.12 Solid and liquid mass fractions for an alloy in the freezing range $(T_I^b - T_I^e)$.

cating that latent heat would dominate; this feature is common to most alloys.

The one-dimensional temperature field corresponding to this model is shown in Fig. 4.13 which suggests that the transition temperature limits T_I^e and T_I^b may be used to define the boundaries Y_I^e and Y_I^b, respectively, of a freezing zone, frequently called the *mushy* zone. At first glance, it appears that Stefan-type approximations, for which $Ste \ll 1$, might provide simple solutions to this transient heat conduction problem. However, the high melting points of most metals may preclude this possibility. A solution must then be obtained using the method of Neumann described in Section 4.3. Error functions for the temperature profiles in each of the three regions shown in Fig. 4.13 may thus be used to determine the location of the interfacial planes at Y_I^e and Y_I^b. Hence, we obtain

$$\left.\begin{aligned} Y_I^e &= \beta_e(\kappa_S t)^{1/2} \\ \text{and} \qquad Y_I^b &= \beta_b(\kappa_m t)^{1/2} \end{aligned}\right\} \tag{4.41}$$

where $\kappa_m = k_m/\rho_m c_p'$ is the mushy zone thermal diffusivity with c_p' being determined from eqn (4.40). Since the latent heat is now released throughout the mushy zone and not at the interfaces Y_I^e and Y_I^b, the heat flux is no longer discontinuous on these planes. The corresponding interface equations are thus simplified by the suppression of the planar latent heat term.

When the temperature of the original melt T^i approaches the liquidus temperature T_I^b, the mushy zone becomes infinitely thick, theoretically. The problem then reduces to the determination of β_e alone. For example. the interface equation at Y_I^e for a large alloy mass cast in a large mould at T_∞, yields

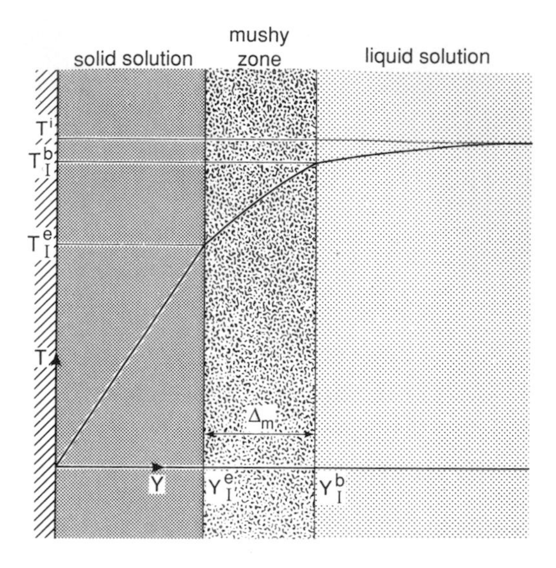

Fig. 4.13 Temperature profiles in an alloy with a freezing temperature range $(T_I^b - T_I^e)$.

$$\frac{k_{mo}\kappa_S^{1/2}\exp[(\kappa_S - \kappa_m)\beta_e^2/\kappa_m]\operatorname{erfc}[\beta_e(\kappa_S/\kappa_m)^{1/2}]}{k_S\kappa_{mo}^{1/2} + k_{mo}\kappa_S^{1/2}\operatorname{erf}\beta_e} = \frac{(T_I^b - T_I^e)k_m\kappa_S^{1/2}}{(T_I^e - T_\infty)k_S\kappa_m^{1/2}}, \tag{4.42}$$

where the subscript *mo* designates the mould. This provides β_e in terms of the material properties and the specified temperatures T_I^b, T_I^e, and T_∞. Movement of the interface Y_I^e is then found from the first of eqns (4.41). Problems Q4.8 and 4.9 in Chapter 8 illustrate this situation.

4.5.2 Casting

Pure and constitutional supercooling commonly result in castings which are neither homogeneous nor isotropic. This situation is common in ingots which are later worked into bars, strips, etc. Figure 4.14 provides a schematic representation of crystal size and orientation within an ingot. Near the mould surface is a thin chill zone characterized by numerous small, equiaxed crystals randomly orientated; these are caused by heterogeneous nucleation in the presence of pure supercooling. Some crystals attach to the mould surface and some do not. Pure and constitutional supercooling, acting together or alone, also give rise to a columnar zone populated by long crystals emanating from the chill zone. Beyond the columnar zone, in the ingot interior, is a core zone which is usually superheated before its formation. The core crystals are equiaxed and randomly orientated, but

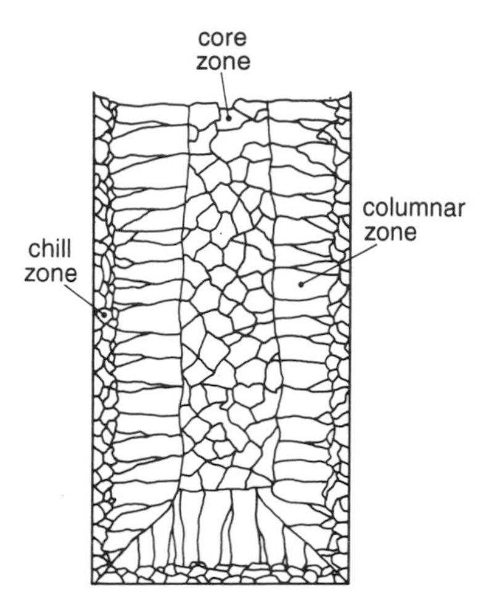

Fig. 4.14 Grain distribution in an ingot showing the three characteristic zones.

they tend to be larger than those found in the chill zone. They grow over a longer period from nuclei formed firstly during the initial chill and later through dendrite melting and free surface seeding; these nuclei are re-distributed by natural convection.

These basic features occur to a greater or lesser extent in any casting. They may be controlled to achieve the desired crystallographic structure. In most ingots and castings, for example, a fine-grained, equiaxed structure is desired, thus demanding the suppression of the columnar zone as far as possible. This may be accomplished by pouring close to the liquidus temperature, thereby permitting nucleation and crystal dispersal assisted by natural convection in the melt. Pure supercooling may be inhibited by lowering the conductance of the mould. Freeze-shrinkage gaps at the mould surface naturally assist in the reduction of heat loss; their effect may be substantial, as Table 4.1 indicates. A simple calculation shows that a 0.1 mm air gap gives a conductance of about $400\,\mathrm{W\,m^{-2}\,K^{-1}}$ at 500 K. Mould conductance is explored in Q4.4, 4.7, and 4.9 in Chapter 8.

On the other hand, when a columnar structure is desired, the opposite strategy is employed: high pour temperatures and high conductance moulds free of shrinkage gaps must be used. These measures promote pure super-cooling in the chill zone and inhibit nucleation in the core zone. A columnar structure is sought, for example, with magnetic alloys which demand crystal-lographic alignment. In the casting of turbine blades, the production of a single (columnar) crystal growing along the length of the blade eliminates potential failure associated with grain boundaries.

Table 4.1 Gap Conductances (source: Poirier and Poirier 1992)

Casting situation	Conductance ($W\ m^{-2}\ K^{-1}$)
Ductile iron in cast iron mould	1700
Steel in cast iron mould	1020
Aluminium alloy in small copper mould	2300
Steel chilled by steel mould	
before gap forms	680
after gap forms	400
Aluminium die castings	
before gap forms	3400
after gap forms	400

4.5.3 Soldering, brazing, and welding

Turning from conventional casting processes, let us consider various thermal joining methods; notably soldering, brazing, and welding. In each of these, molten alloy is applied to the common edge of two or more solid metal components in contact with each other. The component metal substrate invariably exhibits a very high conductance, thereby giving rise to pure supercooling which may be expected to create a dominant columnar zone; direct growth (without nucleation) on the metal substrate essentially eliminates the chill zone. The joining process is thus characterized by a mushy zone which grows rapidly away from the metal substrate.

The phase diagram for the tin–lead system (solder) is shown in Fig. 4.15. This is seen to be a combination of the types given earlier in Figs 4.8 and 4.11. With tin compositions in the range 20–62 per cent, the solidus and liquidus lines are reminiscent of Fig. 4.8 whereas for compositions less than 20 per cent they resemble Fig. 4.11. A eutectic point is evident. Although the liquidus temperature is kept low, thus minimizing thermal stresses, there is a significant freezing range. Plumbers' solder, for example, has a composition of about 33 per cent tin with a liquidus temperature near 240°C but a freezing range extending down to 183°C (the eutectic temperature); the range thus provides time for the joint to be 'wiped'. For a higher strength joint, brass (copper–zinc alloy) is used in a similar way. The process, then known as brazing, requires higher temperatures; the melting point of pure copper is 1083°C, while that of pure zinc is 420°C. The addition of silver to the copper–zinc alloy creates silver solder with an intermediate range of melt temperatures.

If soldering and brazing are thought of as casting processes in miniature, fusion welding may be thought of as a continuous casting process in which a mushy zone follows behind the welding rod as it moves along the joint. The highly conductive substrate again promotes columnar growth while

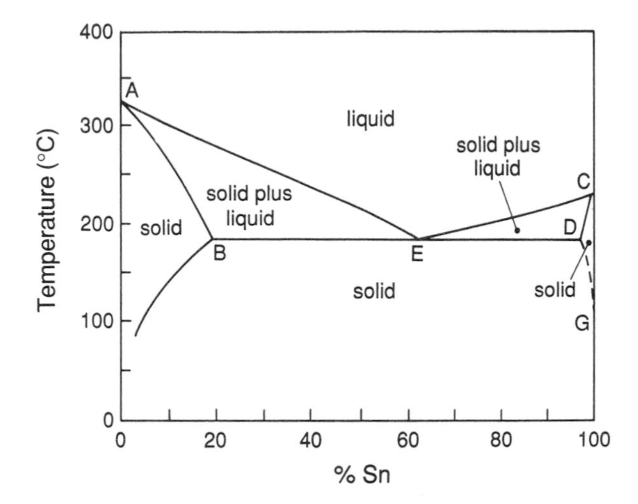

Fig. 4.15 Phase diagram for tin–lead alloy.

melting of the substrate material again eliminates the chill zone. The equiaxed core zone is also suppressed because, even though strong circulation in the weld pool is sufficient to ensure the effective redistribution of naturally occurring nuclei, the temperature is too high for their survival. It should be added, however, that the artificial stimulation of nucleation on the weld pool surface can create equiaxed crystals.

Typically, columnar crystals in the weld grow rapidly from the substrate, their growth rate and direction depending upon the magnitude and direction of the temperature gradient at the changing isotherm bounding their tips. Although this growth does not take place under stable, equilibrium conditions, the shape and movement of the liquidus isotherm provides a useful indication of behaviour. If the heat source were stationary, the liquidus and solidus isotherms would form a pair of concentric circles. Motion of the source elongates these circles in the direction of travel, as indicated in Fig. 4.16. Treating the event as a heat conduction problem created by the steady movement of a point heat source along the joint, e.g. a straight seam, the temperature distribution around the weld pool may be found using the moving coordinate transformation employed in Section 4.3.2; the isotherms in the figure have a spatial distribution which is fixed relative to the heat source.

Figure 4.16 suggests that as the heat source passes by any particular location on the seam, columnar grains will begin growing at the liquidus isotherm more or less perpendicular to the line of travel of the heat source. However, the grain tips must somehow turn and follow the liquidus isotherm as it moves away with the heat source. At first thought, it might

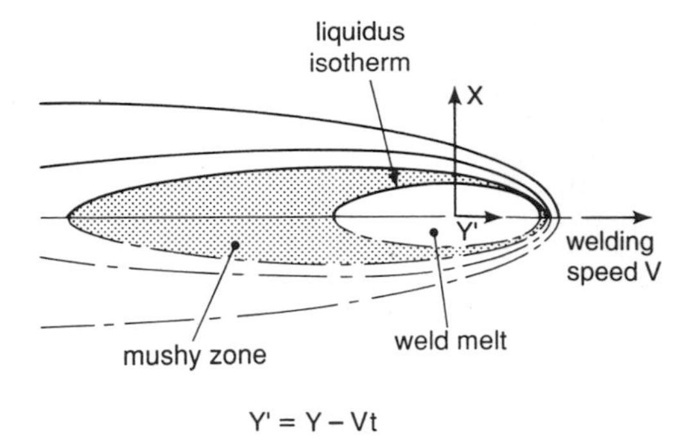

Fig. 4.16 Moving isotherms lying slightly beneath the plane of travel of a point heat source simulating a fusion seam weld.

appear that the grain tips would simply change direction in continuous curves which lie normal to the liquidus isotherm as it sweeps past. But crystallographic structure makes this impossible unless conditions at the grain tips permit periodic adjustments in the growth direction: that is, unless growth proceeds in a series of linear bursts. The inability of the crystals to renucleate and reorientate easily, particularly at higher weld speeds, may induce a structural weakness along the line of travel of the weld.

4.6 Ablation, deposition, and accretion

We now complete our study of fluidification and solidification with a description of important latent heat transfer phenomena involving the vapour phase. The discussion will cover a wide range of situations including the frosting of heat exchanger coils, the ablation of glaciers and meteors, and the icing of ships and aircraft.

4.6.1 Sublimation and deposition

Sublimation is described in Fig. 4.17 using the example of ice in the presence of air temperatures less than the H_2O triple point temperature. The loss of ice at the interface implies that water vapour is diffusing down the vapour pressure gradient, thus moving away from the interface. In these circumstances, the vapour pressure P_I at the interface is less than the triple point pressure P_T and hence the vapour mass fraction is very small: $m_V \leqslant P_T \mathcal{M}_V / P_m \mathcal{M}_m \simeq 4 \times 10^{-3} \ll 1$. Heat and mass transfer may therefore be treated as purely diffusive phenomena (see Section 5.4).

At the interface,

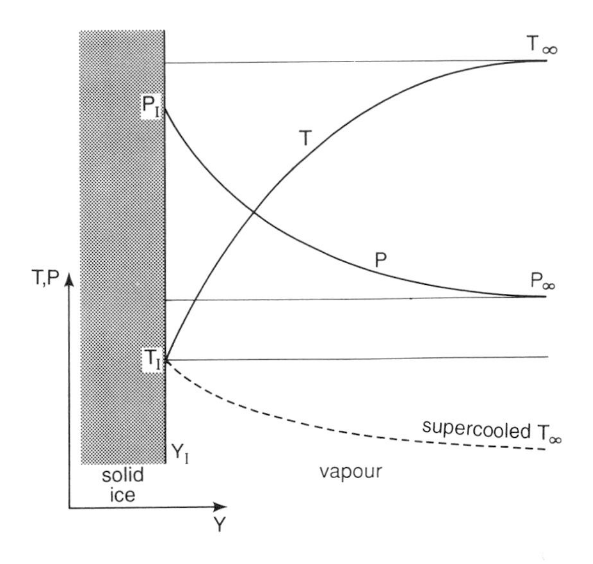

Fig. 4.17 Temperature and pressure profiles during sublimation.

$$\left.\begin{aligned} j_{\mathrm{MI}} &= -\rho_{\mathrm{m}} \mathfrak{D}_{Vg} \left(\frac{\partial m_{\mathrm{V}}}{\partial Y}\right)_{\mathrm{I}} = h_{\mathrm{M}}(m_{\mathrm{I}} - m_{\infty}) \\ \text{and} \qquad j_{\mathrm{QI}} &= -k \left(\frac{\partial T}{\partial Y}\right)_{\mathrm{I}} = -h_{\mathrm{Q}}(T_{\infty} - T_{\mathrm{I}}) = -\lambda_{\mathrm{SV}} j_{\mathrm{MI}} \end{aligned}\right\} \tag{4.43}$$

The interface equation may thus be written

$$\rho_{\mathrm{S}} \lambda_{\mathrm{SV}} \frac{\mathrm{d}Y_{\mathrm{I}}}{\mathrm{d}t} = -h_{\mathrm{Q}}(T_{\infty} - T_{\mathrm{I}}) \tag{4.44}$$

in which the local heat transfer coefficient h_{Q} is determined from the local air flow. After a short-lived transient, $\mathrm{d}Y_{\mathrm{I}}/\mathrm{d}t = -V_{\mathrm{I}}$ becomes constant. Hence

$$V_{\mathrm{I}} = \frac{h_{\mathrm{Q}}(T_{\infty} - T_{\mathrm{I}})}{\rho_{\mathrm{S}} \lambda_{\mathrm{SV}}} \tag{4.45}$$

gives the steady ablation rate. As we shall see in later chapters, T_{I} is the wet bulb temperature T_{wb}. Its determination is illustrated in Q7.13 in Chapter 8.

Sublimation usually takes place with the bulk vapour temperature T_{∞} greater than the interface temperature T_{I}. The interior solid temperature $T_{\mathrm{S}} \leqslant T_{\mathrm{I}}$, but the net heat flux toward the interface is positive. In general, however, the latent heat of sublimation may be provided from either side of the interface. Hence if $T_S > T_I$, it is possible for sublimation to occur into the supercooled vapour suggested by the dashed curve in Fig. 4.17. In either event, the receding interface is stable and smooth.

Deposition is the opposite of sublimation. Again using the example of ice, the situation is described in Fig. 4.18. Equations (4.43) once more apply with the direction of the vapour flux being reversed. As with sublimation, the interface equation reveals that it is the net heat flux which controls the direction of interface motion. Deposition may therefore occur from a supercooled vapour, e.g. during ice crystal growth in cold air, or may be caused by a highly cooled substrate situated in a superheated vapour, e.g. during frosting of a tube containing a refrigerant.

In the presence of a supercooled vapour, a planar interface is unstable; any protuberance being exposed to a lower local ambient temperature, grows faster than the neighbouring area of the interface. This condition, suggested by the dashed curve in Fig. 4.18, gives rise to winter hoarfrost with its characteristic beauty; the temperature of any substrate is then very close to T_I. Ice formed in this way is dendritic and produces a porous matrix with a density which decreases from the substrate to the dendrite tips. Vapour not only diffuses from the supercooled bulk towards the nominal interface at Y_I but also enters the ice matrix where it deposits on the side faces and branches of the dendrites.

It is surprising to find that a porous ice matrix may also form when the substrate is surrounded by superheated vapour. In macroscopic terms, a

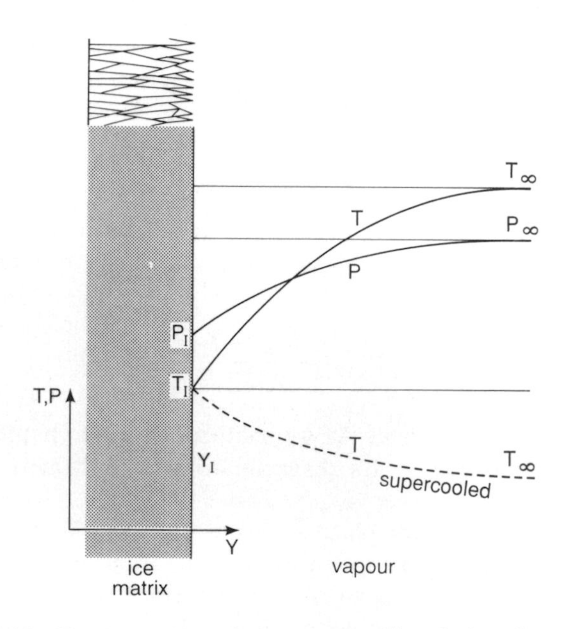

Fig. 4.18 Temperature and pressure profiles during deposition.

planar interface is stable in the presence of warmer vapour, but a highly cooled substrate may induce a high degree of initial supercooling before nucleation takes place. The initial dendrites are evidently capable of growing until the ice matrix is several millimetres thick, by which time dendritic branching has effectively closed off many 'pore' spaces within the matrix (Hayashi *et al.* 1977). Densification of the matrix is thus inhibited while tip growth is not. If densification is completed, the simple analysis given above may be used to calculate the deposition rate. Before this occurs, however, the porous matrix must be treated in much the same way as the freezing or mushy zones encountered in the previous sections. The porous matrix is always present when the vapour is supercooled.

4.6.2 Ablation and accretion

Ablation may be defined as the process in which fluidification produced by the addition of latent heat alters the contours of a solid body or surface. Sublimation is thus a simple form of ablation. This process is represented by the line ab on the *PT* diagram shown in Fig. 4.19. It is a *dry* process, taking place below the triple point.

When the vapour temperature is raised above the triple point value, the situation changes significantly. The corresponding ablation process is represented by the line cde in Fig. 4.19 from which it is evident that this is a *wet* process. Between the solid at c and the vapour bulk at e lies a liquid film. The physical situation is described in Fig. 4.20 in which the solid–liquid interface is designated by Y_I while the liquid–vapour interface is designated by

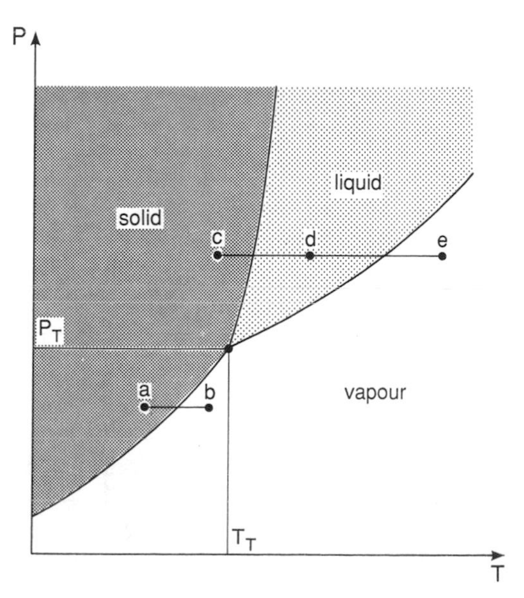

Fig. 4.19 Dry and wet ablation processes.

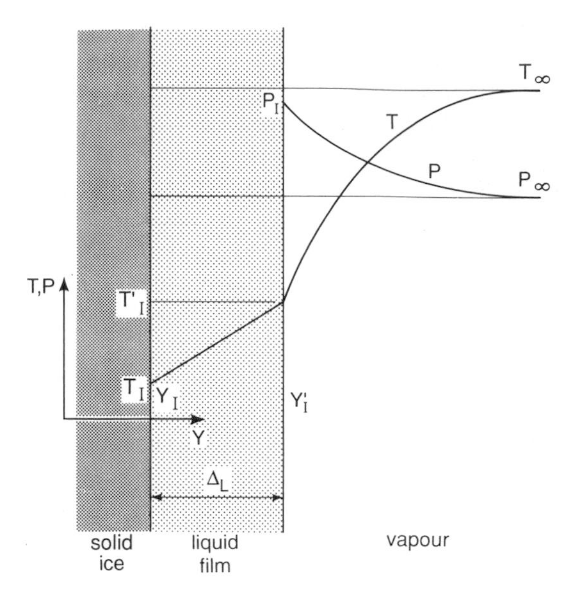

Fig. 4.20 Temperature and pressure profiles during wet ablation.

Y_I'; the corresponding interface temperatures are T_I and T_I', respectively.

Assuming the Stefan number $c_{pL}(T_I' - T_I)/\lambda_{LV} \ll 1$, the two interface equations may be combined to yield the steady ablation rate

$$V_I = -\frac{dY_I}{dt} \simeq \frac{h_Q(T_\infty - T_I')}{\rho_S \lambda_{SV}} \tag{4.46}$$

which has the same form as eqn (4.45). This result may be used, for example, to estimate the ablation rate of hail descending in warm air. For a 1 cm diameter hailstone falling at 15 m s^{-1} ($h_Q \simeq 170$ W m^{-2} K^{-1}) through 10°C air, the rate is 0.7 μm s^{-1}. Such a low value illustrates why ice particles formed at high elevations are capable of reaching and destroying summer crops.

The use of eqn (4.46), and the interface equations on which it is based, is subject to two important qualifications. Firstly, if T_∞ greatly exceeds the triple point temperature the vapour mass fraction may become large enough to preclude the use of the diffusive relations given in eqns (4.43). The bulk mass flux must then be incorporated, as outlined in Sections 2.1 and (5.4). Secondly, unless the liquid film is very thin, an increasing vapour velocity will ripple its surface, as discussed in Section 3.3. Surface ripples create periodic variations in the local heat transfer rate and introduce a roughness which adds to the influence of vapour shear; surface waves enlarge these effects. At high vapour velocities, the film may shift around the body surface and re-freeze to create the fascinating shapes found in hailstones and meteorites.

In distinction to ablation, accretion is the process in which mass is added to the body surface, usually in the form of liquid drops. The widespread occurrence of supercooled water drops in the atmosphere provides us with many examples: from cloud icing of aircraft to spray icing of ships; from the freezing of rain on roads to the formation of hailstones. An important feature of the accretion process is the way in which the drops approach and impact on the icing surface. This was discussed in Section 3.5.2 where Fig. 3.12 describes the drop trajectories. The inertia of drops congregated close to the stagnation streamline may prevent them from veering past the body. Drops further out can avoid impact by responding sufficiently to the lateral viscous forces created by the veering air. The forward surface of the body is thus subjected to a spray extending from the forward stagnation point to the location where drops approach the body (or ice) surface tangentially. These features are incorporated in problems Q4.10 to Q4.13 in Chapter 8.

Supercooled drops are metastable in flight but freeze after impact. This freezing process is governed principally by the wind temperature and speed which control the rate of heat loss from the impacted drops. The lower the wind temperature and speed, and the smaller the drops, the more likely that each drop will freeze completely as a distinct sphere without being spread. These conditions produce a porous accretion, usually known as *rime*, and is typified by the protruding nose illustrated in Fig. (4.21(a); it is a dry accretion. With higher wind temperatures and speeds, and larger drop diameters, surplus water remains to percolate through and over the accretion, creating solid ice known as *glaze*; needless to say, it is a wet accretion. The mobility of this water, coupled with its tendency to freeze

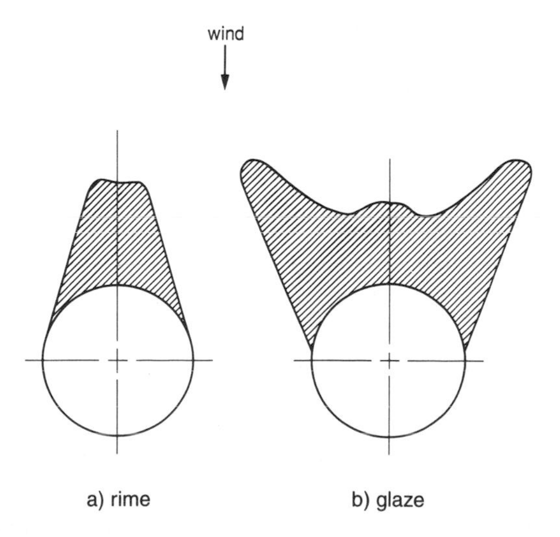

Fig. 4.21 Typical ice shapes growing into the wind on a cylinder.

beyond the point of drop impact, creates accretion shapes which differ markedly from those characteristic of rime. The twin-lobe configuration illustrated in Fig. 4.21(b) is typical.

Any form of accretion may dramatically alter the aerodynamic characteristics of the surface on which it forms. It will also add weight, as demonstrated in the important example of an overhead cable. In many situations it is sufficient to calculate the icing rate using the collision efficiency η defined as the *colliding* fraction of the spray mass approaching through the forward projected area of the body. If all of this colliding material is retained, the icing rate $\dot{M}_i = \eta UA(LWC)$, where U is the body speed, A is its forward projected area and LWC is the liquid water content of the spray.

Selected bibliography

Armstrong, T., Roberts, B., and Swithinbank, C. (1973). *Illustrated glossary of snow and ice*. Scott Polar Research Institute, Cambridge.

Ashton, G. D. (ed.) (1986). *River and lake ice engineering*. Water Resources Publications, Littleton, Colorado.

Berry, J. T. and Dantzig, J. (eds.) (1984). *Modelling of casting and welding processes*. Engineering Foundation, New York.

Davies, G. J. (1973). *Solidification and casting*. Applied Science, London.

Easterling, K. (1983). *Introduction to the physical metallurgy of welding*. Butterworth, London.

Flemings, M. C. (1974). *Solidification processing*. McGraw-Hill, New York.

Gilpin, R. R. (1978). The effects of dendritic ice formation in water pipes. *International Journal of Heat Mass Transfer*, **20**, 693–99.

Gray, D. M. and Male, D. H. (eds.) (1981). *Handbook of snow*. Pergamon, Toronto.

Hayashi, Y., Aoki, K., and Yuhara, H. (1977). Study of frost formation based on a theoretical model of the frost layer. *Heat Transfer—Japanese Res.*, **6**(3), 79–94.

Hobbs, P. V. (1974). *Ice physics*. Oxford University Press, Oxford.

Lock, G. S. H. (1974). The growth and decay of ice. *Proceedings of the 5th International Conference Tokyo*. Japan Society of Mechanical Engineers, Tokyo.

Lock, G. S. H. (1990). *The growth and decay of ice*. Cambridge University Press, Cambridge.

Lunardini, V. J. (1991). *Heat transfer with freezing and thawing*. Elsevier Science Publishers, Amsterdam.

Minkoff, I. (1986). *Solidification and cast structure*. Wiley, New York.

Poirier, D. R. and Poirier, I. J. (1992). *Heat transfer fundamentals for metal casting*. The Minerals, Metals and Materials Society, Warrendale, Pennsylvania.

Williams, P. J. and Smith, M. W. (1989). *The frozen earth*. Cambridge University Press, Cambridge.

Exercises

1. Aluminium is poured at its melting point (660°C) into a large open mould. If nucleation begins immediately on its free upper surface

where it loses heat by natural convention (h_A) to the surrounding air at 20°C, estimate the upper surface temperature T_0 and the Stefan number after 1 cm of solid has formed. Take $h_A = 20\ \mathrm{W\,m^{-2}\,K^{-1}}$, $c_{pS} = 896\ \mathrm{J\,kg^{-1}\,K^{-1}}$, $\lambda_{SL} = 400 \times 10^3\ \mathrm{J\,kg^{-1}}$ and $k_S = 220\ \mathrm{W\,m^{-1}\,K^{-1}}$. (Ans: $T_0 = 659.4$°C, $Ste_S = 1.34 \times 10^{-3}$.)

2. On the side walls of the above mould, $h_A = 4000\ \mathrm{W\,m^{-2}\,K^{-1}}$. Estimate the surface temperature and Stefan number along the side walls when 1 cm of aluminium has frozen.

   (Ans: $T_0 = 561.5$°C, $Ste_S = 0.22$.)

3. Refrigerant at a bulk temperature T_c flows with a heat transfer coefficient h_c through a pipe of diameter D. Neglecting the thermal resistance of the pipe wall, show that the radius of ice formed in surrounding water at its freezing point is given by

$$t = \frac{\rho_S \lambda_{SL} D^2}{8k_S(T_I - T_C)} \left[\left(\frac{1}{Bi} - \frac{1}{2}\right)(r_I^2 - 1) + r_I^2 \, ln \, r_I \right]$$

   in which $Bi = Dh_c/2k_S$.

4. Warm water discharge beneath an ice cover Y^i thick creates a heat transfer coefficient h_w. If the bulk water temperature T_w and the ice upper surface temperature T_0 are both fixed, show that the ice thickness Y_I is given by

$$t = \frac{(Y^i - Y_I)}{B} + \frac{A}{B^2} \ln \left(\frac{A - BY^i}{A - BY_I} \right),$$

   where

$$A = \frac{k_S(T_I - T_0)}{\rho_S \lambda_{SL}} \quad \text{and} \quad B = \frac{h_w(T_w - T_I)}{\rho_S \lambda_{SL}}.$$

5. A thin metal wall at temperature T_0 separates freezing water from deep space. If the radiative heat loss from the wall is given by $j_Q = \sigma T_0^4$, where σ is the Stefan–Boltzmann constant, show that the ice thickness is given by

$$Y_I \simeq \frac{\sigma T_I^4 t}{\rho_S \lambda_{SL}},$$

6. Derive eqn (4.36).

7. Consider a simple, planar Stefan problem in which $k_S(T)$ is the temperature-dependent thermal conductivity. Defining

$$\hat{k} = \int_{T_R}^{T} k_S(T)\, \mathrm{d}T$$

where T_R is an arbitrary reference temperature, show that the frozen solid thickness is given by

$$Y_I^2 = \frac{2}{\rho_S \lambda_{SL}} [\hat{k}(T_I) - \hat{k}(T_0)]t$$

where T_0 is the temperature at $Y = 0$.

8. Using the results developed in Q4.4 and 4.5 in Chapter 8, find the mould surface temperature.
   (Ans: $T_0 = 654.4°C$.)
9. Derive eqn (4.42).
10. In a chill mould, the mould surface temperature T_0 is maintained at a much lower value than the metal pour temperature, but $Y_I = 2\beta(\kappa_S t)^{1/2}$. Show that for an alloy poured at the liquidus temperature T_2,

$$\frac{\exp[(\kappa_S - \kappa_m)\beta^2/\kappa_m]\text{erfc}\,\beta(\kappa_S/\kappa_m)^{1/2}}{\text{erf}\,\beta} = \left(\frac{T_2 - T_1}{T_1 - T_0}\right) \frac{(k\rho c_p)_m^{1/2}}{(k\rho c_p)_S^{1/2}}$$

   if T_1 is the solidus temperature.
11. Find the icing rate per unit length of cable under the circumstances described in Q4.10 and Q4.11 in Chapter 8.
   (Ans: $\dot{m}_i = 7.55 \times 10^{-6}\ \text{kg m}^{-1}\text{s}^{-1}$.)
12. The icing conditions described in Q4.10–Q4.12 of Chapter 8 create added weight on the cable. If the cable density is $8.9 \times 10^3\ \text{kg m}^{-3}$, calculate the weight of ice added in a day and decide if the situation is dangerous.
   (Ans: $w_i = 6.42\ \text{N m}^{-1}$, an addition of 23.4 per cent.)

Solidification and fluidification projects

1. Review and assess the thermophysical processes in freeze drying.
2. Compare and contrast two methods for continuous casting.
3. Develop a physical and mathematical model of freezing and thawing of a plant cell.
4. Compare die casting and investment (lost wax) casting.
5. Review and assess the freezing of polymers.
6. Devise an anti-icing technique for an aircraft wing.
7. Develop a physical and mathematical model of hailstone growth.
8. Review and assess freezing and thawing in moist soil.
9. Devise a de-icing technique for a ship's mast.
10. Review and assess freezing in strong solutions.

5
CONDENSATION

We now enter the first of two chapters dealing exclusively with vapour–liquid transformations: condensation and evaporation. In an engineering context, condensation frequently occurs in the form of a liquid film, and consequently most of the discussion below is devoted to the dynamic behaviour of this film and its influence on condensation rates. Once a basic heat transfer model has been developed in Sections 5.1 and 5.2, the treatment is extended to include the important effects of surface geometry, vapour composition, and vapour motion, all referring to film condensation on an external surface in typical circumstances. Condensation inside tubes, which makes full use of the discussion of two-phase flow in Chapter 3, is treated separately. However, before attempting to discuss any of these situations, it is first necessary to explain the circumstances which lead to the creation of drops and films.

5.1 The formation of drops and films

5.1.1 *Free drops*

In the bulk of a vapour, the appearance of a cloud of drops is often described as fogging. Its occurrence in the atmosphere is well known but it may also occur in industrial equipment, especially with high molecular weight vapours. Cluster theory discussed in Section 2.7 suggests that liquid embryos begin growing spontaneously from pure vapour when the cluster Gibbs function reaches a critical value $\mathcal{G}^* = 16\pi\sigma^3/3(\rho_L RT \ln S^*)^2$. This threshold level corresponds to a critical supersaturation ratio

$$S^*(T) = \frac{P_V^*}{P_{sat}},$$

where P_V^* and P_{sat} are, respectively, the critical vapour pressure and its saturation value at the same temperature. Now the rate of droplet formation J in the cloud depends upon the size distribution which takes the Boltzmann form. Hence, at the threshold,

$$J = K \exp\left(\frac{-\mathcal{G}^*}{kT}\right),$$

where k is the Boltzmann constant and $K = O(10^{31})$ droplets m^{-3} s^{-1} for water. This rate is found to be an extremely steep function of S^*, as may be demonstrated by calculating the time to generate 1 droplet per second in each cubic centimetre. Using water vapour as an example, this time is

$O(10^{12})$s when $S^* = 3$ but $O(10^{-8})$s when $S^* = 5$. It thus follows that $S^* \simeq 4$ is the critical value for pure water vapour. However, the critical supersaturation (or supercooling) depends strongly on the presence of foreign substances and particulates, as noted in Section 2.7. In a typical industrial situation, the critical ratio is much closer to 1.0.

No matter what the origin of a drop, once formed it will grow rapidly in the presence of supercooled vapour. Figure 5.1 describes the situation. Throughout the drop, the temperature remains equal to the saturation temperature T_I while the temperature at a great distance $T_\infty < T_I$. The vapour is generally part of a mixture, e.g. water vapour in air, and diffuses towards the interface at a rate j_{MI} given by

$$j_{MI} = h_M(m_{V\infty} - m_{VI}), \tag{5.1}$$

where m_{VI} and $m_{V\infty}$ are the vapour mass fractions at the interface and far from it, respectively; h_M is the mass transfer coefficient. Heat diffuses away from the interface at a rate given by

$$j_{QI} = h_Q(T_I - T_\infty), \tag{5.2}$$

where h_Q is the heat transfer coefficient. As we shall see later, these diffusive fluxes are accompanied by a bulk (inward) vapour flux which may exert a strong influence on the rate of condensation.

There is a striking similarity between the above process and particle solidification from a supercooled fluid. The growing phase remains at the saturation temperature, above the ambient temperature, and the interface is potentially unstable. However, there is an important difference. During the formation of crystalline solids, intermolecular forces are strong enough to ensure the orderly construction of a regular lattice structure usually in dendritic shapes. Liquids, on the other hand, are usually incapable of withstanding non-uniform internal stresses induced by surface tension; any emergent dendrites are rapidly rounded off and submerged beneath the interface. A growing spherical drop may thus be expected to remain spherical, except perhaps near the triple point.

5.1.2 Drops on a substrate

In many engineering applications, drops do not form throughout the bulk of the vapour but instead appear on a chilled surface, e.g. a condenser tube, as the result of heterogeneous nucleation. Once again, they may grow outwards into supercooled vapour but the presence of a cold substrate removes the restriction that $T_\infty < T_I$; it is more common that $T_\infty > T_I$. Figure 5.2 describes conditions in which a hemispherical drop grows from *superheated* vapour; heat removed through the drop exceeds that supplied by the vapour.

The precise origin of drops on a chilled surface is still being debated, although it is clearly connected with a high surface tension. There are two

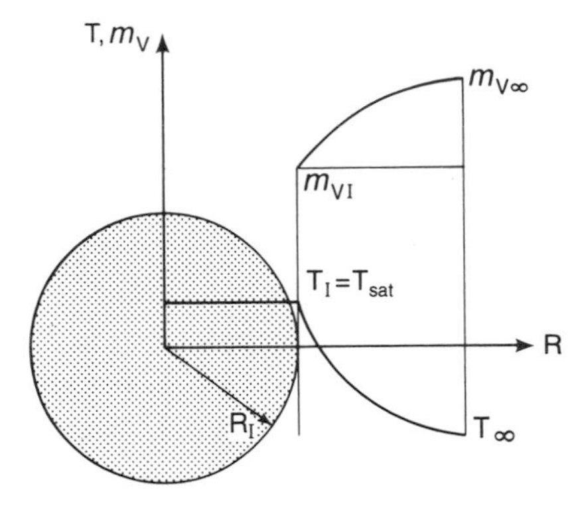

Fig. 5.1 Temperature and mass fraction profiles near a liquid drop growing through condensation of a supercooled vapour.

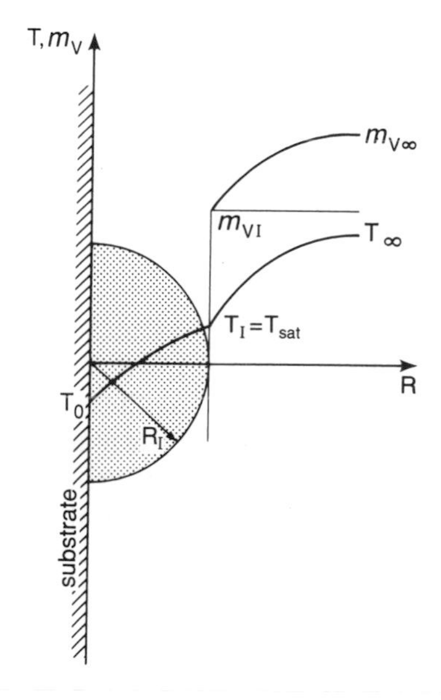

Fig. 5.2 Conditions during condensation on a surface drop in a superheated vapour. The substrate temperature T_0 must be low enough for the complete removal of latent and sensible heat from the vapour.

schools of thought, both of which have found experimental support. In one, it is postulated that condensation begins as heterogeneous nucleation but only in microcavities distributed over the surface as a consequence of manufacturing and other treatment processes. These cavities, which may number in excess of $10^8\,\text{cm}^{-2}$, facilitate the formation and initial growth of microscopic clusters which subsequently become drops. Elsewhere on the surface, intermolecular attraction between vapour and substrate is not great enough to permit condensation, at least while the temperature difference $T_I - T_0$ is small; growth is thus limited by condensation directly on the drop surfaces.

The other theory assumes that the attraction between vapour and substrate molecules is great enough to create a very thin liquid film spread uniformly across the surface. As this microfilm grows, surface tension renders it increasingly unstable until, before it is 1 μm thick, it breaks up into a set of more or less regularly spaced bumps which then grow as discrete drops. Further condensation takes place directly on the drop surfaces, as for the microcavity drops, but also on the remaining free surface which provides an extended catchment area supplying the drops through a rapid, radially focusing liquid flow mechanism.

Whether drops originate at randomly distributed sites, or regularly through surface tension instability, is usually unimportant in practice, except at very low condensation rates. For most engineering applications, it is their subsequent growth which is more important, especially when they exceed 0.1 mm in diameter. As each drop grows it enters increasingly into competition with its neighbours. This is particularly evident when drops encroach close enough to each other to create a sudden coalescence. Larger drops then grow at the expense of smaller drops. On a vertical surface, this process is enhanced by gravity drainage; on a horizontal surface it leads to a continuous film.

Drop condensation heat transfer coefficients are among the highest available in thermal equipment, e.g. 50–500 kW m^{-2} K^{-1}. They reflect the high conductances of the smallest drops, as illustrated in problem Q6.2 of Chapter 8. To maintain such a high performance it is essential that the liquid does not wet the substrate, but this is difficult to achieve in practice for an extended period of time because of gradual surface degradation, even with hydrophobic coatings. Moreover, as condensation rates increase the number of drops, their growth rate and mobility also increase with the result that the substrate is effectively covered by a continuous liquid film in which the only natural mechanism for liquid removal is continuous drainage. Condensation is then limited to the relatively passive surface of the film; heat transfer coefficients are reduced to lower, though still substantial, levels, e.g. <10 kW m^{-2} K^{-1}. Very high heat loads are thus in conflict with high condensation efficiency. For practical and economic reasons, surface condensation is often modelled conservatively as though surface tension were unimportant: that is, by assuming the liquid wets the substrate uniformly to create a continuous film which flows under the influence of gravity. Even in the absence of gravity, drop condensation from a stagnant vapour eventually poses a film condensation problem analogous to deposition or freezing, i.e. a Stefan problem.

5.2 Scale analysis and the Nusselt problem

5.2.1 Scales

The Nusselt model of steady, film condensation is described in Fig. 5.3. A solid substrate of vertical length L and surface temperature $T_0 < T_I$, the pure vapour saturation temperature, removes heat by conduction through the liquid film. This heat transfer consists almost entirely of latent heat released at the interface located at $Y_I(X)$. The film itself is driven downward by Archimedean buoyancy, as discussed in Section 3.1. The downward liquid velocity thus increases from zero at the substrate to a maximum on or before reaching the interface where vapour and liquid velocities are equal and the shear stress is given by

$$\tau_I = \mu_L \left(\frac{\partial U_L}{\partial Y}\right)_I = \mu_V \left(\frac{\partial U_V}{\partial Y}\right)_I . \tag{5.3}$$

When the film thickness is small in relation to the substrate length, i.e. $Y_I \ll L$, the laminar equation of motion for the liquid reduces to

$$U\frac{\partial U}{\partial X} + V\frac{\partial U}{\partial Y} = \frac{g\Delta\rho}{\rho} + \nu\frac{\partial^2 U}{\partial Y^2}, \tag{5.4}$$

according to eqn (3.7): $\Delta\rho = \rho_L - \rho_V$, and the subscript L has been omitted for clarity. The corresponding form of the energy equation in the film is

$$U\frac{\partial \theta}{\partial X} + V\frac{\partial \theta}{\partial Y} = \kappa\frac{\partial^2 \theta}{\partial Y^2}, \tag{5.5}$$

where $\theta = T - T_0$, while the continuity equation is given by

$$\frac{\partial U}{\partial X} + \frac{\partial V}{\partial Y} = 0. \tag{5.6}$$

Finally, the energy balance at the interface may be written

$$\lambda_{LV}\dot{m} \simeq k\nabla T \simeq k\left(\frac{\partial T}{\partial Y}\right)_I, \tag{5.7}$$

where $\dot{m}$ is the interfacial condensation rate; under steady conditions this condensation is balanced by liquid being carried away in the film. Following the procedure developed in Chapter 1, we now determine the natural scale of each variable, denoting it by the superscript c: length scales X^c and Y^c; velocity scales U^c and V^c; and temperature scale θ^c.

Two of these scales are prescribed: $\theta^c = T_I - T_0$, and $X^c = L$. To determine the others we introduce the normalized variables

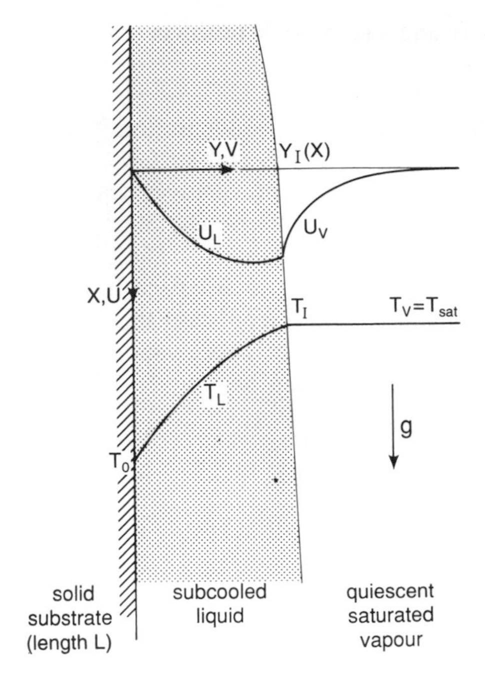

Fig. 5.3 Temperature and velocity profile during film condensation of a pure, quiescent, saturated vapour on a vertical surface: the Nusselt problem.

$$x=\frac{X}{X^{\mathrm{c}}},\quad y=\frac{Y}{Y^{\mathrm{c}}},\quad u=\frac{U}{U^{\mathrm{c}}},\quad v=\frac{V}{V^{\mathrm{c}}},\quad \text{and}\quad \phi=\frac{\theta}{\theta^{\mathrm{c}}}. \tag{5.8}$$

Substituting these into the continuity eqn (5.6) we obtain

$$\left[\frac{U^{\mathrm{c}}\,Y^{\mathrm{c}}}{X^{\mathrm{c}}\,V^{\mathrm{c}}}\right]\frac{\partial u}{\partial x}+\frac{\partial v}{\partial y}=0. \tag{5.9}$$

Thus,

$$\frac{U^{\mathrm{c}}\,Y^{\mathrm{c}}}{X^{\mathrm{c}}\,V^{\mathrm{c}}}=O(1) \tag{5.10}$$

To find the three unknown scales U^{c}, X^{c}, and V^{c}, two more relations containing them are required. The first of these expresses continuity across the interface by combining a statement for the conservation of mass

$$\dot{m}=O(\rho_{\mathrm{L}}\,V^{\mathrm{c}}),$$

with a statement for conservation of energy, eqn (5.7), to obtain

$$V^{\mathrm{c}}=O(k\theta^{\mathrm{c}}/\rho\,Y^{\mathrm{c}}\lambda_{\mathrm{LV}}). \tag{5.11}$$

The second relation may be found from the equation of motion in the normalized form

$$\left[\frac{\rho(U^c)^2}{X^c g\Delta\rho}\right]\left(u\frac{\partial u}{\partial x}+v\frac{\partial u}{\partial y}\right)=1+\left[\frac{\mu U^c}{(Y^c)^2 g\Delta\rho}\right]\frac{\partial^2 u}{\partial y^2}. \tag{5.12}$$

When the buoyancy force (the first term on the right-hand side) is balanced principally by the viscous force (the second term),

$$\frac{\mu U^C}{(Y^c)^2 g\Delta\rho}=O(1). \tag{5.13}$$

Using eqns (5.10), (5.11), and (5.13) we find that

$$Y^c=L\left(\frac{Ja}{ArPr}\right)^{1/4},\quad U^c=\frac{\kappa Ja}{L}\left(\frac{ArPr}{Ja}\right)^{1/2},\quad \text{and}\quad \frac{V^c}{U^c}=\left(\frac{Ja}{ArPr}\right)^{1/4}, \tag{5.14}$$

bearing in mind that as a matter of choice we are free to set $O(1)=1.0$ exactly.

The dominant non-dimensional group

$$\frac{ArPr}{Ja}=\frac{\lambda_{LV}L^3 g\Delta\rho}{\nu k\theta^c} \tag{5.15}$$

is seen to consist of three other distinct groups: $Ar=gL^3\Delta\rho/\rho\nu^2$ is the Archimedes number which determines the driving buoyancy force; $Pr=\nu/\kappa$ is the Prandtl number which measures the magnitude of the momentum diffusivity relative to the thermal diffusivity; and $Ja=c_p\theta^c/\lambda_{LV}$ is the Jakob number which, like the Stefan number, measures the importance of liquid sensible heat relative to latent heat. For water vapour condensing at atmospheric pressure on a 1 m plate at 40°C, $ArPr/Ja\simeq 0.2\times 10^{16}$, thus confirming that the film is thin (0.15 mm) and has a lateral velocity scale much less than its longitudinal velocity scale (10^{-4}:1).

5.2.2 *The Nusselt problem*

The velocity and length scales given in eqn (5.14) now permit the energy equation and the equation of motion to be restated as

$$\left.\begin{aligned} Ja\left(u\frac{\partial\phi}{\partial x}+v\frac{\partial\phi}{\partial y}\right)&=\frac{\partial^2\phi}{\partial y^2}\\ \text{and}\qquad \frac{Ja}{Pr}\left(u\frac{\partial u}{\partial x}+v\frac{\partial u}{\partial y}\right)&=1+\frac{\partial^2 u}{\partial y^2}\end{aligned}\right\} \tag{5.16}$$

respectively. The solutions of these equations must satisfy certain boundary conditions. For temperature $\phi(x,y)$ these are: $\phi(0,y)=1$, at the top the substrate, $\phi(x, 0) = 0$, at the substrate surface, and $\phi(x, \Delta) = 1$, at the interface where $\Delta = Y_I/Y^c$.

Typically, sensible heat in the liquid is relatively unimportant and hence $Ja \ll 1$. The energy equation thus reduces to the one-dimensional approximation $\partial^2\phi/\partial y^2 = 0$ with the solution

$$\phi(y) = y/\Delta, \tag{5.17}$$

given the above boundary conditions. This linear form implies pure conduction between the interface and the substrate, as in the simple Stefan problem discussed in the previous chapter. A similar argument suggests that the equation of motion may be reduced to

$$\frac{\partial^2 u}{\partial y^2} = -1, \tag{5.18}$$

at least for non-metallic liquids ($Pr > 1$). At the substrate surface the no-slip condition requires that $u = 0$. At the interface, the quiescent vapour may be expected to exert a negligible shear stress. The solution of eqn (5.18) is then

$$u = y\Delta - \frac{y^2}{2}. \tag{5.19}$$

The above analysis has demonstrated that laminar film condensation of a pure quiescent, saturated vapour has certain basic features. Heat transfer through the film may be approximated by pure conduction if the Jakob number is much smaller than unity. At the same time, the parabolic velocity profile within the liquid film reflects a balance between the driving Archimedean buoyancy force and the retarding viscous force, provided the Prandtl number is not much smaller than unity. Situations of this type are appropriately described as *Nusselt problems*, recognizing the pioneer analysis of Wilhelm Nusselt. They are common during condensation (or evaporation) over external surfaces.

The heat transfer rate in these circumstances is easily estimated because, at the substrate, the heat flux density $\dot{q}_0$ is given by

$$\dot{q}_0 = k_L\left(\frac{\partial T}{\partial Y}\right)_0 = \left(\frac{k_L\theta^c}{Y^c}\right)\left(\frac{\partial\phi}{\partial y}\right)_0 = O\left(\frac{k_L\theta^c}{Y^c}\right). \tag{5.20}$$

Hence

$$Nu = \frac{\dot{q}_0 L}{\theta^c k_L} = 0\left(\frac{L}{Y^c}\right) = O\left(\frac{ArPr}{Ja}\right)^{1/4}, \tag{5.21}$$

using eqn (5.14). The analysis has thus revealed the central importance of the non-dimensional group $ArPr/Ja$ which sets the scales of the velocity and film thickness, and succinctly determines the overall heat transfer rate. The power law form of the heat transfer relation has also emerged. Some idea of heat transfer coefficient magnitudes may be obtained using the earlier example of pure water vapour condensing at atmospheric pressure on a 1 m plate at 40°C. For this, $h \simeq 4500\,\mathrm{W\,m^{-2}\,K^{-1}}$ which corresponds to a heat flux density of $270\,\mathrm{kW\,m^{-2}}$.

5.3 External condensation on smooth films

The simple Nusselt problem is a useful introduction to film condensation on the external surfaces of plates and tubes. We now build on this essential base by extending the coverage to a wider range of practical circumstances. Consideration will therefore be given to several specific questions. What if the liquid sensible heat is not negligible? What if the Prandtl number is small? What is the effect of surface geometry; and how does turbulence alter the picture?

5.3.1 The effects of Jakob number and Prandtl number

Under steady, laminar conditions, film behaviour is expressed principally through eqns (5.16). Since these contain the parameters *Ja* and *Pr*, it follows that their general solutions must be in the form

$$\phi = \phi(x, y, Ja, Pr), \quad u = u(x, y, Ja, Pr), \quad \text{and} \quad v = v(x, y, Ja, Pr). \qquad (5.22)$$

This suggests that the heat transfer relation given by eqn (5.21) has the more general form

$$Nu = C_v(Ja, Pr)\left(\frac{ArPr}{Ja}\right)^{1/4}. \qquad (5.23)$$

The coefficient $C_v(Ja, Pr)$ thus contains the effects of liquid sensible heat and Prandtl number.

A succinct approach to the determination of $C_v(Ja, Pr)$ is to cast the problem in a boundary layer similarity form, as suggested by Sparrow and Gregg (1959). This may be done by using the structure of the film thickness scale Y^c given in eqn (5.14) to define a similarity variable $\eta = Y/Y^c = (\lambda g \Delta\rho / \nu k \theta^c)^{1/4}\, Y/X^{1/4}$ which enables the partial differential eqns (5.16) to be transformed into ordinary differential equations for which the solutions then take the form $\phi(\eta, Ja, Pr)$, etc. The results of these solutions are presented in Fig. 5.4 which shows the function

$$f(Ja, Pr) = Nu\left(\frac{4Ja}{PrAr}\right)^{1/4}$$

plotted against the Jakob number for each of a series of Prandtl numbers. This provides a quantitative confirmation of the qualitative analysis of eqns (5.16) presented earlier. The simple Nusselt prediction $C_v = 1$ is evidently valid for the conditions under which it was derived; namely $Ja \leqslant 0.1$ for

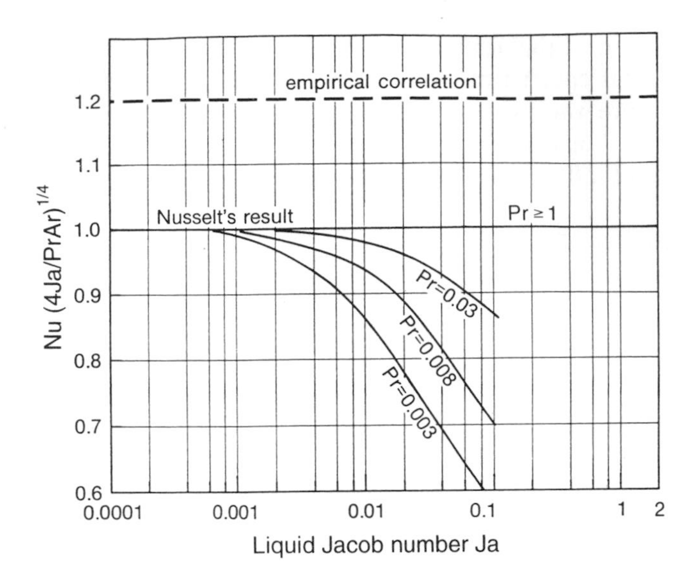

Fig. 5.4 Prediction of laminar film condensation heat transfer rate on a vertical surface in the presence of a pure, quiescent, saturated vapour (based on Sparrow and Gregg, 1959).

all liquids except the metals. Even so, the predictions are consistently lower than empirical findings ($C_v \simeq 1.2$) for reasons discussed in Section 5.5.

5.3.2 The effects of surface inclination and geometry

In general, heat transfer and condensation rates also depend upon the orientation and shape of the cold substrate. If the substrate surface remains planar but is tilted backward (facing up) at an angle α to the horizontal, the structure of eqn (5.23) is retained for laminar conditions, except that g is replaced by $g \sin \alpha$ in the Archimedes number. This situation is explored in problem Q5.1 in Chapter 8. When α is close to zero, however, the situation changes substantially. For a horizontal substrate facing downward, a Bondian instability (Section 3.3.3) occurs in the film; continuing condensation then replaces the liquid leaving periodically as drops. On the other hand, for a horizontal substrate facing upward, the equation of motion parallel to the substrate reduces to

$$0 = -\frac{\partial P}{\partial X} + \mu \frac{\partial^2 U}{\partial Y^2} \tag{5.25}$$

if inertia is ignored. The corresponding equation normal to the substrate is

$$0 = -\frac{\partial P}{\partial Y} - \rho_L g. \tag{5.26}$$

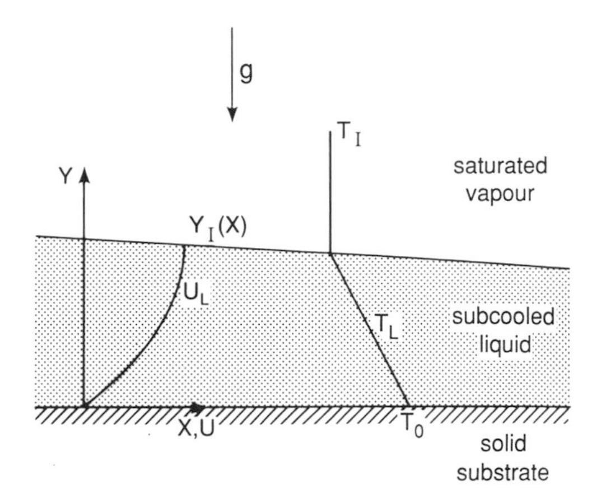

Fig. 5.5 Film condensation above a horizontal substrate: the liquid is driven by variations in film thickness.

These equations apply to the liquid film depicted in Fig. 5.5. We face a Nusselt problem.

Equation (5.26) integrates to give

$$P = -\rho_L g Y + \rho_L g Y_I + P_I(Y_I), \tag{5.27}$$

where $P_I(Y_I)$, is the vapour pressure exerted at the liquid–vapour interface. Hence

$$\frac{\partial P}{\partial X} = \rho_L g \frac{dY_I}{dX} + \frac{\partial P_I}{\partial X}$$

or, since $\partial P_I/\partial X = (\partial P_V/\partial X)_I = -\rho_V g\, dY_I/dX$,

$$\frac{\partial P}{\partial X} = g\Delta\rho \frac{dY_I}{dX}.$$

Equation (5.25) may therefore be rewritten

$$0 = -g\Delta\rho \frac{dY_I}{dX} + \mu \frac{\partial^2 U}{\partial Y^2}$$

or, using normalized variables,

$$0 = -\left[\frac{g\Delta\rho (Y^c)^3}{\mu U^c X^c}\right] \frac{dy_I}{dx} + \frac{\partial^2 u}{\partial y^2}.$$

Given that the driving 'hydrostatic' force, the first term on the right-hand side, is balanced by the viscous force, the second term, then

$$\frac{g\Delta\rho(Y^c)^3}{\mu U^c X^c} = O(1) \tag{5.28}$$

relates the unknown scales Y^c and U^c. The interface and continuity equations together provide a second relation, as described in the previous section. Thus we find that

$$Y^c = L\left(\frac{Ja}{ArPr}\right)^{1/5} \quad \text{and} \quad U^c = \frac{\kappa Ja}{L}\left(\frac{ArPr}{Ja}\right)^{2/5}, \tag{5.29}$$

where L is the substrate length in the direction of drainage.

The heat transfer rate during laminar film condensation on a horizontal substrate is thus represented by

$$Nu = \frac{\dot{q}_0 L}{k\theta^c} = O\left(\frac{L}{Y^c}\right) = O\left[\frac{ArPr}{Ja}\right]^{1/5}. \tag{5.30}$$

Bearing in mind the earlier findings for a vertical substrate, this may be generalized to

$$Nu = C_h(Ja, Pr)\left(\frac{ArPr}{Ja}\right)^{1/5}, \tag{5.31}$$

where C_h (*Ja, Pr*) must be determined from exact solutions or experiments. Note again the central importance of the group *ArPr/Ja*. This result is not too different from eqn (5.23), for a vertical substrate, and thus suggests that the relation

$$Nu = C(Ja, Pr)\left(\frac{ArPr}{Ja}\right)^{m}, \tag{5.32}$$

where $0.20 \leqslant m \leqslant 0.25$, is a useful correlating form for laminar film condensation on the outside of most common substrates. In particular, it may often by applied to circular tubes, as illustrated in Q5.2, 5.3 and 5.5.

Vertical tubes may sometimes be regarded as flat plates wrapped into a cylindrical form. Equation (5.23) then applies directly unless the tube diameter is very small, in which case the corresponding Nusselt problem must be cast in cylindrical coordinates to accommodate the surface curvature. When the tubes are horizontal, the more general form of eqn (5.32) may be used, but only if the tubes are independent of each other. In a typical tube bank, this is untrue, as Fig. 5.6 illustrates. For most tube layouts, some tubes are beneath others thus giving rise to the dripping of condensate from one tube to another. The condensate film therefore tends to be thicker on successively lower tubes. This feature, together with the cumulative subcooling of the condensate, leads to lower heat transfer coefficients on successively lower tubes, but the reduction is partly offset by turbulence and condensation on the drops which originate on the underside of each tube, their size and spacing being governed by the

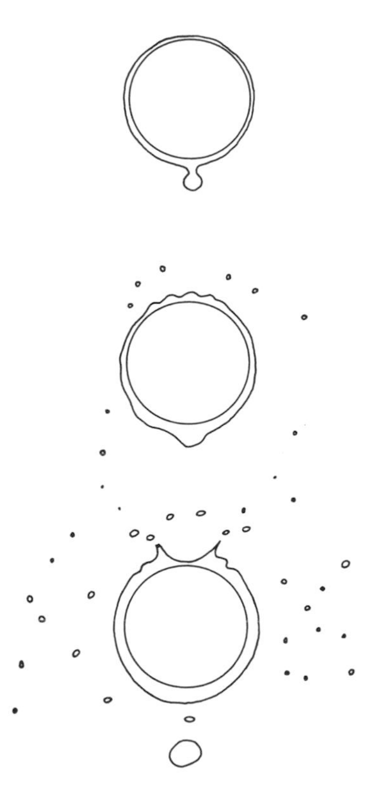

Fig. 5.6 Film condensation on a column of horizontal tubes in a tube bundle.

Bondian stability criterion discussed in Section 3.3. Depending on the vertical separation between the tubes, the drops may promote splashing and vigorous convection upon impact. The overall result is to create an average heat transfer coefficient $\bar{h}_n$ for n tube rows given by

$$\bar{h}_n = h_1/n^{1/6}, \tag{5.33}$$

where h_1 refers to the first row which obeys eqn (5.32). The exponent of n is empirically based. In a shallow condenser, e.g. $n=5$, the average heat transfer coefficient is about 76 per cent of h_1; in a deep condenser, e.g. $n=30$, the average is reduced to 57 per cent. The message is clear: tubes should be staggered and bundles kept shallow. Problems Q.5.4, 5.5, and 5.6 in Chapter 8 provide numerical illustrations for a tubular condenser.

5.3.3 Transitional and turbulent flow

Turbulence in the film arises naturally when the film Reynolds number

$$Re_f = \frac{\bar{U}\Delta}{\nu} = \frac{\Gamma}{\mu} \tag{5.34}$$

exceeds a critical value: Γ is the mass flow rate per unit span. Using the length and velocity scales given in eqn (5.14), $Re_f \simeq Ar^{1/4}(Ja/Pr)^{3/4}$, and hence the average Nusselt number for a falling film may be written

$$\bar{N}u_f = \frac{\bar{h}}{k_L}\left(\frac{\mu^2}{\rho_L g \Delta\rho}\right)^{1/3} = \frac{C_1}{Re_f^{1/3}} \tag{5.35}$$

in which $(\mu^2/\rho_L g\Delta\rho)^{1/3}$ is a length scale corresponding to $Ar = 1.0$. This result, and a corresponding result for a horizontal surface, reveals that an increase in Jakob number produces an increase in Reynolds number and thus leads to a decrease in heat transfer coefficient.

Figure 5.7 suggests that an inverse relation between $\bar{N}u_f$ and Re_f holds until $Re_f \simeq 400$ at which point turbulence is noticeable. The origin of the transition may be attributed to the growth and disintegration of Tollmien–Schlichting waves in the liquid but, because of the significant role played by fluid properties in the laminar sublayer, the transition itself is gradual, spanning more than an order of magnitude in the Reynolds number. The presence of capillary waves adds a further complexity to the transitional régime. The Nusselt solution for most fluids extends only up to $Re_f \simeq 10$, as Fig. 5.7 indicates. Beyond this point, eqn (5.35) must be modified to account for interfacial waves which enlarge as the Reynolds number is increased. A detailed discussion of the effect of waves on heat transfer is deferred until Section 5.5.

In the fully turbulent region, the film may be treated in the same way

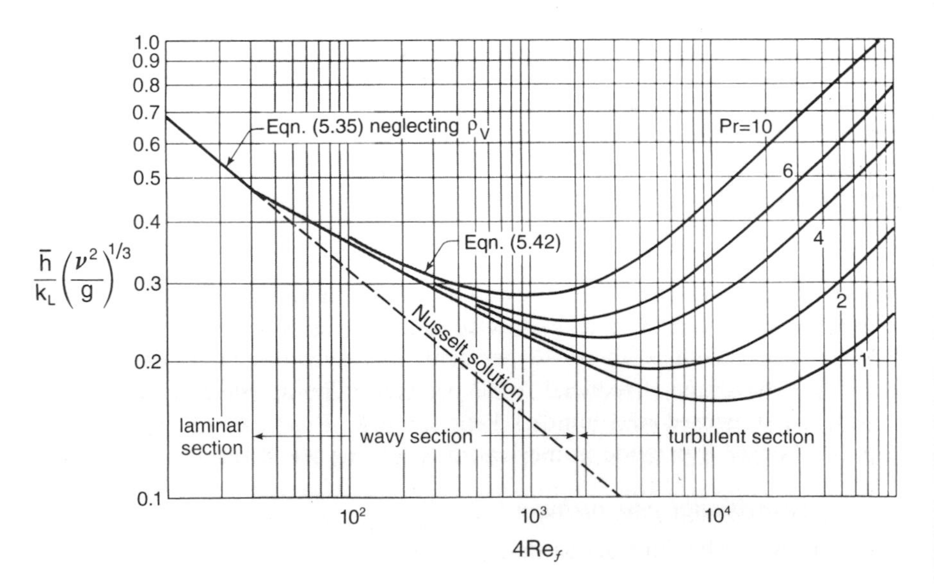

Fig. 5.7 Empirical curves showing the effect of Reynolds number and Prandtl number on film condensation over a vertical surface (after Chen *et al.* 1987).

as the planar turbulent film discussed in Section 3.2. The heat transfer rate may thus be determined from

$$\dot{q} = -(k_L + \rho_L c_{pL} \varepsilon_T) \frac{dT}{dY} \tag{5.36}$$

in which, by analogy, the thermal eddy diffusivity ε_T is equated to the momentum diffusivity ε described in Chapter 3. Defining the nondimensional temperature by

$$\theta^+ = (T_w - T)c_{pL} \frac{(\tau \rho_L)^{1/2}}{\dot{q}}, \tag{5.37}$$

eqn (5.36) may be rewritten

$$\frac{d\theta^+}{dy^+} = \frac{1}{\varepsilon^+ + 1/Pr}$$

in which $\varepsilon^+ = \rho_L \varepsilon/\mu_L$. This equation may be integrated over the three regions within the condensate film. Thus,

$$y^+ \leqslant 5: \qquad \theta^+ = Pr\, y^+ \tag{5.39}$$

$$5 \leqslant y^+ \leqslant 30: \qquad \theta^+ = 5[Pr + \ln\{1 - Pr(1 - y^+/5))] \tag{5.40}$$

$$30 \leqslant y^+: \qquad \theta^+ = 2.5 \ln \left[\frac{y^+ + 2.5(1/Pr - 1)}{30 + 2.5(1/Pr - 1)} \right] + 5[Pr + \ln(1 + 5Pr)]. \tag{5.41}$$

The above equations, and other similar expressions, have been used to estimate turbulent film heat transfer rates. By and large, the various predictions show the same general trends evident in Fig. 5.7 which provides an empirical correlation spanning the transitional régime (Chen *et al.* 1987). The correlation has the advantage that it may be stated in the concise form

$$\bar{N}u_f = [(4Re_f)^{-0.44} + 5.82 \times 10^{-6}(4Re_f)^{0.8} Pr^{0.33}]^{1/2} \tag{5.42}$$

which is, however, restricted to non-metallic fluids. Problems Q5.12, 5.13, and 5.14 in Chapter 8 provide numerical illustrations for a water condensate film.

When $Pr \ll 1$, the characteristically higher thermal conductivity tends to restore the linear temperature profile found in the Nusselt solution, but inertial effects then become important (see Fig. 5.4). This suggests turbulent metal condensate films would produce heat transfer results much closer to the dashed extrapolation of eqn (5.35), and even beneath it. Under these conditions, it is unlikely that interfacial waves would exert much influence.

5.4 The role of the vapour

Thus far, we have assumed that the vapour condenses unimpeded on a smooth interface but otherwise exerts no influence. This assumption is frequently valid under saturated conditions when the vapour is pure and quiescent. If the vapour is superheated, impure, and flows over the interface, the behaviour of the system may be expected to change. We now examine each of these modifying factors in turn.

5.4.1 The effects of superheat in a pure vapour

Interfacial thermal resistance in the vapour becomes evident in the presence of a superheat $T_V - T_I$. The condensation rate $\dot{m}_{cond}$ associated with this superheat was derived in Section 2.3.3 and may be used to define the interfacial heat transfer coefficient h_I as

$$h_I = \lambda_{LV}\dot{m}_{cond}/(T_V - T_I) = 2\rho_V\lambda_{LV}^2/T(2\pi RT)^{1/2}. \tag{5.43}$$

With the exception of metallic vapours at low pressures, this interfacial heat transfer coefficient is usually large enough (e.g. $10^7\,\mathrm{W\,m^{-2}\,K^{-1}}$) to permit the neglect of thermal resistance between a pure, superheated vapour and the interface. Problem Q7.2 in Chapter 8 provides an illustration.

Under these circumstances, condensation may be modelled as the simultaneous removal of both latent and sensible heat at the interface. The interface equation may therefore be restated as

$$\begin{aligned} k_L\left(\frac{\partial T}{\partial Y}\right)_I &= \dot{m}[\lambda_{LV} + c_{PV}(T_V - T_I)] \\ &= \dot{m}\lambda_{LV}(1 + Ja_V), \end{aligned} \tag{5.44}$$

where $Ja_V = c_{pV}(T_V - T_I)/\lambda_{LV}$ is the vapour Jakob number. In essence, this alteration produces no change in the earlier analyses except that λ_{LV} is replaced by $\lambda_{LV}(1 + Ja_V)$. The structure of the Nusselt problem is retained and the results follow automatically. For laminar conditions, the removal of a given heat flux requires that

$$h_{sup} = h_{sat}(1 + Ja_V)^m, \tag{5.45}$$

where $1/5 \leqslant m \leqslant 1/4$, depending on the substrate geometry, and h_{sup} and h_{sat} are the heat transfer coefficients with and without superheat, respectively. When $Ja_V \ll 1$, the effect of superheat on the condensation rate of a pure vapour may be neglected. For example, 200°C steam at atmospheric pressure modifies the heat transfer coefficient by only 2 per cent. Numerical illustrations are also provided in Q5.7 in Chapter 8.

5.4.2 Vapour mixtures: the basic features

When the vapour is a mixture, condensation becomes more complex. Before condensing, the vapour must first diffuse towards the interface. Simultaneous diffusion of heat and mass therefore takes place. Here we will limit ourselves to the practical situation of a binary mixture. In particular, we will concentrate on a condensable vapour V mixed with an incondensable gas g. The situation is described in Fig. 5.8.

At the outset, it is important to recognize that the thermal and mass fluxes are no longer simply diffusive. As noted in Section 2.1.2, this requires us to take

$$\dot{m}_V = j_{MV} + \mathfrak{m}_V G \tag{5.46}$$

and

$$\dot{m}_g = j_{Mg} + \mathfrak{m}_g G \tag{5.47}$$

as the vapour and gas mass flux densities, respectively. The diffusive mass flux density $j_M = -\rho\mathfrak{D}\,\partial\mathfrak{m}/\partial Y$, while the total mass flux density is defined by

$$G = \dot{m}_V + \dot{m}_g. \tag{5.48}$$

Likewise, the thermal flux density is taken as

$$\dot{q} = j_Q + hG, \tag{5.49}$$

where $j_Q = -k\partial\theta/\partial Y$ and $h = c_p\theta$, with $\theta = T - T_I$, defines the specific enthalpy: $k = k_V + k_g$ and $c_p = \mathfrak{m}_V c_{pV} + \mathfrak{m}_g c_{pg}$ represent the mixture.

When the gas does not condense, the above expressions simplify. Equation (5.47), then reduces to $\dot{m}_g = 0$, i.e.

$$j_{Mg} = -\rho\mathfrak{D}_{Vg}\partial\mathfrak{m}_g/\partial Y = -\mathfrak{m}_g G,$$

implying that diffusion of the gas in one direction is balanced by its bulk flow in the other. Equation (5.46) thus simplifies to

$$G = j_{MV} + \mathfrak{m}_V G$$

or

$$G = \dot{m}_V = j_{MV}/(1 - \mathfrak{m}_V). \tag{5.50}$$

Hence

$$\dot{m}_V = \frac{-\rho\mathfrak{D}_{Vg}}{(1-\mathfrak{m}_V)}\frac{\partial\mathfrak{m}_V}{\partial Y} = \frac{\rho\mathfrak{D}_{Vg}}{\mathfrak{m}_g}\frac{\partial\mathfrak{m}_g}{\partial Y}, \tag{5.51}$$

since $\mathfrak{m}_V + \mathfrak{m}_g = 1$. The interfacial mass and heat flux densities thus become

$$\dot{m}_{VI} = -\frac{\rho\mathfrak{D}_{Vg}}{1-\mathfrak{m}_{VI}}\left(\frac{\partial\mathfrak{m}_V}{\partial Y}\right)_I \tag{5.52}$$

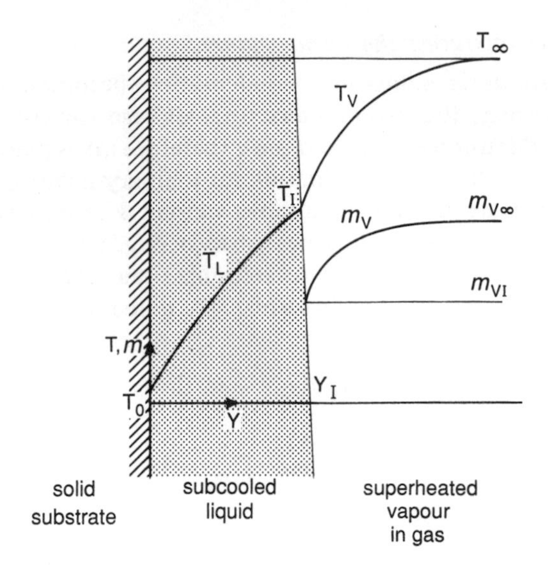

Fig. 5.8 Condensation in the presence of an incondensable gas.

and

$$\dot{q}_I = -k\left(\frac{\partial \theta}{\partial Y}\right)_I, \tag{5.53}$$

respectively.

It is now useful to examine the interface equation

$$k_L\left(\frac{\partial T_L}{\partial Y}\right)_I = -\lambda_{LV}\dot{m}_{VI} + k\left(\frac{\partial \theta_V}{\partial Y}\right)_I,$$

or, in normalized form,

$$\left(\frac{\partial \phi_L}{\partial y}\right)_I = -\frac{\lambda_{LV}\dot{m}_{VI}Y_L^c}{k_L\theta_L^c} + \left(\frac{k}{k_L}\right)\left(\frac{\theta_V^c}{\theta_L^c}\right)\left(\frac{Y_L^c}{Y_V^c}\right)\left(\frac{\partial \phi_V}{\partial y}\right)_I.$$

The significance of vapour sensible heat is evidently determined by the three coefficients of the normalized derivative $(\partial\phi_V/\partial y)_I$. With steam, for example, $k/k_L \simeq 0.05$. Hence, only when steam is very highly superheated $(\theta_V^c \gg \theta_L^c)$ in a thin diffusion layer $(Y_V^c \ll Y_L^c)$ is it necessary to account for steam temperatures higher than the saturation value. It is thus evident again that, except for metals, vapour (mixture) sensible heat plays only a minor role in condensation.

5.4.3 *The effects of vapour concentration and velocity*

Diffusive resistance may play a major role. The reasons for this are developed schematically in Fig. 5.9, which is a more detailed version of Fig. 5.8. The curve T_V^0 represents the vapour temperature profile in the absence of the gas; it terminates at the interface where the saturation temperature is $T_I^0 \leqslant T_\infty$. With the gas present, the vapour pressure P_V decreases towards the interface, reflecting the diffusive resistance. The reduced vapour pressure creates a lower vapour temperature, as indicated by the curve T_V. In particular, the reduced vapour pressure at the interface, following the Clapeyron equation, creates a reduced saturation temperature $T_I < T_I^0$. The rate of heat conduction through the liquid film, along with the condensation rate, is therefore reduced accordingly.

The interface equation describing this situation is given by

$$k_L \left(\frac{\partial T_L}{\partial Y}\right)_I = \lambda_{LV} \dot{m}_{VI} = -\lambda_{LV} \rho \frac{\mathfrak{D}_{Vg}}{m_g} \left(\frac{\partial m_g}{\partial y}\right)_I, \tag{5.54}$$

if vapour sensible heat is neglected. Normalizing, we find that

$$\frac{k_L(T_I - T_0)}{\Delta} = O\left[\lambda_{LV} \rho \frac{\mathfrak{D}_{Vg}}{\delta}\right] \tag{5.55}$$

where Δ is the film thickness and δ is the diffusion layer thickness scale in the vapour. To maintain a high condensation rate, both sides of this equation must be kept as high as possible. The right-hand side indicates that this may be achieved by reducing the diffusion layer thickness. Figure 5.10 provides some predictions for steam at T_∞ condensing in the presence of small concentrations of air m_a. It is seen that even small air concentrations may produce a significant reduction in heat transfer rate, here presented as a fraction of the air-free value $\dot{q}^0$. Also evident is the substantial improvement which may be introduced when natural convection is replaced by forced convection with a thinner diffusion layer.

The above qualitative analysis reveals the essential structure of the mixture condensation problem. A quantitative evaluation must incorporate the details of the vapour concentration and velocity fields in the interface eqn (5.54). The simplest approach is to use the diffusive analogy between heat and mass transfer expressed by

$$(j_Q)_I = -h_Q(T_\infty - T_I) \tag{5.56}$$

and

$$(j_{MV}^0)_I = -h_M^0(m_{V\infty} - m_{VI}) \tag{5.57}$$

in which $h_M^0 = h_Q \rho \mathfrak{D}/k$, based on the analogous relation $Nu(Re, Pr) = Sh(Re, Sc)$, is the limiting mass transfer coefficient created as $m_{V\infty} \rightarrow m_{VI} \rightarrow 0$. This leaves us with two difficulties: the actual diffusive mass flux density is

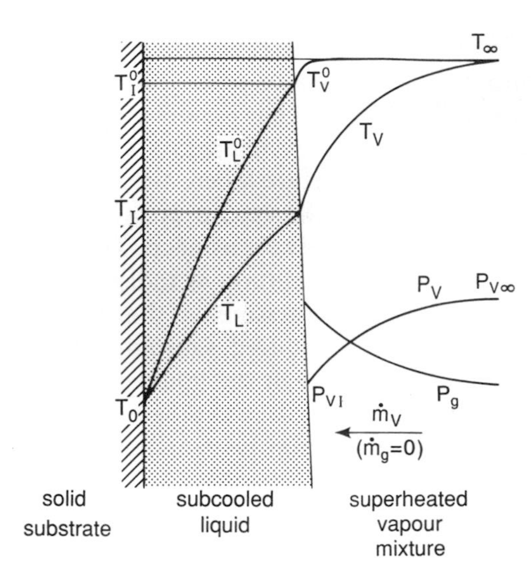

Fig. 5.9 Temperature and pressure profiles during condensation in the presence of an incondensable gas.

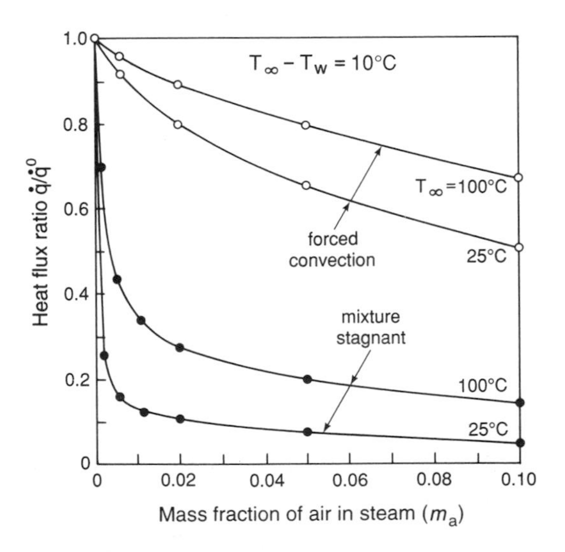

Fig. 5.10 Influence of air concentration on steam condensation (based on Sparrow *et al.* 1967).

$$(j_{MV})_I = -h_M(m_{V\infty} - m_{VI})$$

in which h_M, the actual mass transfer coefficient, is *not* analogous to h_Q; and the interfacial values T_I and m_{VI} are not given.

The first difficulty may be resolved by using eqn (5.50) to write

$$(j_{MV})_I = G_I(1 - m_{VI}) = -h_M(m_{V\infty} - m_{VI}). \tag{5.58}$$

Hence

$$G_I = h_M\left(\frac{m_{V\infty} - m_{VI}}{m_{VI} - 1}\right) = h_M B, \tag{5.59}$$

where B is the blowing parameter: $B > 0$ for evaporation, but $B < 0$ for condensation. Also from eqn (5.50)

$$j_{MV} = -\rho\mathfrak{D}\frac{\partial m_V}{\partial Y} = -G(m_V - 1)$$

which may be integrated across the vapour diffusion layer measured from the interface $(0 \leqslant Y \leqslant \infty)$ to give

$$h_M = \frac{\ln(1+B)}{B}\Bigg/\frac{1}{\rho\mathfrak{D}}\int_0^\infty g(Y)\,\mathrm{d}Y, \tag{5.60}$$

where the normalized flux $g(Y) = G(Y)/G_I$. In particular, as $B \to 0$,

$$h_M^0 = 1\Bigg/\frac{1}{\rho\mathfrak{D}}\int_0^\infty g^0(Y)\,\mathrm{d}Y$$

and hence

$$h_M \simeq h_M^0\frac{\ln(1+B)}{B}, \tag{5.61}$$

providing $\int_0^\infty g^0(Y)\,\mathrm{d}Y \simeq \int_0^\infty g(Y)\,\mathrm{d}Y$, as we might reasonably assume.

The second difficulty is handled by writing the interfacial molar fraction

$$x_{VI} = \frac{P_{VI}}{P} = \exp \lambda_{LV}(T_I - T_\infty)/R_V T_\infty^2,$$

using the Clausius–Clapeyron equation, and noting that

$$m_{VI} = \frac{x_{VI}\mathfrak{M}_V}{x_{VI}\mathfrak{M}_V + (1 - x_{VI})\mathfrak{M}_g}.$$

A guessed value of T_I thus provide x_{VI} and m_{VI}; hence B, h_M, and $G_I(=\dot{m}_{VI})$ may be found for prescribed values of T_∞, $m_{V\infty}$, and h_M^0 (or h_Q). This value of T_I must satisfy the interface eqn (5.54), which includes the heat conducted across the condensate film; otherwise, the guess must be refined iteratively. Problems Q5.15 and 5.16 in Chapter 8 provide a numerical example.

In addition to controlling the thickness of the diffusion layer, vapour motion also creates an interfacial shear stress. If great enough, this stress converts the characteristic parabolic velocity profile of a buoyancy-driven film to the linear profile of a shear-driven film. The film thickness is thereby reduced and an increase in heat transfer and condensation rate occurs. This feature becomes important during condensation inside a tube, as discussed below in Section 5.6. It may also be important during condensation on the outside of a tube, as illustrated in Q5.8 to 5.11.

5.5 The role of the interface

5.5.1 Interfacial architecture

Having discussed the effect of the vapour near a smooth, stable interface we now take a closer look at the interface itself. From the discussion of film hydrodynamics in Chapter 3, it is evident that a condensate film will rarely be perfectly smooth in practice even though surface tension continues to act as a restraining influence on any disturbances tending to produce surface irregularities. But the shaping influence of surface tension becomes more evident whenever σ_{LV} alone does not provide a complete description of interfacial forces.

For a pure vapour, σ_{LV} will act in concert with σ_{SV} when the substrate is exposed: during the overlap between drop and film condensation, for example, or when an established film becomes thin enough to break into rivulets. In both of these situations the shape of the interface is strongly influenced by at least two surface tensions. For a mixture of condensing vapours, the situation is more complex even when the condensate completely covers the substrate. Depending on the spreading coefficient, the condensate may shape itself into various heterogeneous patterns.

Consider, for instance, the condensation of a eutectic vapour mixture into two immiscible liquids. This is illustrated in Fig. 5.11 using an organic-water vapour mixture. When both of the condensates wet the substrate, the water (which has the higher surface tension) is distributed as drops throughout a continuous film of organic liquid, as suggested in Fig. 5.11(a). However, if the water does not wet the substrate, two modes of condensation occur simultaneously, as suggested in Fig. 5.11(b). Drop condensation of water vapour is then accompanied by film condensation of the organic. Should the water vapour condensation rate increase relative to the organic condensation rate, the drop mode will change into a film mode. The condensate film will then consist of water and organic rivulets running more or less in parallel. This is suggested in Fig. 5.11(c) which cannot,

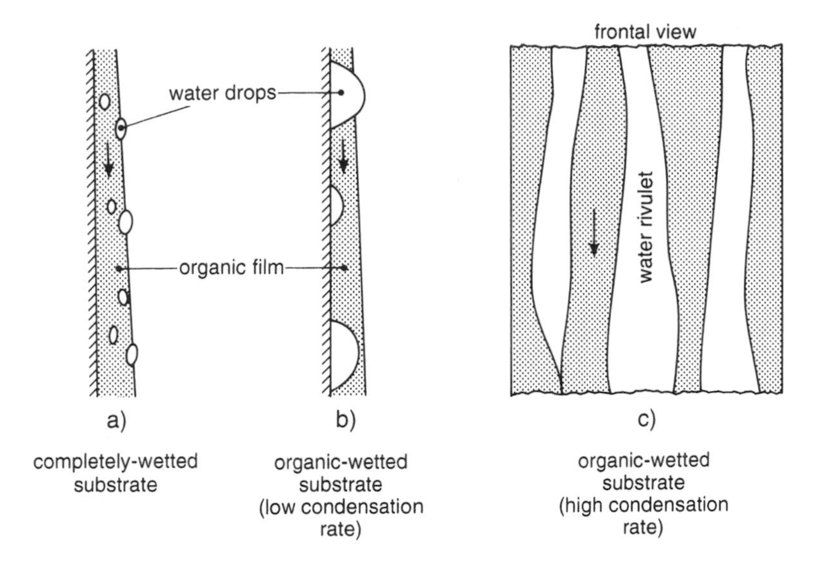

Fig. 5.11 The behaviour of immiscible condensates formed from an organic-water vapour mixture.

however, portray the capricious shifts and alternations so characteristic of rivulet behaviour.

5.5.2 The effects of waves

It has been noted previously that ripples begin to appear on the interface even when the film velocity is small. They have been observed at $Re_f \simeq 10$, and it has been argued that a falling film may be inherently unstable. The origin of these waves is the relative velocity between the vapour and the liquid film. On a vertical substrate the onset of vapour-driven *capillary* waves is associated with a critical value of the film Weber number $We_f = \rho_L U_V^2 \Delta / \sigma$, where Δ is the film thickness. On a horizontal substrate facing upward, vapour-driven *gravity* waves begin appearing when the film Froude number $Fr_f = \rho_L U_L^2 / g(\rho_L - \rho_V)\Delta$ reaches a critical value.

Waviness alters the Nusselt problem discussed earlier, but not in a fundamental way unless the wave amplitudes become very large. It will be recalled from Fig. 5.7 that the Nusselt solution, which neglects waviness, underestimates the heat transfer rate. To explain the observed improvement, consider Fig. 5.12, which depicts a laminar falling film with an interfacial wave described by

$$\Delta = \Delta_0 \left(1 + A \sin \frac{2\pi X}{\Lambda}\right), \tag{5.62}$$

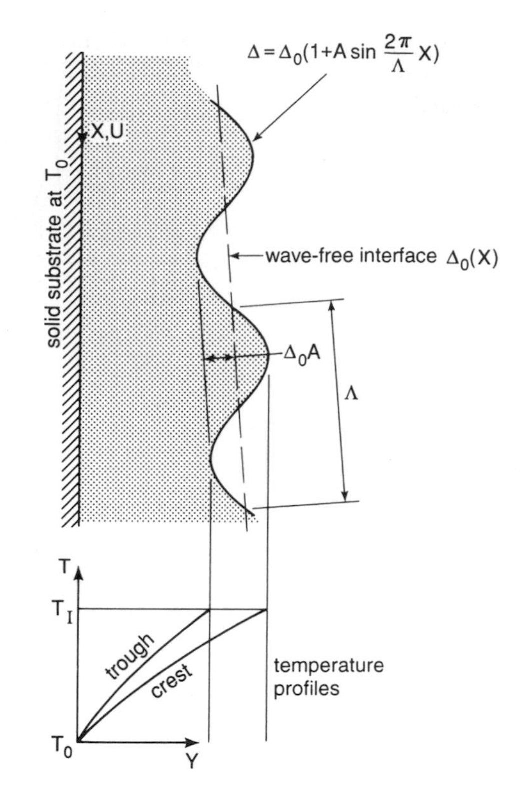

Fig. 5.12 Film condensation with a wavy interface.

where $\Delta_0(X)$ is the mean film thickness at X and $A\Delta_0$ is the wave amplitude about the mean. Since the local heat flux density when $Ja \ll 1$ is given by

$$\dot{q}_0(X) = k\left(\frac{\partial T}{\partial Y}\right)_0 \simeq \frac{k(T_\mathrm{I} - T_0)}{\Delta},$$

where T_I is the uniform temperature on the wavy interface, the average heat transfer rate over a wavelength Δ,

$$\bar{\dot{q}}_{0\Lambda} = \frac{1}{\Lambda}\int_0^\Lambda \dot{q}_0(X)\,\mathrm{d}X,$$

takes the form

$$\bar{\dot{q}}_{0\Lambda} = \frac{k(T_\mathrm{I} - T_0)}{\Delta_0(1 - \mathrm{A}^2)^{1/2}} \tag{5.63}$$

in which $\Delta_0(X)$ has been held constant during the integration. This reveals that the net effect of cyclical variations in the temperature gradient, from a maximum in the wave trough to a minimum at the crest, is to increase the average heat transfer rate above the wave-free mean value.

The result carries two important implications. Firstly, it is evident that increases in the temperature gradient caused by thinning of the film more than offset decreases caused by thickening, even though the mean film thickness remains the same. This asymmetric feature occurs in many film condensation systems, as will be seen again later. Secondly, it is clear that the average (conduction) heat transfer rate increases as the wave amplitude increases, tending to a theoretical limit of infinity as the amplitude approaches the film thickness. It was noted in Section 5.3 that observed increases may exceed 20 per cent, even with low vapour velocities.

When the wave-induced increase in heat transfer has reached 11 per cent, the wave amplitude has already exceeded 44 per cent of the film thickness. Such waves invariably have a significant curvature which grows as the vapour speed increases. As observed in Section 3.4.1, the flow over each crest then induces separation on the sheltered side of the wave. With a rising vapour velocity, the resulting recirculation creates an undercut near the peak which eventually forms what might be described as a breaking lip. The droplets formed by this lip as it cascades forward are torn away and thus enter the mainstream in the process known as *entrainment*. Thus, as the vapour speed increases, the disturbing inertial force of the vapour, causing the waves to grow, gradually rises until it exceeds the restraining force and generates spray.

The equality of the Froude and Weber numbers, which signals entrainment, implies that the inertial, buoyancy, and surface tension forces are substantial and roughly equal. By contrast, the interfacial shear force usually remains relatively small. Thus, even though the shear at lower vapour velocities may help give rise to surface waves, just as surface shear over a stretched plastic film may cause buckling, the waves themselves grow principally in response to inertial forces. When the inertial forces have risen to dominate, the vapour may be capable of blowing the entire film off the substrate. As we shall see, actual behaviour is dependent on the circumstances and, in particular, on whether or not the vapour and film are flowing in the same direction. The entire process of wave development, from the first appearance of tiny ripples to the final breakup of spray-generating crested waves, becomes especially important during condensation inside tubes.

5.6 Condensation in tubes

Condensation inside a tube may take any of the forms discussed above: drop or film mode; laminar or turbulent films; forced or natural vapour flow; saturated or superheated vapour; pure or mixed vapour; and vertical or horizontal orientations. The principal difference from external condensation arises from the confining effect of the tube wall. Typically, the vapour is forced through the tube by a pressure gradient; natural convection is usually unimportant. Given these circumstances, we now divide our enquiry into two main categories: vertical tubes and horizontal tubes.

5.6.1 The vertical tube

Drop condensation inside a tube usually differs from external drop condensation in only one minor respect: the vapour velocity tends to be higher. The additional shear stress acting on the drops accelerates their natural progress if the vapour flows downward and impedes it if the vapour flow is upward. These are not common situations and thus deserve no further comment. Internal film condensation, on the other hand, often arises in industrial heat exchangers. It is usually associated with short tubes or high vapour velocities.

The role of the vapour pressure gradient is crucial in annular film condensation, as may be illustrated by considering a downward through-flow. The situation is illustrated in Fig. 5.13, which shows a condensate film running over the inside surface of a tube of diameter D. The imposed pressure gradient induces an interfacial shear stress given by

$$\tau_{\mathrm{I}} \simeq -\frac{D}{4}\frac{\mathrm{d}P_{\mathrm{V}}}{\mathrm{d}X} \tag{5.64}$$

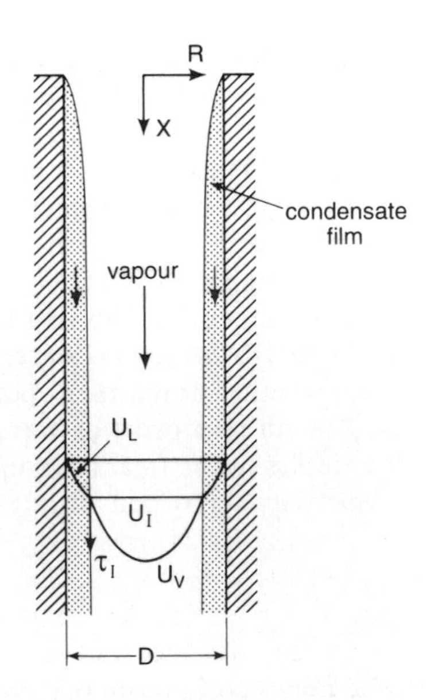

Fig. 5.13 Annular, through-flow film condensation in a vertical tube.

if the mean film thickness $\bar{\Delta} \ll D$. It is thus assumed that the film is thin enough for the tube surface to be treated as a vertical plane wrapped into a cylindrical shape. This strategy has the advantage that film behaviour may then be anticipated from the previous discussion. However, the heat transfer rates presented earlier in Fig. 5.7, for example, are limited to $\tau_I = 0$. When the film is driven by shear, $\tau_I \simeq \tau_w \simeq \mu_L U_I/\bar{\Delta}$ if the condensate flow is laminar. Under these circumstances, the triangular relation $\bar{\Delta} = \bar{\Delta}(\tau_I, \Gamma)$ takes the simple, explicit form

$$\bar{\Delta}^2 = \frac{2\nu_L \Gamma}{\tau_I} = 2\rho_L \nu_L^2 \frac{Re_f}{\tau_I}, \tag{5.65}$$

from eqn (3.17). Since the average heat transfer rate is then given by $\bar{\dot{q}}_w = k_L(T_I - T_w)/\bar{\Delta}$, if $Ja_L \ll 1$, we find that

$$\overline{Nu}_f = \left(\frac{\alpha}{Re_f}\right)^{1/2}, \tag{5.66}$$

where $\alpha = \tau_I / 2\rho_L(\nu_L g)^{2/3}$. This represents a family of shear curves lying well above the Nusselt number equation (5.35) in Fig. 5.7.

The above result may be used for a laminar film whenever the pressure gradient in eqn (5.64) is specified. Near the tube entry, we may adopt an alternative approach based on the neglect of sensible heat. thus

$$\bar{\dot{q}} = \left[\frac{k_L(T_I - T_0)}{\bar{\Delta}}\right] = O(\lambda \bar{\dot{m}})$$

from a film heat balance averaged over an entry length L, while

$$\bar{\dot{m}} = O\left(\frac{\rho_L U_I \bar{\Delta}}{2L}\right)$$

from a corresponding mass balance. From these it is evident that

$$\bar{\Delta} = O\left[\frac{k_L(T_I - T_0)L}{\lambda \rho_L U_I}\right]^{1/2} \tag{5.67}$$

and hence

$$\overline{Nu}_D = \frac{\bar{\dot{q}} D}{k_L(T_I - T_0)} = C\left(\frac{Gz_D}{Ja_L}\right)^{1/2} \tag{5.68}$$

where $C = O(1)$ and $Gz_D = D^2 U_I / L\kappa_L$ is Graetz number. Further from the tube inlet the cumulative effect of condensation reduces the vapour Reynolds number but increases the film Reynolds number. Additional information is therefore required, as noted in Q5.8 in Chapter 8.

When the annular condensate film is turbulent, we must return to the universal velocity profiles given by eqn (3.28). Integration of these profiles to satisfy the triangular relations is now more complex, and the process becomes iterative. Problems Q3.2–3.4 in Chapter 8 illustrate the technique. The film thickness Δ thus obtained (along with Γ and τ_I), may then be used with eqns (5.39)–(5.41) to determine θ^+ and hence $\bar{q}_w$. Problem Q5.17 in Chapter 8 provides an illustration. The strategy is identical to that used under laminar conditions, but the relations are implicit. Only in the simplest, laminar entry situation may Δ and τ_I be written explicitly. Alternatively, the annular condensate may be treated empirically (Travis *et al.* 1973), as illustrated by eqns (6.27)–(6.30).

A turbulent annular film is most often associated with a turbulent vapour core. The friction factor is therefore given by $f = 0.079/Re_V^{1/4}$ if the interface is smooth. However, the data in Fig. 5.7 indicate that even a falling film is seldom, if ever, smooth. Waviness introduces an equivalent roughness at the interface. It has been suggested by Whalley (1987) that this effect may be represented by a relation of the form

$$f = f_s(1 + \beta\Delta/D), \tag{5.69}$$

where f_s refers to smooth interfacial conditions and β is an empirical roughness coefficient. The effect of waviness may be substantial, as illustrated in Q3.4 in Chapter 8.

Near the tube inlet, the condensate film is laminar and smooth, whether the vapour is laminar or turbulent. In either event, the condensate thickens as it moves downstream, thus tending to increase the film Weber number. At the same time, the corresponding decrease in vapour velocity tends to reduce the Weber number. In any event, the vapour velocity may be high enough to destabilize the film and the interface downstream then becomes covered in steepening waves. Eventually, these reach across the tube to enclose large pockets of the vapour which by then has slowed to a velocity comparable with the interface velocity. This situation marks the end of the annular through-flow régime.

Interfacial waves are common in downward annular flows where their effect ranges from the improvement in heat transfer described in the previous section to a complete change in régime. This change is particularly dramatic when the vapour flows upward, as in a tubular thermosyphon. Reflux condensation is illustrated in Fig. 5.14, which shows vapour entering at the bottom of a vertical tube in which annular film condensation is occurring. When the film is much thinner than the tube radius, this situation is not unlike that shown in Fig. 5.13. Although the vapour no longer flows through, its dynamic effect on the condensate film continues to be exerted through the interfacial shear stress and pressure distribution. This effect is slight when $\Delta \ll D/2$.

As the waves grow with increases in the heat transfer, circulation, and condensation rate, the influence of interfacial shear becomes *relatively* unimportant; the effect of vapour inertia (or pressure gradient) eventually

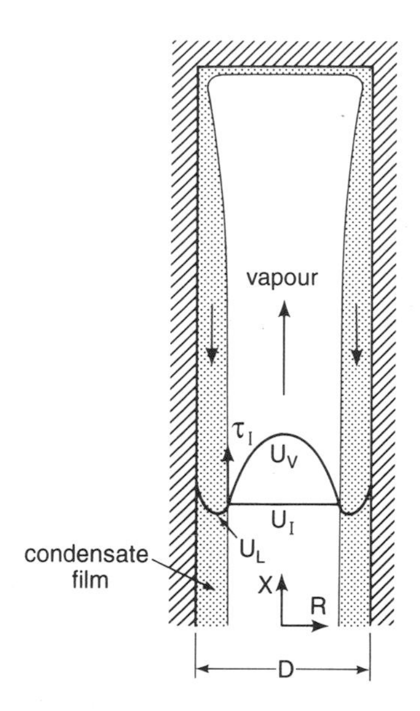

Fig. 5.14 Reflux condensation in a vertical tube.

dominates. As these limiting conditions are approached, the adverse pressure gradient in the vapour tends to slow, and ultimately halt, the descent of the liquid film, as described in Chapter 3. During this process, breakup on the wave crests creates large amounts of spray, drops being both entrained in the vapour and re-deposited in the film later. The transfer of momentum from the vapour to the film gradually causes the surface layers in the latter to flow in the reverse direction.

The condition at which condensate flowing downward (near the wall) is balanced by that flowing upwards (near the interface) is usually described as *flooding*. Waves then fill most of the tube cross section thus producing a performance limit. Attempts to increase the heat transfer (circulation) rate beyond this point lead to a complete reversal of the film and the condition known as *holdup*. This creates an unstable situation in which vapour entry is stifled, condensation ceases, and the liquid held up then falls slug-like out of the tube mouth, thus paving the way for re-establishment of the annular film. Such limiting behaviour is characterized by vigorous flow oscillations which can be reduced by inserting a concentric smaller bore tube to confine the vapour and help keep the opposed flows separate.

5.6.2 The horizontal tube

Turning now to condensation in horizontal tubes it is necessary to recall the two-phase, through-flow régimes discussed in Chapter 3. When the condensation rate is low, behaviour is not unlike that described earlier for condensation on the tube outside. A thin annular film then flows circumferentially around the inside of the tube wall as suggested in Fig. 5.15. The principal difference with external condensation is attributable to the rivulet which drains the two films flowing in from both sides. As long as the depth of this rivulet remains much smaller than the tube diameter, the problem is simply the internal version of the Nusselt problem for external condensation, providing the axial vapour velocity is not too great. Integrating around the tube surface to account for circumferential variations in the component of gravity, the heat transfer coefficient is given by

$$\bar{h} \simeq 0.725\left(1-\frac{\gamma}{\pi}\right)\left[\frac{\lambda_{LV} g k_L^3(\rho_L-\rho_V)}{\nu_L D(T_I-T_w)}\right]^{1/4} \tag{5.70}$$

in which $\gamma/\pi \ll 1$. This neglects condensation on the rivulet and applies so long as the vapour Reynolds number is less than 3×10^4.

As the vapour velocity increases, it converts the two-dimensional, circumferential film flow into a three-dimensional spiral flow in which turbulent vapour shear progressively thins the film and thus increases both the heat transfer rate and the condensation rate. Condensate flow is then more complex. However, with further increases in vapour speed, the flow eventually reverts to the axisymmetric form found in the vertical tubes discussed above. That is, an annular model incorporating a circumferentially averaged film thickness provides an appropriate asymptote at high vapour velocities regardless of tube orientation. For purely laminar conditons, eqns (5.68) and (5.70) represent limiting forms which, taken together, suggest the more general interpolation

$$\bar{Nu}_D = C\left[\frac{Gz_D}{Ja_L} + a\left(\frac{Ar_D Pr}{Ja}\right)^{1/2}\right]^{1/2}, \tag{5.71}$$

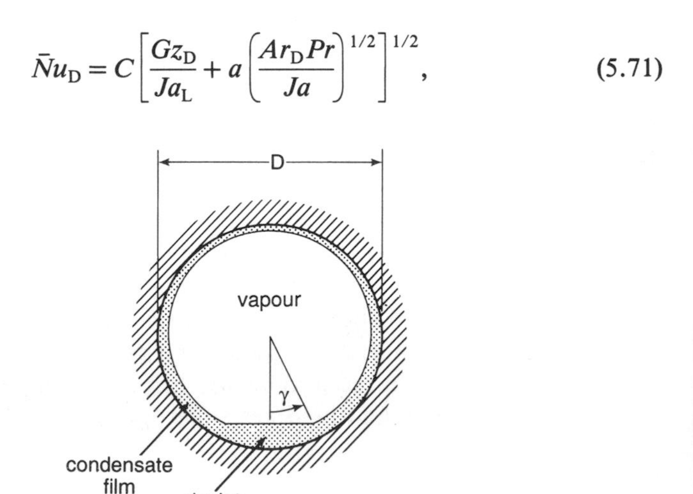

Fig. 5.15 Schematic representation of annular film condensation and drainage in a horizontal tube.

where C and a are empirical coefficients. Unfortunately, this low velocity entry length equation has yet to be tested. Interpolation may also be used when the flow at the upper annular limit is completely turbulent (Butterworth, 1986).

As with vertical tubes, the above discussion assumes that the condensate film is thin, on average. It also assumes that interfacial ripples or waves have little effect on behaviour beyond creating an equivalent roughness which alters the interfacial shear stress. In addition, the rivulet running over the bottom of the tube is taken to be the passive occupant of a small surface area. Any attempt to lower the exit quality of the liquid–vapour mixture by lengthening the tube alters the situation significantly. The rivulet deepens, eventually producing the stratified flow schematically illustrated in Fig. 5.16. Equation (5.70) then provides only a rough estimate with $\gamma \geqslant \pi/2$. Well before this, interfacial waves have grown substantially under what are usually turbulent conditions.

Figure 5.17 provides a useful interpretation of events using the régime map from Chapter 3. A typical quality curve is shown superimposed. Some idea of the corresponding flow conditions may be obtained from the schematic representation in Fig. 5.18. In this example, the vapour flow at entry quickly establishes a rippled surface on both the rivulet and the condensate film. Further into the tube, a thickening annular film produces waves large enough to create a spray of drops; these are entrained in the superheated vapour which is thereby slowed and further cooled by evaporation. Eventually, the wave amplitudes reach across the tube, enclosing sporadic pockets of vapour which become slugs moving near the upper surface. These in turn tend to merge into a single vapour space, producing the stratified flow illustrated in Fig. 5.16.

The quality curve on Fig. 5.17 indicates that the liquid mass flow rate increases rapidly near the tube entry, without much effect on the vapour mass flow rate; the void fraction $\varepsilon \simeq 1$ and the vapour velocity remains fairly high. This behaviour continues well through the annular régime.

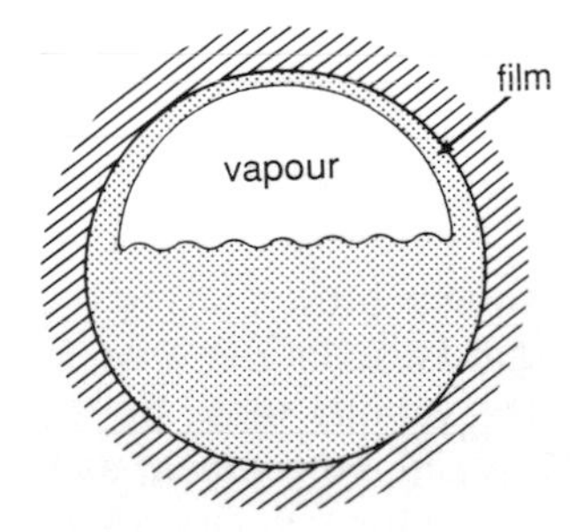

Fig. 5.16 Schematic representation of stratified film condensation in horizontal tube.

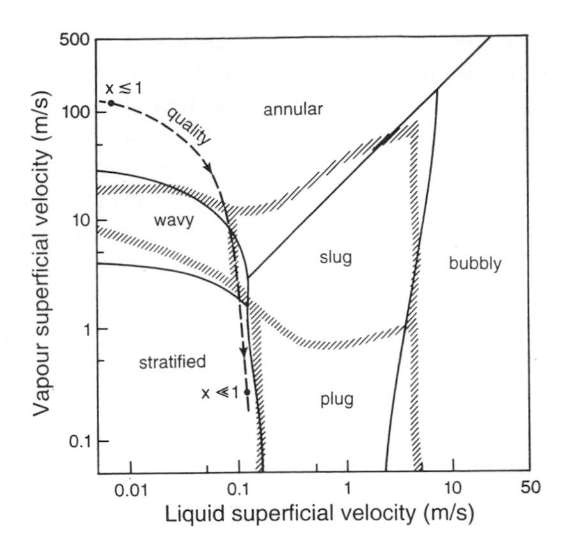

Fig. 5.17 Condensation path superimposed on régime map for air–water flow in a horizontal tube.

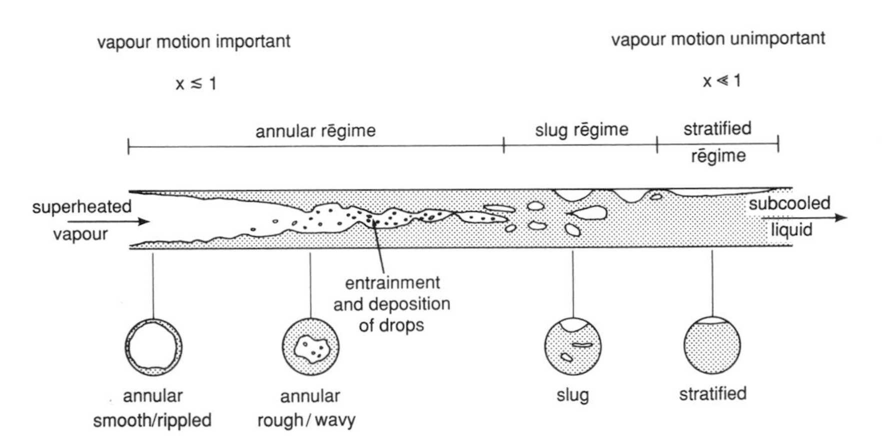

Fig. 5.18 Schematic representation of complete condensation in a horizontal tube (based on Collier 1972).

Predictions based on an annular model are fairly reliable up to the beginning of the slug régime: that is, over the greater part of the tube length. Beyond the slug régime, the vapour mass flow rate falls rapidly without altering the liquid mass flow rate very much, as Fig. 5.17 indicates. Under these latter conditions, the vapour velocity approaches the liquid velocity and the shearing effect at the interface is much less important; the stratified flow model applies with $\gamma \geqslant \pi/2$ in eqn (5.70). In the transitional wavy/slug region, condensation and heat transfer rates are difficult to predict accurately but are usually a minor contribution. A higher mass flow rate shifts the quality curve further to the right and upward; transition then occurs in the intermittent slug régime which separates annular (entry) flow from bubbly (exit) flow.

Selected bibliography

Butterworth, D. (1986). Condensers: basic heat transfer and fluid flow. In *Heat exchanger sourcebook* (ed. J. W. Palen), pp. 389–413. Hemisphere, Washington.

Chen, S. L., Gerner, F. M, and Tien, C. L. (1987). General film condensation correlations. *Experimental Heat Transfer*, **1**, 93–107.

Collier, J. G. (1972). *Convective boiling and condensation.* McGraw-Hill, New York.

Collier, J. G. (1981). Heat transfer in condensation. In *Two-phase flow and heat transfer in the power and process industries* (eds A. E. Bergles, J. G. Collier, J. M. Delhaye, G. F. Hewitt, and F. Mayinger), pp. 330–65. Hemisphere, Washington.

Fujii, T. (1991). *Theory of laminar film condensation.* Springer, New York.

Griffith, P. (1982). Condensation. In *Handbook of multiphase systems* (ed. G. Hetsroni). McGraw-Hill Book Company, New York.

Hetsroni, G. (ed.) (1982). *Handbook of multiphase systems.* McGraw-Hill, New York.

Hewitt, G. F. and Hall-Taylor, N. S. (1971). *Annular two-phase flow.* Pergamon, Oxford.

Lock, G. S. H. (1992). *The tubular thermosyphon.* Oxford University Press, Oxford.

Palen, J. W. (ed.) (1986). *Heat exchanger sourcebook.* Hemisphere, Washington.

Sparrow, E. M. and Gregg, J. L. (1959). A boundary layer treatment of laminar film condensation. *Journal of Heat Transfer*, **81C**, 13–18.

Sparrow, E. M., Minkowycz, W. J., and Saddy, M. (1967). Forced convection condensation in the presence of non-condensibles and interfacial resistance. *International Journal of Heat Mass Transfer*, **10**, 1829–45

Traviss, D. P., Rohsenow, W. M., and Baron, A. B. (1973). Forced convection condensation inside tubes: a heat transfer equation for condenser design. *ASHRAE Transactions*, Pt. 1, 157–65.

Whalley, P. B. (1987). *Boiling, condensation and gas-liquid flow.* Oxford University Press, Oxford.

Xin, Mingdao (ed.) (1989). *Advances in phase change heat transfer.* Proceedings of the International Symposium on Phase Change Heat Transfer, Chongquing, China. International Academic Publishers/Pergamon Press, Oxford.

Exercises

1. Neglecting inertia, show that the average velocity in a freely falling condensate film of thickness Δ is given by

$$\bar{U}_L = \frac{g\Delta^2(\rho_L - \rho_V)}{3\mu_L}.$$

Hence find the mass flux $\Gamma = \rho_L \bar{U}_L \Delta$ and the film Reynolds number for a water film of thickness 0.25 mm.

(Ans: $\Gamma = 29.2 \times 10^{-3}\,\text{kg m}^{-1}\,\text{s}^{-1}$, $Re_f = 16.7$.)

2. Show that the wall shear stress acting on a freely falling condensate film is given by

$$\tau_w^* \simeq 11\ \rho_L(\nu_L g)^{2/3}$$

at the laminar-turbulent transition point ($\text{Re}_f \simeq 400$), and hence find τ_w^* for a water film.

(Ans: $\tau_w^* \simeq 7\,\text{N m}^{-2}$.)

3. Compare the structures of the three non-dimensional buoyancy groups: $ArPr/Ja$, Grashof number, and Rayleigh number.

4. Show that for laminar film condensation of a saturated vapour on a vertical surface the average heat transfer coefficient over a height L is given by

$$\bar{h}_L = C\left[\frac{g\lambda k_L^3(\rho_L - \rho_V)}{\nu_L(T_I - T_w)L}\right]^{1/4}$$

where $C = 0.943$ for a smooth interface and $C \simeq 1.13$ for a wavy interface. Hence find $\bar{h}_L$ for condensation of saturated steam at 1 bar on a 10 cm plate when $ArPr/Ja = 2 \times 10^{12}$.

(Ans: $\bar{h}_L = 9138\,\text{W m}^{-2}\,\text{K}^{-1}$.)

5. Consider laminar film condensation of a saturated vapour on the curved surface of a short cylinder. Find the length–diameter ratio such that the condensation rate is the same whether the cylinder is vertical or horizontal. Which way should condenser tubes be oriented?

(Ans: $L/D = 5.9$. (empirical), 2.9 (Nusselt))

6. Derive eqn (5.35) and the corresponding equation for a horizontal surface.

7. Derive eqn (5.55) from eqn (5.54).

8. Derive eqns (5.65) and (5.66). Hence calculate the forced film Nusselt number $\bar{Nu}_f$ at $Re_f = 10$ when $\tau_I = 10\ \tau_w^*$ (see question 2) and compare with the value of $\bar{Nu}_f$ for a freely falling film at the same Reynolds number.

(Ans: $\bar{Nu}_f$ increases by about 5.6:1.)

9. Derive eqns (5.68) and (5.71).

10. The water heat transfer coefficient inside a horizontal steam condenser tube is $3114\,\text{W m}^{-2}\,\text{K}^{-1}$ and the tube wall temperature is 41°C. Calculate the heat flux density, the external condensation heat transfer coefficient, and the condensate film thickness if saturated steam is

admitted at 10 kPa and the bulk cooling water temperature is 23°C. (Ans: $\bar{q} = 56\,\mathrm{kW\,m^{-2}}$, $\bar{h} = 1.22 \times 10^4\,\mathrm{W\,m^{-2}\,K^{-1}}$ and $\bar{\Delta} = 53\,\mu\mathrm{m}$.)

11. Show that for annular film condensation

$$\frac{k_L(T_I - T_0)}{\lambda} = \frac{\Delta\,\mathrm{d}\Gamma}{\mathrm{d}X}$$

where X is measured in the direction of flow, T_0 is the wall temperature, Δ is the film thickness, and $\Gamma = \rho_L \bar{U}_L \Delta$ is the mass flow per unit perimeter. Also show that if the film is vapour driven, the average film velocity is given by

$$\bar{U}_L = \tau\Delta/2\mu_L$$

where τ is the uniform shear stress in the film.

12. Steam superheated by 20 K enters a condenser where the pressure is 25 kPa. Calculate the vapour Jakob number and determine the additional length (ΔH) a vertical condensing surface must have to maintain the same (total) laminar condensation rate as when the steam is saturated.
(Ans: $Ja_V = \Delta H/H = 1.7$ per cent.)

Condensation projects

1. Review and assess condensation of metals.
2. Review and assess reflux condensation in vertical and horizontal tubes.
3. Develop a physical and mathematical model for condensation in very low gravity conditions.
4. Review and assess the laminar-turbulent transitions during through-flow condensation in vertical and horizontal tubes.
5. Compare the theory and practice of thermal design for a small condenser.
6. Review and assess condensation of non-eutectic vapour mixtures.
7. Develop a physical and mathematical model for through-flow condensation in inclined tubes.
8. Review and assess the effect of vapour Reynolds number on annular, through-flow condensation in a horizontal tube.
9. Develop a physical and mathematical model for annular, through-flow condensation in a tube rotating about its own axis at high speed.
10. Review and assess the drop-to-film condensation transition and its reversal.

6
EVAPORATION

In this chapter we will consider various forms of evaporation, covering both bipartitioned and dispersed systems. We may easily anticipate some of the non-dimensional groups which affect evaporation rates: for example, the ratio of sensible to latent heat, the Jakob number. To begin with, we limit ourselves to situations in which evaporation is essentially the inverse of condensation, thus permitting the use of ideas and results presented in the previous chapter. In particular, we will consider the classical Nusselt problem with condensation replaced by evaporation at the interface.

6.1 Inverse Nusselt problems

6.1.1 Film evaporation

Evaporation in a falling film evaporator is represented schematically in Fig. 6.1. A mass flux of Γ_L per unit width (normal to the page) is shown entering at the top of a vertical surface where $X = L$. Evaporation from the free surface of the descending film gradually reduces the thickness Δ, here shown with an enlarged scale, until it vanishes completely at a point which has been chosen as the origin of the X, Y coordinate system. Throughout its overall length L, the film cools the surface by virtue of conduction towards, and evaporation from, the interface. In the simplest situation, the vapour is pure and saturated at temperature T_I. More generally, the bulk vapour temperature T_∞ is less than T_I if the vapour is mixed with gas through which it must diffuse on leaving the interface; T_I then corresponds to the saturation vapour pressure at the interface.

Under typical conditions, the results of the previous chapter may be applied immediately. Thus, for example, the film thickness and longitudinal velocity scales for a pure fluid under laminar conditions are given by

$$Y^c = \Delta = L\left(\frac{Ja}{ArPr}\right)^{1/4} \quad \text{and} \quad U^c = \frac{\kappa}{L} Ja \left(\frac{ArPr}{Ja}\right)^{1/2}.$$

Hence the heat transfer rate may be written

$$Nu = C_v \left(\frac{ArPr}{Ja}\right)^{1/4}, \tag{6.1}$$

provided $Ja \ll 1$ and $Pr \gg 1$. Similarly, if the film flows over a horizontal surface

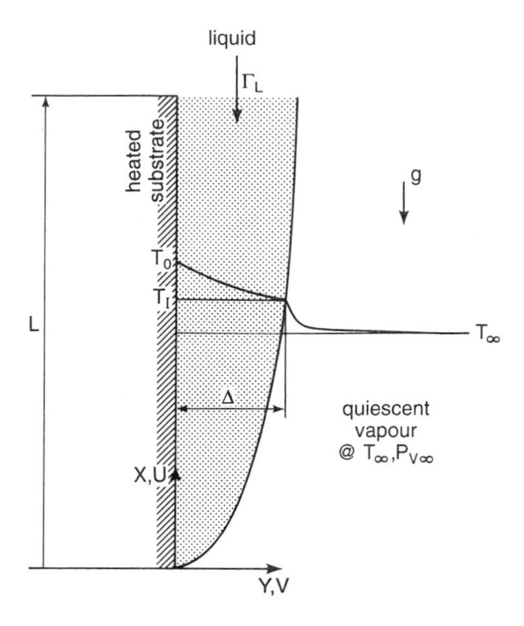

Fig. 6.1 Schematic representation of evaporation from a falling liquid film in a quiescent vapour.

$$Nu = C_h \left(\frac{ArPr}{Ja} \right)^{1/5}. \tag{6.2}$$

Equations (6.1) and (6.2) suggest that a more general relation of the form

$$Nu = C \left(\frac{ArPr}{Ja} \right)^{m} \tag{6.3}$$

would be suitable for film evaporation over most external surface geometries and would accommodate either laminar or turbulent conditions. As with condensation, if the vapour is not saturated, the latent heat term in the above expressions must be modified to include the vapour sensible heat; in practice, $Ja_V \ll 1$ so that the effect may often be ignored. In any event, heat transfer coefficients are comparable with those found in film condensation.

In the discussion of the classical Nusselt problem of film condensation it was noted that the film has a tendency to develop surface waves. This type of Weberian instability may not occur in the evaporative equivalent shown in Fig. 6.1 unless the vertical vapour velocity is high or the waves are already present in the liquid injected in the upper reaches. The Nusselt solution may therefore apply up to the turbulent transition point ($Re_f \simeq 400$). However, the thinning of the film may give rise to another type of Weberian instability, also discussed in Chapter 3. As the inertial forces fall with decreasing elevation, the comparative rise in surface tension forces may eventually cause the film to break up into rivulets. In practice, these tend to meander over the substrate surface.

Many of the above remarks also apply to the *vapour* film depicted in Fig. 6.2. As we shall see, this is often described as film boiling. Archimedean buoyancy now acts upward to cause the vapour to ascend, starting at the origin shown. The superheat $(T_0 - T_I)$ is usually much greater than in Fig. 6.1; typically $k_V \ll k_L$. Once again we may use the form of eqn (6.3) to represent laminar heat transfer providing it is understood that fluid properties, ν and κ in particular, then refer to the vapour phase: see problems Q6.6 and 6.7 in Chapter 8 for illustrations. The hydrodynamic boundary conditions must also be re-phrased to accommodate a quiescent liquid; thus we take $U_\Delta = 0$ but $(\partial U/\partial Y)_\Delta \neq 0$. Given that vapour film Reynolds numbers are typically greater than liquid film Reynolds numbers, the vapour flow will often be turbulent and the liquid–vapour interface will be more susceptible to a Weberian wave instability, especially if the substrate is vertical. A relation having the form of eqn (5.42) may then be more appropriate.

For horizontal tubes, on the other hand, eqn (6.3) has wide applicability in the form

$$\bar{h}_{NC} = C_{NC}\left[\frac{\lambda g k_V^3 \Delta\rho}{\nu_V D(T_0 - T_I)}\right]^{1/4} \tag{6.4}$$

In which $C_{NC} \simeq 0.62$, in comparison with the value of 0.725 appropriate to film condensation. The parallel with film condensation is also seen under forced flow conditions. Corresponding to eqn (5.67), for example, we find that

$$\bar{h}_{FC} = C_{FC}\left[\frac{\lambda \rho_V U_\infty k_V}{D(T_0 - T_I)}\right]^{1/2} \tag{6.5}$$

in which $C_{FC} \simeq 2.7$. The ratio of these results,

$$\frac{\bar{h}_{FC}}{\bar{h}_{NC}} = 4.35\left[\left(\frac{Re_D Pr_V}{Ja_V}\right)^2\left(\frac{Ja_V}{Ar_D Pr_V}\right)\right]^{1/4},$$

reveals that the relative importance of forced convection is dictated by the ratio $(Re_D Pr_V/Ja_V)^2/(Ar_D Pr_V/Ja_V)$ which is equivalent to the ratio Re_D^2/Gr_D in single phase convention.

6.1.2 *Surface spray cooling*

Another type of inverse Nusselt problem arises when cold drops are sprayed on to a hot substrate where they become superheated and evaporate. This occurs, for example, during the cooling of hot metal strip or on evapora-

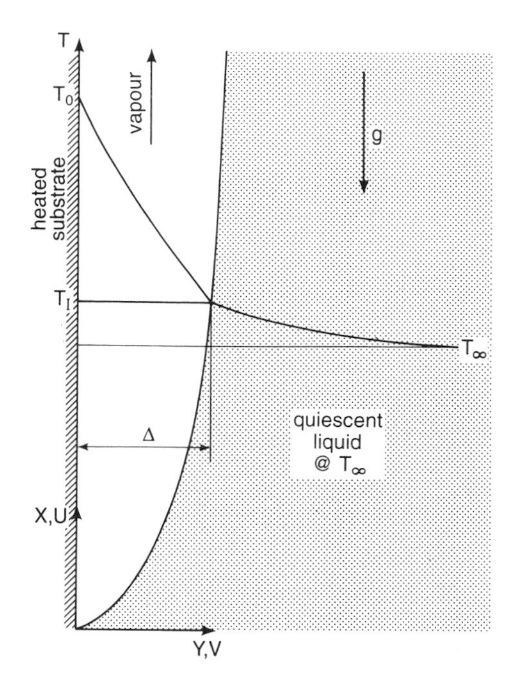

Fig. 6.2 Schematic representation of evaporation into a vapour film ascending in a quiescent liquid: film boiling.

tively cooled heat exchanger tubes. Problems Q6.1–6.5 in Chapter 8 illustrate the situation which is reminiscent of drop condensation in reverse; it may produce very high heat transfer rates, as revealed in Q.6.2. Figure 6.3 provides a schematic representation of a hemispherical drop evaporating into a vapour–gas mixture with a bulk temperature $T_\infty < T_I$ and a corresponding vapour mass fraction $m_{V\infty}$.

This description suggests that the interface equation may be written

$$\rho_L \lambda_{LV} \frac{dR_I}{dt} = k_L \left(\frac{\partial T_L}{\partial R}\right)_I - h(T_\infty - T_I),$$

where h is the heat transfer coefficient attributable to bulk motion of the vapour–gas mixture. When evaporation is largely balanced by conduction through the drop, external convection may be neglected. Normalization then reveals that the evaporation time may be estimated from

$$t_f = 0\left[\frac{\rho_L \lambda_{LV} R_i^2}{k_L(T_0 - T_\infty)}\right],$$

where R_i is the initial drop radius. Alternatively, since the substrate heat flux *density* beneath the drop is represented by

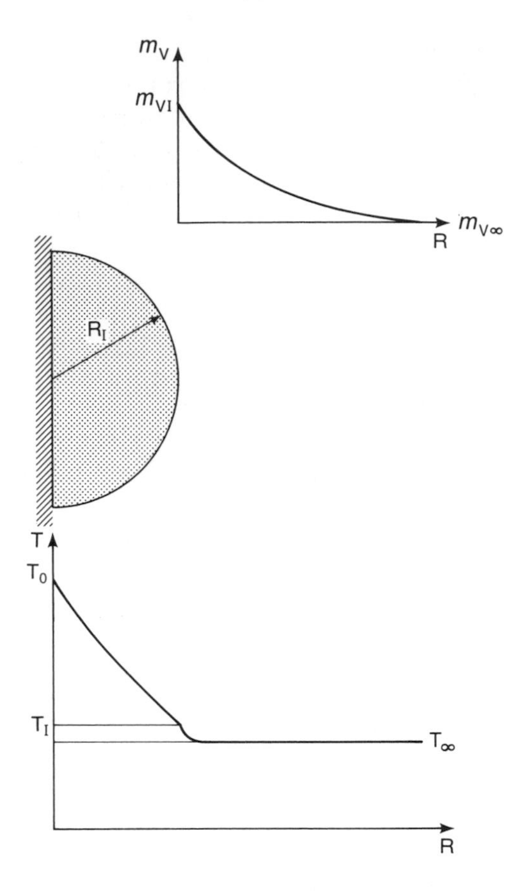

Fig. 6.3 Schematic representation of evaporation from a hemispherical drop on a substrate: spray cooling.

$$\dot{q}_0 = 0\left[\frac{k_L(T_0 - T_\infty)}{R_I}\right],$$

the evaporation time may be written

$$t_f = O\left(\frac{\rho_L \lambda_{LV} R_i}{\dot{q}_0}\right).$$

It is evident that the smaller the drop the greater its cooling effect. With $\dot{q}_0 = 1\,\mathrm{MWm^{-2}}$, $t_f \simeq 2\,\mathrm{s}$ for a 1 mm water drop and 2 ms for a 1 μm drop. Problem 6.4 in Chapter 8 provides a further illustration.

In general, the impact of the drop itself will influence heat transfer and evaporation rates. Higher impact velocities produce a flatter drop, ultimately leading to a shape which is more like a disc than a hemisphere. Such a shape creates a shorter, wider conduction path in the liquid and introduces liquid convection parallel to the substrate surface. The evaporation rate thus improves but the imbalances between inertial and surface tension forces complicate behaviour by creating various bouncing, splashing, and breakup phenomena. These are not easily modelled. When the spray is sufficiently intense, the substrate is completely covered by liquid and behaviour reverts to that of a film.

With the substrate temperature not too much greater than the saturation (interface) temperature, evaporation remains a well-defined, if complex, process. When the substrate temperature becomes too high, however, very different events may occur. For example, if $T_0 - T_I$ is great enough, the liquid superheat may be sufficient to promote vigorous nucleate boiling, as described later; heterogenous nucleation has been created. On the other hand, if T_0 exceeds the *Leidenfrost* temperature, ebullition ceases and a film of vapour then separates the drop from the substrate. The familiar sight of a water drop skidding about over a very hot horizontal plate is another example of an inverse Nusselt problem. Evaporation from the underside of such a drop is caused by conduction and radiation across the thin vapour film separating it from the hot plate.

The Leidenfrost temperature represents a thermodynamic limit above which no liquid may remain in contact with the substrate, at least not for very long. The existence of film boiling mentioned in the previous subsection, for example, requires that this limit be exceeded. The limit may be defined simply as the substrate temperature at which the vapour film collapses and sudden re-wetting occurs. In practice, however, this temperature is difficult to predict because it is sensitive to variations in substrate contamination, including the oxidation caused by local overheating in the presence of the vapour film itself.

6.2 Pool boiling

The rate of heat loss from a solid substrate to most liquids is limited if it is attributable solely to conduction. The single-phase diffusive and convective mechanisms of heat transfer normal to the substrate are not capable of producing and sustaining very high levels of heat flux. However, the appearance of vapour bubbles dramatically alters the picture. At the very least, their rapid growth and detachment will introduce a vigorous stirring action. Equally important is the transfer of latent heat which suggests a further improvement in heat transfer rate. In this section, we consider boiling within an enclosed volume of liquid which is otherwise quiescent. It will become evident that the process, commonly known as *pool* boiling, is both complex and variable.

6.2.1 *Bubble formation*

It will be recalled from Chapter 2 that nucleation of vapour in a liquid phase requires the formation of a molecular cluster of critical size. For this to occur in the bulk of the liquid, i.e. as homogeneous nucleation at T^*, demands superheats which are usually very large except near the critical point; typically, $T^* \simeq 0.9T_C$ and increases towards T_C at the critical point. It is observed, however, that superheats in boiling are commonly of the order of a few degrees. This suggests that heterogeneous nucleation takes place, and implies that the chemical composition of the substrate and the liquid are important. Experiments confirm this and suggest that the micro-geometry of the substrate surface is also important. Even on apparently smooth surfaces it is inferred that microcavities, e.g. 0.1–10 μm in diameter, are randomly distributed in densities as high as $10^8\,\text{cm}^{-2}$.

Before boiling begins, and especially during the initial flooding of the substrate, many of the microcavities trap small pockets of gas, usually air. On warming the substrate, gas may be release from solution thereby creating expanded pockets which are easily seen protruding beyond the microcavity mouth during the run up to boiling. The boiling process itself requires the production of vapour bubbles in three stages: initiation, growth, and detachment. Let us examine these in detail.

Figure 6.4 provides a schematic representation of gas pockets trapped in a microcavity, here idealized as a slender cone. The degree of wetting m was defined in Chapter 2 by

$$m = \cos\theta = \frac{\sigma_{SV} - \sigma_{SL}}{\sigma_{LV}},$$

where θ is the static contact angle. For a highly wetting liquid (Fig. 6.4(a)), θ is small and $\sigma_{LV} \simeq (\sigma_{SV} - \sigma_{SL})$ is also small. On the other hand, for a

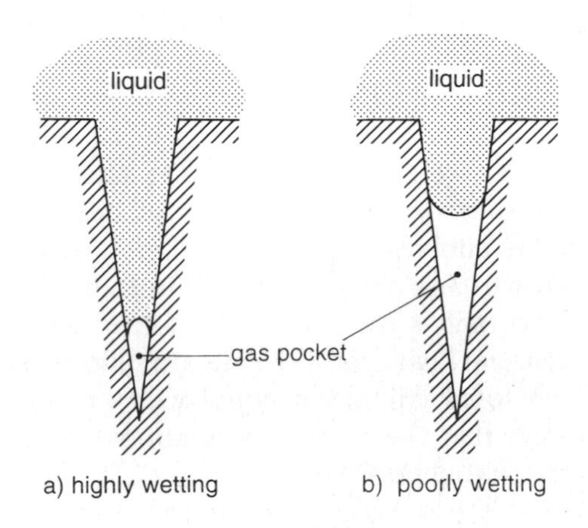

Fig. 6.4 Gas pockets trapped in microcavities prior to nucleation.

poorly wetting liquid (Fig. 6.4(b)), θ and σ_{LV} are not small. As illustrated, highly wetting liquids create smaller gas pockets bounded by a concave liquid surface, while poorly wetting liquids create larger pockets having a convex liquid surface. The higher gas pressure in the highly wetted microcavity may cause gas diffusion into the liquid, thus reducing the pocket size further. The lower gas pressure in the poorly wetted microcavity has the opposite effect.

Initiation of a vapour bubble is marked by sudden and rapid evaporation from the curved liquid surface into the microcavity. Strictly, this is not a nucleation event because a stable vapour-gas pocket already exists. Even so, the embryonic vapour bubble cannot grow until it has reached a critical size corresponding to the local temperature. Following from Chapter 2, this threshold size may be estimated from the Clausius–Clapeyron eqn (2.44) and the static balance

$$P_{act} - P_{sat} = \frac{2\sigma_{LV}}{r_b} \tag{6.6}$$

in which P_{act} is the bubble activation pressure for a pocket interface of mean radius of curvature r_b, less than the microcavity mouth radius; the gas pressure may be ignored for simplicity. The resulting activation temperature T_{act} is then given by

$$T_{act} - T_{sat} = \frac{2RT_{sat}^2}{\lambda_{LV} P_{sat}} \left(\frac{\sigma_{LV}}{r_b} \right). \tag{6.7}$$

Equations (6.6) and (6.7) define, respectively, the bubble activation isobar and isotherm shown in Fig. 6.5. For a given fluid at a given pressure P_{sat}, the activation superheat $T_{act} - T_{sat}$ depends specifically on σ_{LV}/r_b which turns out to be a monotonically *decreasing* function of σ_{LV}. This feature has the important consequence that highly wetting liquids (small σ_{LV}) require higher activation superheats than poorly wetting liquids. The effect of microcavity size may be estimated from eqn (6.7) by assuming that the mouth radius is of the same order as r_b. In Chapter 8, Q6.6 provides a numerical illustration. For water at atmospheric pressure, the equation reduces to

$$T_{act} - T_{sat} \simeq \frac{3 \times 10^{-5}}{r_b}.$$

Thus for the size range $0.1\ \mu m < r_b < 10\ \mu m$, the required superheat lies in the corresponding range $300\ K > T_{act} - T_{sat} > 3\ K$. This suggests that surface roughness may be used to decrease the minimum superheat, as confirmed by experiment. Surface ageing may have the opposite effect on some materials, e.g. copper, which develops an oxide coating that swells and thus shrinks the effective cavity size.

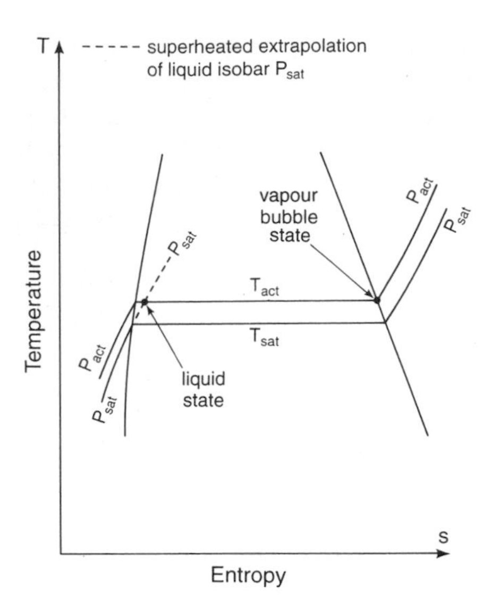

Fig. 6.5 Vapour and liquid states during bubble initiation.

Once the substrate superheat has exceeded the value given in eqn (6.7), bubble growth is spontaneous. Evaporation ensures a concave liquid–vapour interface and a liquid–solid contact line which travels towards the mouth of the cavity, as illustrated in Fig. 6.6. This is a schematic representation in which the scale of the microcavity has been greatly enlarged for clarity. For low heat fluxes, when the liquid temperature gradient normal to the substrate is small and the growth period relatively long, a quasi-static balance between surface tension and buoyancy forces tends to create small spherical bubbles as suggested in Fig. 6.6a; the (low) growth rate decreases with time. For higher heat fluxes, increases in both the substrate temperature and the liquid temperature gradient reduce the growth period and thereby increase the importance of inertial and viscous forces; the bubble base thins into the *microlayer* depicted in Fig. 6.6b. The bubble shape is then more like a hemisphere and its (higher) growth rate is constant.

6.2.2 *Bubble detachment and subcooled boiling*

The growth of a protruding, or detached, bubble may be described by the interface equation

$$\rho_V \lambda_{LV} \frac{dR_I}{dt} = k_L \left(\frac{\partial T}{\partial R}\right)_I = h(T_L - T_I), \tag{6.8}$$

where h is the heat transfer coefficient in the surrounding liquid at T_L, and $T_I > T_{sat}$ is the interfacial temperature. Vapour temperature is essen-

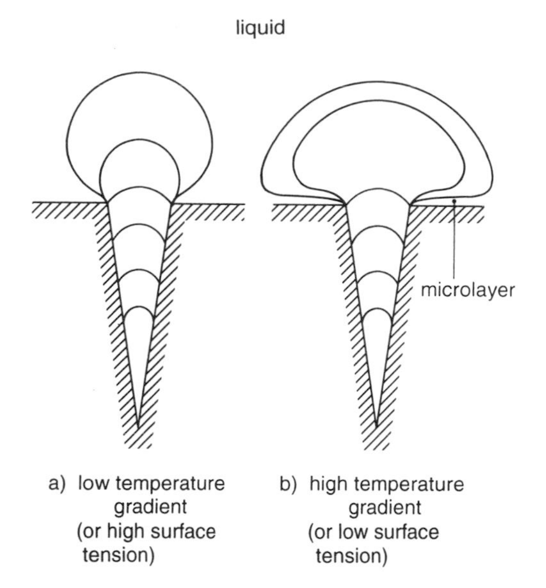

Fig. 6.6 Schematic representation of vapour bubble growth within and beyond a microcavity which has been drawn greatly enlarged for clarity.

tially uniform throughout the bubble because the vapour is generated at, or near, T_{I}. Whenever the liquid surrounding the bubble is subcooled, $T_{\mathrm{I}} > T_{\mathrm{sat}} > T_{\mathrm{L}}$ and the bubble will shrink according to eqn (6.8). This condensation may begin before detachment if the bubble is large enough to protrude beyond the superheated layer immediately adjacent to the substrate; within the superheated layer, $T_{\mathrm{L}} > T_{\mathrm{I}}$. Following detachment, the bubble will eventually collapse in the subcooled liquid. In any event, the bubble gives up latent heat to the enclosed liquid bulk thereby raising T_{L} towards T_{sat}.

For the low heat fluxes which characterize the onset of boiling, detachment is marked by a pinch-off process in which the restraining force of surface tension and the driving force of Archimedean buoyancy strike a rough balance. At the neck of the pinching bubble shown in Fig. 6.7, $P_{\mathrm{B}} \geqslant P_{\mathrm{A}}$ under quasi-static conditions. Relative to the level D in liquid,

$$P_{\mathrm{B}} = O(2R_{\mathrm{b}} g \rho_{\mathrm{L}}),$$

whereas

$$P_{\mathrm{A}} = O(2R_{\mathrm{b}} g \rho_{\mathrm{V}} + 2\sigma_{\mathrm{LV}}/R_{\mathrm{b}})$$

if the principal radii of the curvature just above the neck are roughly equal and opposite. Hence

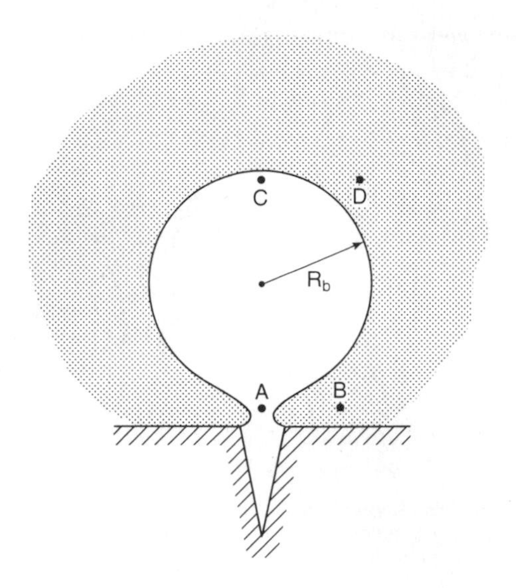

Fig. 6.7 Schematic representation of incipient pinch off for a spherical bubble. (The microcavity is again enlarged for clarity.)

$$\frac{R_b^2 g \Delta\rho}{\sigma_{LV}} = Bo_b \geqslant O(1), \tag{6.9}$$

where R_b is the detached bubble radius and Bo_b is the bubble Bond number, i.e. the ratio of the bouyancy and surface tension forces. This provides us with a quasi-static estimate of bubble size: for water boiling at atmospheric pressure, $R_b = 0$ (2.5 mm).

Pinch-off ensures that all of the vapour previously generated does not escape into the bubble. A residual vapour volume is an essential condition for the formation of another bubble. As noted above, however, the condition is necessary but not sufficient. Adequate superheat must also be provided. The absorption of latent heat during the growth of one bubble usually produces sufficient cooling around the microcavity wall to cause a delay in production of the next bubble; this is particularly true for a highly wetting liquid.

Re-flooding of the microcavity to re-create the vapour pocket differs from the emptying process in two ways. Firstly, the vapour pressure increases as the pocket is compressed; this is in contrast to the decrease during emptying which is much like inflating a balloon, although there is a local maximum (minimum radius) shortly after the contact line reaches the cavity mouth (see Fig. 6.6(a)). Secondly, the contact angle during filling is greater than under static conditions; the reverse is true while emptying. It is generally believed that re-flooding leaves the microcavity largely void of liquid. For highly wetting liquids, this implies a vapour pocket which is much larger than that required to initiate spontaneous bubble growth. The superheat to *maintain* nucleate boiling of highly wetting liquids is therefore much less than that for *initiation*. The difference is much smaller for poorly wetting liquids.

Subcooled boiling is preceded by natural convection with its characteristically low heat transfer rates. The effect of early bubble growth and detachment is therefore to increase the heat transfer rate through convection and evaporation and thus steepen the lower reaches of the *pool boiling curve* shown plotted as $\ln \dot{q}$ versus $\ln(T_0 - T_L)$ in Fig. 6.8. This effect continues progressively (b–c) until the temperature of the enclosed liquid T_L has reached T_{sat} and saturated boiling begins. When the surface tension is large, and the liquid does not wet the microcavities, the transition from natural convection (a–b) to subcooled boiling (at b) and then to saturated boiling (at c) is smooth, and is characterized by a reversible path; superheats are small. However, when the surface tension is small, the superheat required for bubble initiation is much greater than that required for repetitive bubbling. The boiling curve may then exhibit a reversible overshoot from the natural convection regime to the bubble initiation point b″ before falling back irreversibly on to the subcooled boiling curve at b′. This creates a hysteresis effect.

6.2.3 *Saturated, supersaturated, and pre-transitional boiling*

Under the relatively low superheats commonly associated with subcooled pool boiling, eqn (6.7) reveals that only the larger microcavities are activated; sites are widely separated and the bubbles are isolated from each other. Bubble interaction is minimal. As the substrate temperature rises, and superheat in the layer of liquid immediately adjacent to the substrate becomes both more intensive and more extensive, the number of bubble sites increases; smaller microcavities are activated. Once the bulk liquid temperature has reached T_{sat}, and *saturated pool boiling* begins, bubble behaviour becomes very important. It is revealed in the growth patterns of individual bubbles and through bubble interactions.

The bubble growth history of an isolated bubble may be described by the interface equation

$$\rho_V \lambda_{LV} \frac{dR_I}{dt} = h(T_L - T_{sat}) = \dot{q}_I \tag{6.10}$$

in which the bulk liquid temperature T_L lies between T_{sat} and T_0, the substrate temperature. For rapid (or early) growth, $\dot{q}_I$ maintains a high, fixed value so that $R_I^c \propto t^c$; for slower (or later) growth, $\dot{q}_I = 0[k_L(T_L - T_{sat})/R_I^c]$ and hence $R_I^c \propto (t^c)^{1/2}$. Overall, the bubble growth rate increases with liquid superheat. This suggests that the bubble frequency scale

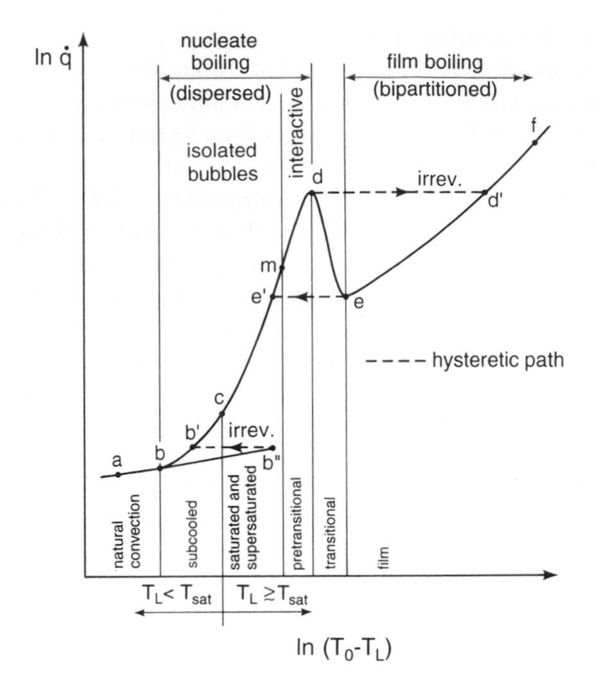

Fig. 6.8 The pool boiling curve: heat flux controlled.

$$f = \frac{1}{t^c} = O\left[\frac{h(T_0 - T_{sat})}{\rho_V \lambda_{LV} R_b}\right], \tag{6.11}$$

estimated using eqn (6.10), also increases with superheat, and keeping in mind that there is a waiting period between successive bubbles.

Growth and departure follow a general sequence of events. After nucleation, bubble growth is very rapid. Evaporative cooling quickly reduces the local substrate temperature, especially beneath the disc-like microlayer which may vanish in the process. Bubble growth then continues at a reduced rate while the substrate temperature begins to rise fairly rapidly, until buoyancy lifts the bubble off the substrate which is then flooded with new liquid. A slower rise in substrate temperature then leads to nucleation of the next bubble after a waiting period. This process is repeated at many sites with the result that the natural plumes attributable to thermal buoyancy in the liquid are re-organized and enhanced by the Archimedean buoyancy of bubble columns. Liquid circulation near the base of these columns greatly increases and, in the presence of a high temperature gradient near the bubble base, is augmented by thermocapillarity.

As the rate of heat transfer increases during saturated and supersaturated boiling, the spacing of active nucleation sites decreases. Once this spacing has been reduced to less than about two bubble diameters, the bubbles (and the sites) begin to interact with each other; we enter the pre-transitional régime (m–d). In general, bubble interaction takes two forms: longitudinal and lateral. Bubble chains at any given site organize and enhance natural convection near the site. With an increased heat transfer rate, chains above a horizontal substrate may thus evolve into discrete upwellings separated by downward returns. Lateral interaction between the bubbles begins before detachment. Bubbles touch and often coalesce to produce a larger vapour pocket than would otherwise occur. It is thus apparent that as the heat flux increases, a model of single bubbles growing and detaching independently must be replaced by one in which larger vapour pockets are created by high-frequency, interactive bubble chains. Groupings of these vapour pockets eventually become separated from the substrate by a liquid film, usually described as the *macrolayer*.

The point at which larger vapour pockets begin appearing is difficult to detect from the boiling curve. The point m in Fig. 6.8 is a representative location. There is no dramatic discontinuity in heat transfer and evaporation rate despite observable shifts in both the bubble interaction and the vapour fraction profile near the substrate. Before these occur, any increase in liquid superheat increases both the bubble frequency f and the number of sites n_A per unit substrate area. Many attempts have been made to build upon the elementary heat balance describing the surface heat flux *density* by

$$\dot{q}_0 = f n_A \left(\frac{4}{3} \pi R_b^3 \rho_V \lambda_{LV} \right), \tag{6.12}$$

where R_b is the bubble radius which may be estimated from equation (6.9); more precisely, the radius first increases with superheat and then decreases through interaction in the chain. The magnitudes of n_A and f are hidden in the superheat, the fluid properties and the substrate micro-geometry and composition. Examining eqns (6.7) and (6.11), it is evident that n_A and f both increase monotonically with $T_0 - T_{sat}$, but their individual forms are difficult to predict. Their product, however, is found to vary roughly as $(T_0 - T_{sat})^m$, where $m \geqslant 1$. Typically, eqn (6.12) assumes the simple power law form

$$\dot{q}_0 = K(T_0 - T_{sat})^3, \tag{6.13}$$

where K, and the precise value of m, are empirical functions of fluid and surface properties; Mikic and Rohsenow (1969) suggest that

$$K = 1.89 \times 10^{-14} \frac{g^{1/2} \lambda_{LV}^{1/8} k_L^{1/2} \rho_L^{17/8} c_{pL}^{19/8} \rho_V^{1/8}}{\sigma_{LV}^{9/8} (\rho_L - \rho_V)^{5/8} T_{sat}^{1/8}}. \tag{6.14}$$

Problem Q6.6 in Chapter 8 illustrates the use of this empirical correlation.

Equation (6.13) applies over most of the range c–d in Fig. 6.8. While this includes the range m–d, it does so fortuitously because of the fundamental change in conditions at m brought about by bubble interaction. It has been observed by several investigators (Dhir and Bergles 1992) that under these interactive conditions the vapour fraction reaches a maximum of about 1.0 at a certain distance from the substrate δ_{max} which thus defines the extent of the macrolayer. Heat transfer from the substrate produces macrolayer evaporation in two ways: at the extensive outer surface shared with large vapour pockets; and through high frequency bubble chains or 'jets' issuing from active sites. Individual bubbles which grow until they intersect with the free surface of the macrolayer burst in a process that is rapid enough, e.g. $O(1\ \mu s)$, to prevent microlayer dry-out and thus create very high local heat transfer rates. Despite these bubble punctuations, however, a significant fraction of the evaporation may occur at the free surface of the macrolayer, especially when the macrolayer thins out and ebullition ceases.

A model of pre-transitional behaviour thus emerges. This is illustrated in Fig. 6.9, for a small horizontal surface, and in Fig. 6.10, for a larger vertical surface. The macrolayer is created by heat flux densities high enough to ensure that bubble coalescence generates large vapour pockets. While this process may be partly offset by subcooling, which limits bubble growth, it eventually impedes the access of replenishing liquid. As Fig. 6.9

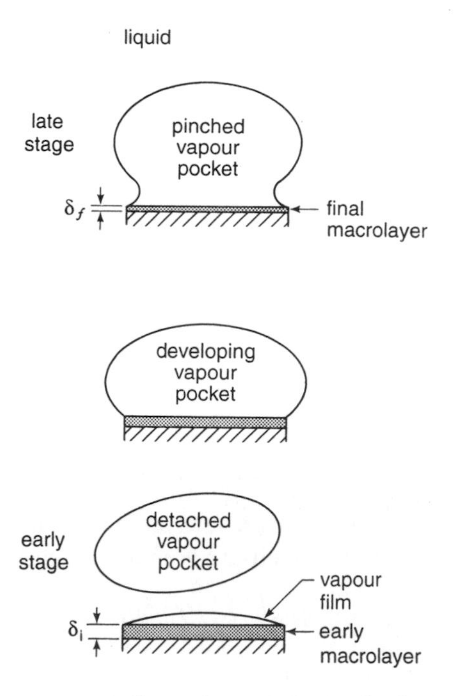

Fig. 6.9 Schematic representation of macrolayer development above a small, horizontal surface.

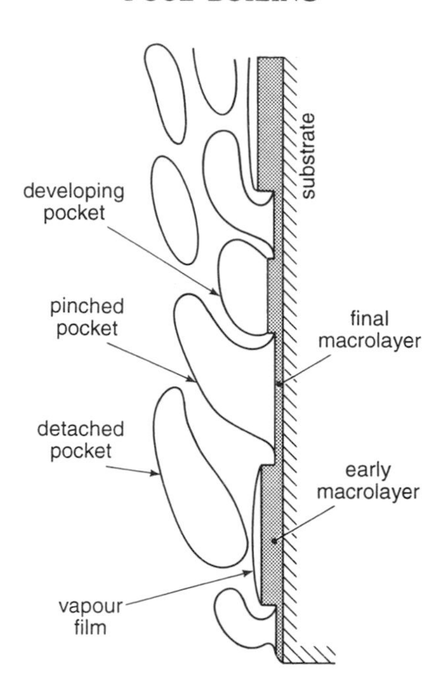

Fig. 6.10 Schematic representation of macrolayer development on a vertical surface.

illustrates, a pocket will grow and mushroom over a small horizontal surface until, like a more conventional bubble, it becomes sufficiently buoyant to be pinched off. During growth, the vapour pocket is fed by boiling in, and evaporation from, the macrolayer which therefore decreases in thickness from an initial value δ_i to the final value δ_f when the pocket detaches. At that point, re-flooding and nucleate boiling rapidly create a new macrolayer and the process is repeated. In general, the local macrolayer thickness will vary from time to time and from place to place. Figure 6.10 illustrates the corresponding situation on a vertical surface.

It has been noted by several investigators (Dhir and Bergles 1992) that bubble size and macrolayer thickness follow essentially the same inverse relation with respect to the heat flux density; for water at atmospheric pressure, this is given approximately by

$$\delta_i = O\left(\frac{10^5}{\dot{q}_0^{3/2}}\right). \tag{6.15}$$

Thus, during pre-transitional boiling, when $2 \times 10^5\,\mathrm{Wm}^{-2} \leqslant \dot{q}_0 \leqslant 2 \times 10^6\,\mathrm{Wm}^{-2}$, the initial macrolayer thickness lies in the range $35\,\mu\mathrm{m} \leqslant \delta_i \leqslant 1.1\,\mathrm{mm}$. If these heat transfer rates represented pure conduction through the macrolayer, its mean thickness $\bar{\delta}$ would be given by $\bar{\delta} = O[k_L(T_0 - T_{sat})/\dot{q}_0]$.

Again considering water, this yields 3.4 μm $\leqslant \bar{\delta} \leqslant$ 34 μm, a range below that of δ_i which is, however, much greater than the final thickness δ_f. Ebullition therefore continues to be important in the macrolayer except when δ_f approaches zero at the critical heat flux and a new régime begins at d on Fig. 6.8.

6.2.4 *Film boiling*

The limiting heat flux at d may be approached from beneath along a *reversible* path. We will return to this local maximum later. It is now instructive to jump to the film boiling régime (see Section 6.1) recognizing only that it is frequently entered by means of an *irreversible* shift from d to d′, as shown in Fig. 6.8. For this to occur, the change in superheat must be large enough for the substrate temperature to exceed the Leidenfrost temperature, marked roughly by the point e on the boiling curve. A bipartitioned system now replaces the dispersed system characteristic of ebullition. Once again behaviour is reversible: that is, along any stretch of the curve to the right of the minimum heat flux at e there is no hysteresis.

At very high substrate temperatures, e.g. at point f, heat transfer across the vapour film is by radiation as well as conduction. To a rough approximation, the two contributions may be added to obtain their overall effect. Strictly, this is incorrect because radiation thickens the vapour film. If the substrate temperature falls, the radiative contribution becomes less important and the problem reverts to the inverse Nusselt form discussed in the previous section. Under these circumstances, however, the vapour superheat is usually large and therefore the Jakob number may not be small. Convection and inertial effects then influence the heat transfer rate.

As noted in Section 6.1, higher vapour velocities often produce a turbulent film with a wavy interface. In extreme circumstances, the wave crests shed drops which simply bounce off the high temperature substrate and may be re-deposited at the interface. The film itself is driven by Archimedean buoyancy, and hence the orientation of the substrate influences the vapour velocity unless the surface is small, e.g. horizontal wires. Equally important is the manner in which the vapour leaves the film and drives the surrounding liquid.

6.2.5 *Transitional boiling*

From Fig. 6.8 it may be seen that if the heat flux is used to control pool boiling behaviour, the attainment of the *critical heat flux* (d), leads to an irreversible jump upwards (d–d′) to the film boiling régime. Likewise, at the lower limit of film boiling, the *minimum heat flux* (e), can lead to a jump downwards (e–e′) to the nucleate boiling régime. Under such circumstances, the *transitional boiling* régime (d–e) is circumvented by the hysteresis loop d–d′–e–e′–d. The transitional régime is then inaccessible. However, when the substrate temperature is used to control behaviour the path d–e is reversible. The entire path e′–d–e–d′ is then reversible.

Stable behaviour of the transitional régime is not well understood. Even so, several physical observations provide valuable insight, particularly during the limiting conditions at d and e. In the lower reaches of the nucleate boiling régime, the buoyancy force B acting on a bubble may be represented by $B \sim (X^c)^3 g\Delta\rho$, where X^c is an appropriate length scale. This force is roughly balanced by the surface tension force T which may be represented by $T \sim \sigma X^c$. The ratio of these forces is the Bond number in which, for nucleate boiling, the appropriate length scale is the bubble radius. In the same circumstances, the vapour inertia force $I = \rho_V U_V^2 (X^c)^2$ is much less than either of the above forces; hence, the Weber number $We = R_b \rho_V U_V^2/\sigma \ll 1$. As the critical heat flux is approached from beneath, the vapour inertial force rises until $I = O(B) = O(T)$. This occurs when the macrolayer covers the substrate, but the appropriate length scale is not then obvious. However, the length scale may be eliminated between the Bond and Weber numbers to yield the vapour velocity as

$$U_V = O\left(\frac{\sigma g \Delta\rho}{\rho_V^2}\right)^{1/4}. \tag{6.16}$$

Hence the critical heat flux corresponding to $Bo = O(We) = O(1)$, may be written

$$\dot{q}_{max} = \lambda_{LV} \rho_V U_V = C \lambda_{LV} \rho_V^{1/2} (\sigma g \Delta\rho)^{1/4} \tag{6.17}$$

This simple prediction is well supported by experimental data: above a large horizontal substrate, $C \simeq 0.15$. However, the argument is based solely on hydrodynamic considerations and takes no account of the substrate. For example, when the substrate length W is not much larger than the equilibrium bubble radius $(\sigma/g\Delta\rho)^{1/2}$, the non-dimensional maximum heat flux must be rewritten

$$\dot{q}_{max}/\lambda_{LV} \rho_V^{1/2} (\sigma g \Delta\rho)^{1/4} = Ku^{1/4} = F(Bo_W)$$

where Ku is the Kutateladze number and $Bo_W = W^2 g\Delta\rho/\sigma$. For $Bo_W \lessdot O(10)$, easier replenishment of liquid near the substrate then raises Ku (and $\dot{q}_{max}$) as Bo_W decreases. Conversely, when $Bo_W \ll O(10^{-1})$ nucleation immediately covers the entire substrate in a vapour film; the local maximum in heat flux no longer exists. This may occur when σ is large, W is very small or with reduced gravity.

Under more common circumstances, the transition from d to e may be attributed to the spread of macrolayer failure, from isolated patches to coverage of the entire substrate. Macrolayer failure is complete when the Leidenfrost temperature is reached at e. Alternatively, the Leidenfrost temperature may be approached from above e. It has been observed that the vapour film above a horizontal substrate, instead of vanishing at the Leidenfrost temperature, first exhibits a Bondian interfacial instability such that $Bo_\Lambda = \Lambda^2 g\Delta\rho/4\pi^2\sigma = O(1)$, where Λ is the disturbance wavelength.

This limit, which defines the limit of macrolayer existence, may be used to show that

$$\dot{q}_{min} = C'\lambda_{LV}\rho_V^{1/2}(\sigma g\Delta\rho)^{1/4}\left(\frac{\rho_V}{\rho_L}\right)^{1/2} \tag{6.18}$$

in which $C' \simeq 0.15$ (Zuber and Tribus 1958). By comparison with eqn (6.17), it is evident that

$$\dot{q}_{min} \simeq \left(\frac{\rho_V}{\rho_L}\right)^{1/2} \dot{q}_{max}. \tag{6.19}$$

Providing the critical pressure is not approached, this reveals that $\dot{q}_{min} \ll \dot{q}_{max}$. For water and steam at 100 kPa, for example, $\dot{q}_{min}/\dot{q}_{max} = 0.024$.

The excess temperatures at which the critical and minimum heat fluxes occur in pool boiling may be estimated by using the above results together with the heat transfer relations for saturated and film boiling, eqns (6.13) and (6.3), respectively. Problem Q6.6 in Chapter 8 provides a representative calculation of the critical and minimum conditions while Q6.7 illustrates the changes in substrate temperature which accompany the irreversible paths d–d′ and e–e′. The effect of pressure on $\dot{q}_{max}$ and $\dot{q}_{min}$ is explored in Q6.8 in Chapter 8.

6.3 Boiling in liquid films

6.3.1 Boiling in a horizontal film

Typically, a stagnant liquid layer which is more than a few centimetres deep boils in accordance with the boiling curve discussed above. Once the liquid depth approaches and falls beneath the average bubble size, however, bubble growth and detachment are more constrained, but the heat transfer rate actually increases, as suggested in Fig. 6.11. A larger number of smaller bubbles begin to protrude above the liquid surface where, in bursting, they create smaller, more numerous drops. As these drops fall back into the liquid they entrain vapour which, in breaking into very small bubbles, create an efficient source of secondary nucleation; this seeding effect was discussed in Chapter 3. The tiny seed bubbles tend to form well beneath the upper surface of the film, closer to the substrate, and thus grow rapidly.

Most of the bubbles in horizontal films are not spherical for very long. Immediately after nucleation, whether on the substrate or within the film, the bubbles begin their lives as spheres but conditions peculiar to the film tend to change their shape. The high temperature gradient associated with boiling tends to accelerate evaporation at the bubble base relative to the upper surface, especially when the bubble protrudes above the film surface. Typically, therefore, the bubbles assume a more hemispherical shape with a basal liquid layer which is much thinner than the film. Continued

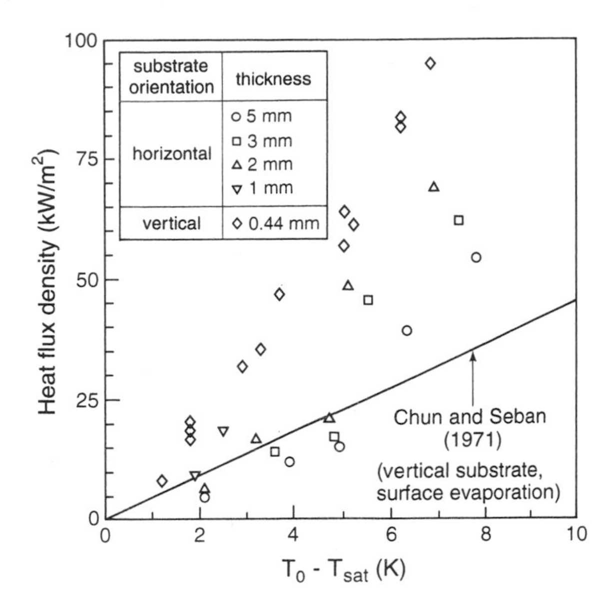

Fig. 6.11 Effect of liquid film thickness on boiling heat transfer rate (after Kim *et al.* 1983).

evaporation from the bubble base may significantly deplete this basal liquid. It is not difficult to see that thin basal films coupled with secondary nucleation can produce high heat transfer and evaporation rates.

Gradual and total evaporation of a horizontal liquid film is not well understood, but the following sequence may be observed in a heated metal pan containing water. Once the layer is thin enough e.g. a few millimetres, it tends to fizz as a result of the numerous bursting bubbles created by secondary nucleation. At some point, primary nucleation on the substrate subsides thus leaving Mesler entrainment as the sole source of nucleation. As the evaporation rate increases with further thinning of the film, the situation is reminiscent of the upper reaches of pool boiling. Bubble coalescence creates fewer and larger bubbles which are separated from the substrate by a macrolayer. Conduction across this basal macrolayer, and evaporation into the bubbles, creates very high heat transfer rates, but as it reduces the number of bubbles it eventually stifles nucleation all together. Naturally, this stifling process begins on the outer rim of the film and spreads inwards until a central patch of very large bubbles is all that remains. Finally, even these subside and only the macrolayer remains, gradually shrinking in extent and thickness until it vanishes completely. Splashing the bare hot surface with water drops invariably reveals that the Leidenfrost temperature has been reached.

Given the existence of the macrolayer, it is natural to enquire into

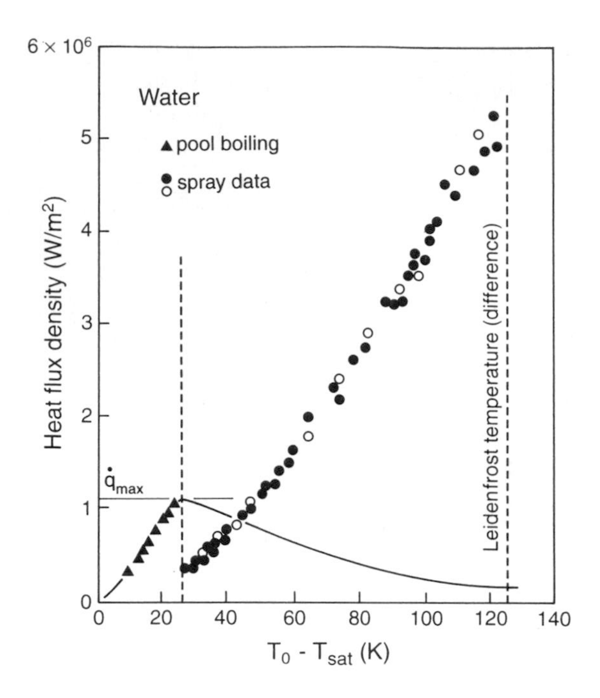

Fig. 6.12 Comparison of boiling heat transfer rates in a pool and a sprayed film (after Kopchikov *et al.* 1969).

situations which would promote its continued existence. Perhaps the most obvious technique is the use of a dense spray. Upon impact, the drops quickly lose their identity and merge into a liquid film within which evaporation behaviour may be significantly different from pool boiling, as Fig. 6.12 illustrates. In general, heat transfer rates within a film may be higher or lower than in pool boiling. The data of Kopchikov *et al.* (1969) indicate pool boiling heat transfer rates that were much higher and rose more steeply with increasing substrate temperature. This suggests that the lower (film) data correspond to a thick convecting film in which boiling was initially restricted, perhaps because of bubble collapse. On the other hand, for excess temperatures ($T_0 - T_{sat}$) higher than the critical value, when the pool macrolayer had finally dried out, the film continued to be effective, thus suggesting that its macrolayer was continually replenished until the substrate reached the Leidenfrost temperature. It is interesting to note that this natural, wetting limit on substrate temperature is not so much the cause of failure as it is the result. Spray-fed films remove the restriction associated with the pool; namely, a fixed value of *Ku*.

6.3.2 Boiling in a falling film

Boiling in a liquid film running over a substrate improves heat transfer and evaporation rates whether secondary nucleation takes place or not. This fact is well illustrated by the falling film. The growth of a vapour bubble in a falling film is shown schematically in Fig. 6.13. If nucleation takes place on the substrate, initial growth is similar to that on a horizontal surface but is gradually modified by the flow of liquid past the site. Eventually, the viscous forces accompanying this flow are enough to detach the growing bubble which is then dragged downwards with the film liquid even though buoyancy opposes the motion. The bubble velocity then exceeds the average film velocity, and is much greater than the liquid velocity near the bubble base. The bubble 'slides' over the thin liquid layer separating it from the substrate. In so doing, it creates a moving local reduction in film thickness and thus becomes a travelling heat sink in which enhanced evaporation takes place.

Most of the bubble growth occurs during the sliding phase which therefore accounts for most of the enhanced heat transfer attributable to boiling. At high heat flux densities, the number of bubbles is greater, their mutual interaction increasing with the evaporation rate. Bursting of the bubbles may also play a role, as the vertical data in Fig. 6.11 suggest. If the substrate is the inside of a tube, the situation may change dramatically. Firstly, drops ejected during bursting may reach across the tube thus allowing each side to impregnate the other with entrained vapour bubbles. Secondly,

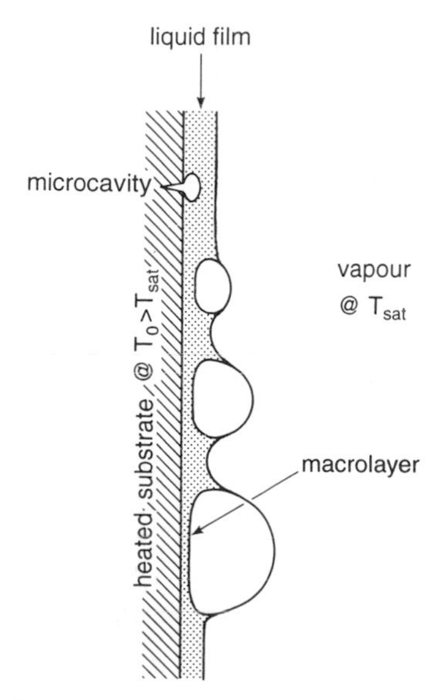

Fig. 6.13 Origin and growth history of an isolated bubble in a falling liquid film (after Cerza and Sernas 1985).

drops may be deposited from the vapour core. Under such conditions, the nucleation rate increases significantly.

Regardless of the source of nucleation, the evaporation rate increases in proportion to it; the liquid is increasingly depleted as the film falls. As noted previously, a thinning film may eventually break into rivulets between which bare patches of substrate appear. Nucleate boiling in the film tends to delay this breakup. Within any rivulets, boiling may continue but with a reduced overall effect. However, the spray thus produced does provide a compensation by dousing the otherwise bare patches with drops which evaporate, at least until the substrate reaches the Leidenfrost temperature.

6.3.3 Quenching

This leads us naturally into the subject of liquid cooling of substrates which have very high temperatures, i.e. above the Leidenfrost temperature. Quenching has long been used as a hardening technique whereby a metal component, heated to a temperature higher than the Leidenfrost temperature, is quickly plunged into a bath of liquid, frequently water or oil. At

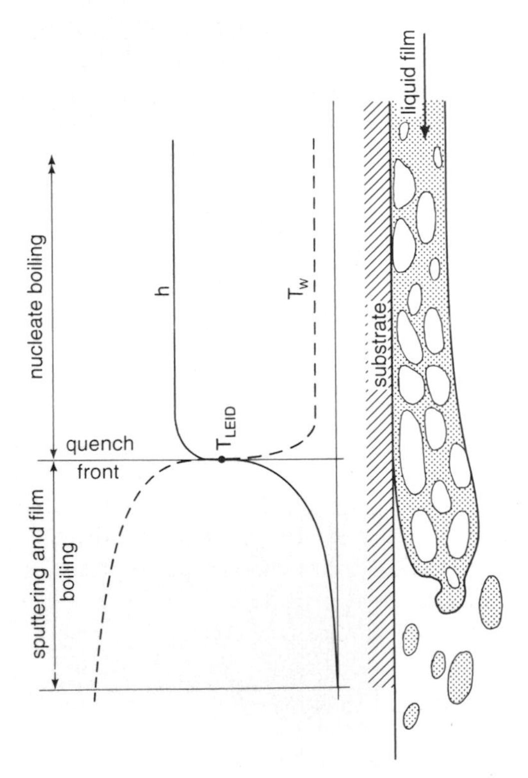

Fig. 6.14 Schematic illustration of quenching action of a falling film on a very hot substrate.

the component surface, the heat flux density tends to traverse the boiling curve more or less in reverse, beginning in the film boiling régime. This produces specific metallurgical changes near the component surface. Another important example of quenching occurs when, in an emergency, an overheated nuclear reactor tube must be suddenly flooded.

Downward flooding of a vertical substrate quenched by a liquid film is illustrated schematically in Fig. 6.14. Well above the quench front, the substrate temperature T_w is beneath the Leidenfrost temperature and nucleate boiling occurs. Approaching the quench front from above, it is found that the vigour of the boiling increases until a critical, and then a minimum, heat flux density is reached. Beyond the latter point, which essentially defines the quench front, a vapour film separates the nose from the heated surface but does not prevent radiant heating. Vigorous boiling may occur in the nose itself, given the attainment of critical conditions immediately upstream. A sputtering effect is caused by the rapid growth and bursting of vapour bubbles in the nose. Some of the drops collide with the substrate where they contribute an additional cooling effect. As Fig. 6.14 suggests, the large boiling heat transfer coefficient h above the nose is suddenly exchanged for a lower and decreasing value in the nose zone, often called the precursory cooling zone. The substrate temperature profile follows the opposite trend.

The above description tacitly assumes that the liquid film is two-dimensional, i.e. the quench front is horizontal. As we have seen previously, such a film may break up into rivulets and, if so, leave bare strips of overheated substrate. Within each rivulet, cooling behaviour is similar to that described immediately above. Between the rivulets, lateral conduction within the substrate coupled with spray cooling tend to even out substrate temperature variations above the Leidenfrost temperature. Equally significant aspects of quenching are the effects of film velocity and initial substrate temperature. As these increase, the conditions necessary for stable nucleate boiling are found further upstream. The vapour film under the nose then extends further upwards, creating an extended nose region completely detached from the substrate. This situation may lead to complete disintegration of the nose into large drops, much like the behaviour of any thin liquid sheet or jet, but the tendency is enhanced by evaporation. Inside a tube, this situation is known as inverted annular evaporation. It occurs when the liquid velocity is high. In-tube evaporation with lower liquid velocities is treated below.

6.4 Boiling in and around tubes

Boiling in a pool or in a liquid film is usually unconstrained geometrically; conditions at any given point distant from the substrate are not influenced by the presence of any other surface. In many industrial situations, however, the proximity of neighbouring surfaces may have a profound effect on evaporation and boiling processes. This is well illustrated by a bundle of tubes comprising a heat exchanger in which boiling may be taking place within each tube or in the spaces between them. We now consider boiling inside a single tube, beginning with a discussion of its onset.

6.4.1 *Incipient flow boiling*

Liquid flow inside a tube exerts a strong influence on the onset of boiling. Immediately adjacent to the wall, e.g. in the viscous sublayer, the liquid temperature profile is essentially linear and may be written in the form

$$\theta = \theta_w + bY, \tag{6.20}$$

where $\theta = T - T_{sat}$, $\theta_w = T_w - T_{sat}$, Y is distance from the wall, and b is the gradient $d\theta/dY$. At the same time, the temperature at which hemispherical bubbles of radius Y are able to grow beyond an active microcavity mouth may be represented by the nucleation curve (see eqn 6.7).

$$\theta = \frac{a}{Y}, \tag{6.21}$$

where $a = 2\sigma T_{sat}/\rho_V \lambda_{LV}$. When the actual liquid temperature profile rises to touch this hyperbolic nucleation curve, bubble growth will occur spontaneously if Y is within the size range of surface microcavities, as it frequently is. This situation is illustrated in Fig. 6.15 by the tangent point P at which θ_P is the same in eqns (6.20) and (6.21), as is the derivative $d\theta/dY$. Using these two equalities, it may be shown that $\theta_p = \theta_w/2$, i.e. $T_p - T_{sat} = (T_w - T_{sat})/2$. Hence the heat flux density at which boiling begins is given by

$$\dot{q}_o = -k_L \frac{dT}{dY} = \frac{\rho_V k_L \lambda_{LV} (T_w - T_{sat})^2}{8\sigma T_{sat}}. \tag{6.22}$$

The flow-controlled temperature gradient at the wall evidently doubles the required superheat for a given wall microcavity to be active in the presence of liquid flow. Forced convection therefore delays nucleate boiling. Q6.9 in Chapter 8 provides a numerical illustration.

6.4.2 *Flow boiling in a vertical tube*

In the vertical riser of a boiler, heat transferred from furnace gases causes a progressive rise in the quality of the mixture with height. This situation is shown schematically in Fig. 6.16. Clearly evident are the bubbly and plug régimes in the lower reaches where the quality x is small. The annular régime is noticeable in the upper reaches where x is greater. Figure 6.17 indicates a representative path ($A \rightarrow B$) on the régime map showing the effects of a gradual change in quality for a given heat flux density $\dot{q}$; the

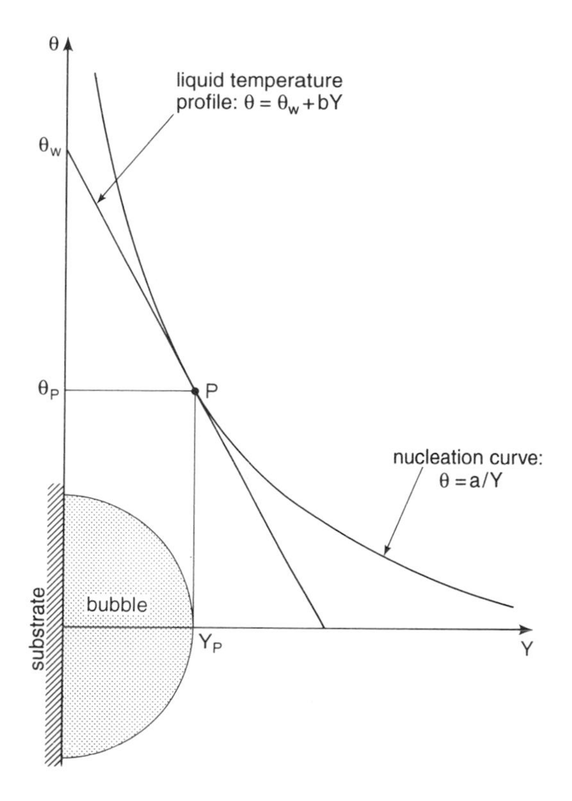

Fig. 6.15 Hyperbolic curve and its tangent representing incipient flow boiling condition.

point A represents the inlet mass flux. In general, the precise contour of this path, which depends upon $\dot{q}$, determines whether or not an excursion into the churn, plug or wispy-annular régimes occurs. The heat flux density thus exerts a substantial influence on flow behaviour.

Assuming the tube to be uniformly heated, and conditions to be more or less axisymmetric, the axial temperature distributions in Fig. 6.16 enable us to relate thermodynamic and hydrodynamic events at each elevation; saturation pressure and temperature are essentially fixed. Below the point N, where the nucleation begins, forced convection heating causes the temperature of the liquid bulk T_L to rise along with the wall temperature T_w. Immediately above N, $T_w > T_{sat}$ but bubbles must grow in the presence of a temperature gradient and many of those released from the wall collapse in the subcooled liquid; $T_L < T_{sat} \simeq T_V$, the vapour temperature (shown dashed). The wall and vapour temperatures stabilize after T_L reaches T_{sat}; saturated boiling then ensues.

The bubbly flow régime, corresponding to subcooled boiling and the

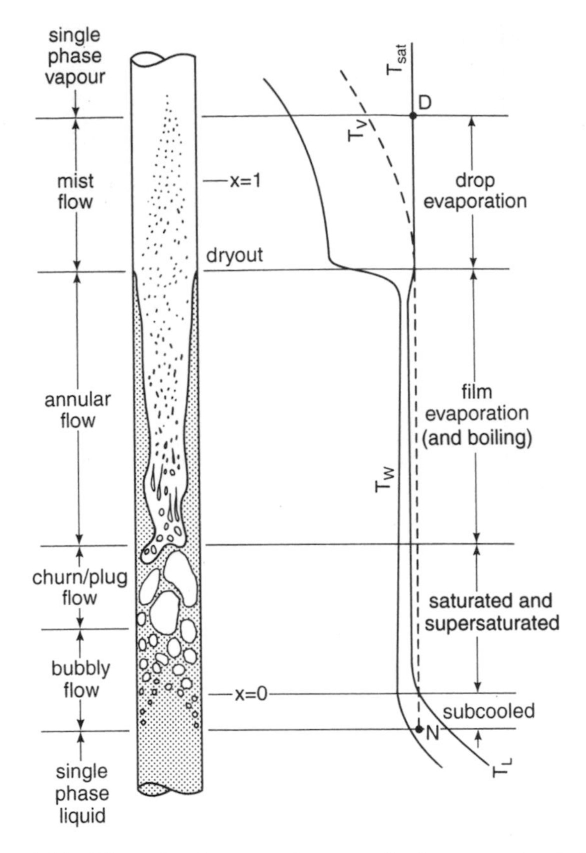

Fig. 6.16 Flow development in a vertical evaporator tube.

lower reaches of saturated boiling, extends above and below the point where the (equilibrium) mixture reaches a saturated liquid condition ($x = 0$). Above this point, vapour production rapidly increases the void ratio to cause a breakthrough into the annular régime. Given that the quality x, the vapour *mass* fraction, is related to the void fraction ε, the vapour *volume* fraction, by the expression $\varepsilon = x/[x + (1 - x)\rho_V U_V/\rho_L U_L]$, this transition is to be found at low qualities, e.g. 5 per cent; typically, $\rho_V U_V/\rho_L U_L \ll x \ll 1$ so that $x \ll \varepsilon \leqslant 1$. Q6.10 in Chapter 8 provides a numerical illustration. The breakthrough is usually accompanied by large amounts of spray, as discussed in Chapter 3. The spray drops are simultaneously deposited and generated at the liquid–vapour surface of the annulus as it develops downstream. As we move vertically upwards, the central core degenerates into a high velocity mist fed by droplet entrainment while the liquid film, although fed by droplet deposition, is gradually depleted by entrainment and evaporation. At lower heat fluxes, evaporation may be accompanied by boiling in which bubble bursting, amplified through secondary nucleation, increases the entrainment and deposition rates.

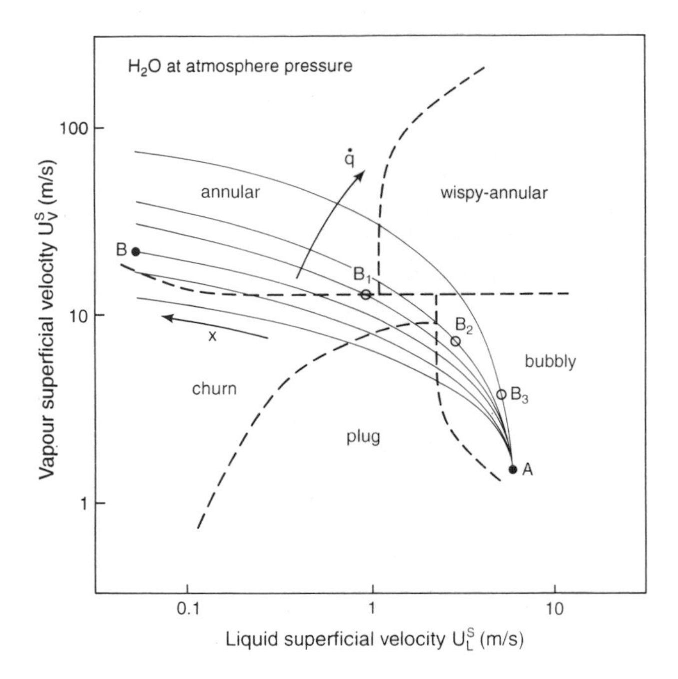

Fig. 6.17 Comparison of flow régime paths for a vertical evaporator tube showing the effect of heat flux density $\dot{q}$.

The annular film eventually vanishes at the *dry-out* point marked by a rapid rise in wall temperature attributable to the lower heat transfer coefficients characteristic of forced convection in a vapour. Immediately upstream of this point the steepening liquid temperature profile falls beneath the nucleation curve in Fig. 6.15; any boiling will then be suppressed in the remnants of a macrolayer. Earlier discussion reminds us that a thinning macrolayer is likely to break into rivulets or patches. Should the local wall temperature exceed the Leidenfrost temperature, however, dry-out is more accurately described by a zone of shifting, sputtering rivulets. Within a short distance downstream of the dry out point, the superheated vapour, along with radiation from the hotter wall, accelerates evaporation of the mist which no longer deposits on the wall. After this evaporation is completed at D, heat transfer is by forced, single phase convection alone. A quality of $x = 1.0$ is first reached below D when the (equilibrium) mixture reaches a dry-saturated condition.

Figure 6.17 indicates that if the heat flux is increased sufficiently, the top of the curve AB is raised until only two flow régimes are present: annular (including wispy-annular) and bubbly. At very high flow rates accompanied by very high wall temperatures the entire curve shifts upward and to the right, but Figs 6.16 and 6.17 no longer offer a comprehensive description of behaviour because the annular flow may then be inverted. This produces a quenching situation in which an annulus of vapour surrounds a jet-like liquid core, not unlike forced convective film boiling. On the other hand, lowering the heat flux sufficiently may eliminate the annular flow régime entirely, implying that a quality of $x = 1$ would not be reached. The latter situation may occur in boiler risers, for example.

6.4.3 Flow boiling in a horizontal tube

The comparable situation for flow boiling in a horizontal tube is shown in Figs 6.18 and 6.19, again assuming a fixed mass (or momentum) flux at

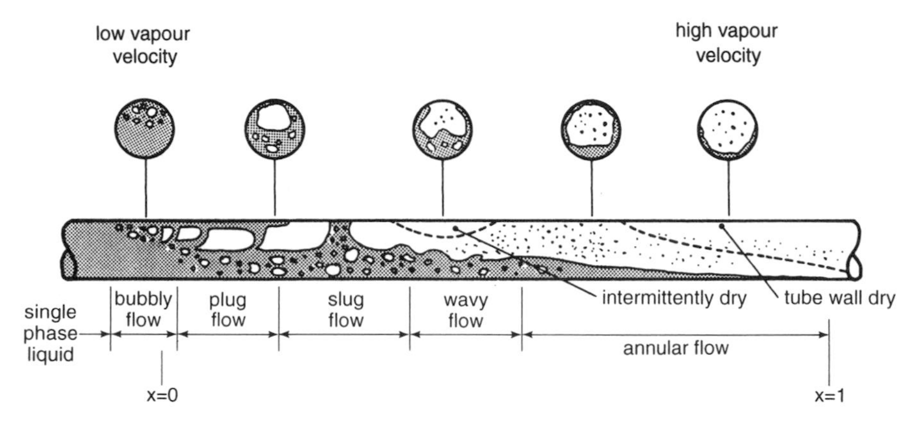

Fig. 6.18 Flow development in horizontal evaporator tube (after Collier 1981).

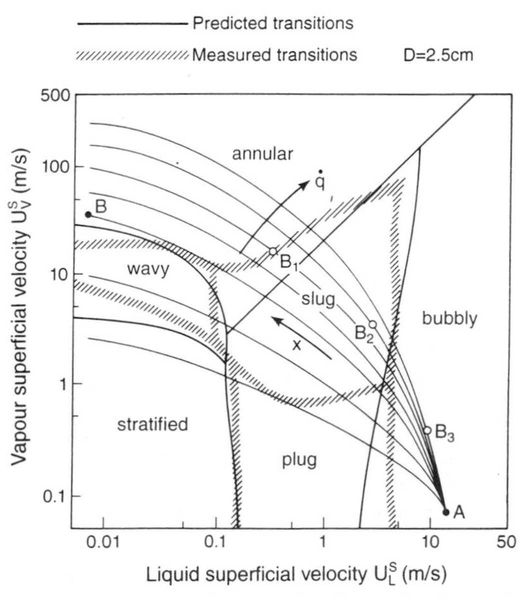

Fig. 6.19 Comparison of flow régime paths for a horizontal evaporator tube showing the effect of heat flux density $\dot{q}$.

inlet; the principal difference is the departure from axisymmetry. The path A → B on Fig. 6.19 corresponds roughly to the representation given in Fig. 6.18 from which five successive régimes may be identified. Nucleation initially creates subcooled boiling followed by saturated boiling during both of which the bubbles tend to collect in the upper half of the tube. Further evaporation enlarges the bubbles which coalesce and thus produce a slug flow which may or may not be preceded by a short-lived plug flow. The vapour velocity is low.

Continued boiling in the lower half of the tube may eventually produce enough vapour with a high enough velocity to give entry to the annular régime but, as Fig. 6.19 illustrates, this is not an invariable occurrence. In any event, at exit from the slug régime the flow is wavy and may therefore be accompanied by intermittent dry-out near the top of the tube. When the vapour velocity is not too high, the succeeding annular flow will differ from that in a vertical tube because Archimedean buoyancy creates a draining effect around the tube wall; this is particularly evident at lower heat fluxes. Before complete dry-out, the circumferential variations in wall temperature may be substantial. The annular régime is again represented by an evaporating film but, as in the vertical tube, nucleate boiling may be present at lower heat flux densities. Increasing the heat transfer rate tends to reduce the departure from axisymmetry in the high quality region by moving the top of the path A → B upward. Conversely, a lower heat transfer rate accentuates the departure and may eliminate the annular régime entirely if evaporation is not completed. Increasing the inlet mass flux again moves the curve up and to the right.

6.4.4 *Heat transfer rates*

It is evident that flow boiling in tubes is both variable and complex. Even so, the heat transfer rate at any given location may be estimated if we first assume that it is divided into two parallel components: single-phase convection in the liquid, usually turbulent; and nucleate boiling. Strictly, these processes are coupled together but as a first approximation they may be treated independently. Thus we take

$$\dot{q} = \dot{q}_{FC} + \dot{q}_{B} \tag{6.23}$$

for the heat flux density constructed from a forced convection component $\dot{q}_{FC}$ and a nucleate boiling component $\dot{q}_B$. The remainder of the analysis is a search for suitable expressions for these components.

To begin with, it is useful to return to the pool boiling curve. Figure 6.20 is a restatement of Fig. 6.8 in the form $\ln \dot{q}$ versus $\ln \theta_w$ where $\theta_w = T_w - T_{sat}$. Superimposed on this figure are single (liquid) phase relations: natural convection at the base of the pool boiling curve, and forced convection higher up; an increasing flow velocity shifts the curves upward. The curve FC, for example, describes single-phase forced convection which, in reality, occurs up to the point i which marks the onset of

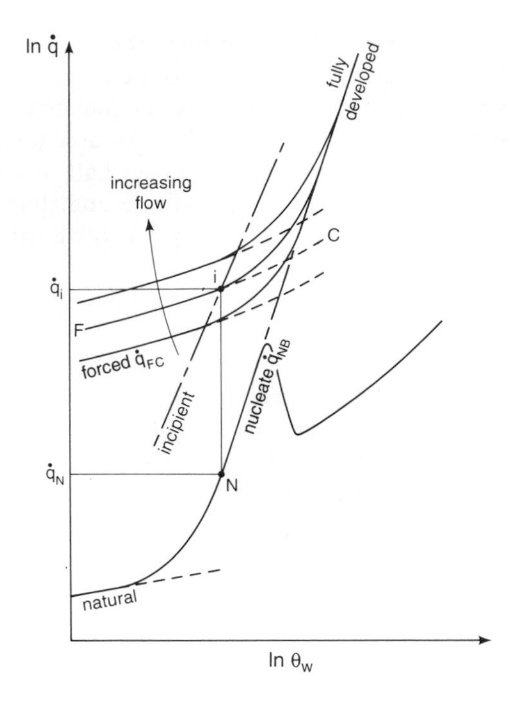

Fig. 6.20 Flow boiling curves superimposed on the pool boiling curve.

nucleate boiling according to eqn (6.22). The slope of the *incipient boiling* curve is usually less than that of the saturated nucleate pool boiling curve, as indicated. To the right of i, boiling increases the heat transfer rate rapidly with further increases in θ_w, the curve eventually becoming asymptotic to a *fully developed boiling* curve. Representative data are shown in Fig. 6.21 for three velocities. It is found that each of the flow boiling curves tends towards this same asymptote which, for most practical purposes, may be taken as an extension of the saturated nucleate pool boiling curve.

In the bubbly flow régime, the heat transfer rate may be estimated from eqn (6.23) as follows. To the left of the incipient boiling curve, $\dot{q} = \dot{q}_{FC} < \dot{q}_i$ in which

$$\dot{q}_{FC} = h(T_w - T_b), \tag{6.24}$$

where T_b is the bulk liquid temperature and

$$h = 0.023 \frac{k_L}{D} Re_L^{4/5} Pr_L^{1/3} \tag{6.25}$$

is a suitable turbulent forced convection correlation for a pipe of diameter D. To the right of the incipient boiling curve, eqn (6.23) requires the addition of $\dot{q}_{FC}$ (now calculated along the extrapolation iC) and $\dot{q}_B$, determined from

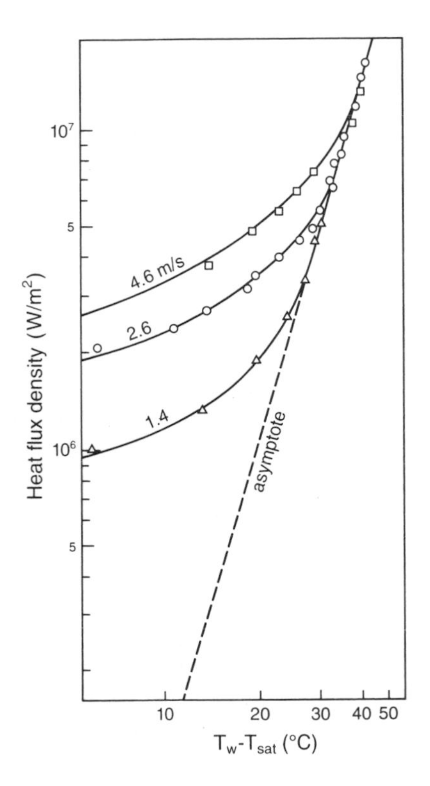

Fig. 6.21 Forced convection boiling in a steel tube (after Bergles and Rohsenow 1964).

$$\dot{q}_{\mathrm{B}} = \dot{q}_{\mathrm{NB}} - \dot{q}_{\mathrm{N}}, \tag{6.26}$$

where $\dot{q}_{\mathrm{NB}}$ is the saturated nucleate pool boiling relation, as represented by eqn (6.13), and $\dot{q}_{\mathrm{N}}$ is the value of $\dot{q}_{\mathrm{NB}}$ when incipient boiling occurs.

Once the void fraction is no longer small, the above approach must be modified to account for the change to annular flow. The boiling component $\dot{q}_{\mathrm{B}}$ may again be determined from eqn (6.13), bearing in mind that if Mesler entrainment is neglected nucleate boiling in an annulus is not unlike that in a filled tube with bubbly flow concentrated near the tube wall. The forced convection component, however, must now accommodate a wide range of void fractions. Equation (6.24) must therefore be restated as

$$\dot{q}_{\mathrm{FC}} = h_{\mathrm{A}}(T_{\mathrm{w}} - T_{\mathrm{sat}})$$

in which T_{b} has been replaced by the more accessible vapour temperature T_{sat}, and h_{A} is obtained from a suitable two-phase flow relation. Traviss *et al.* (1972) suggest that

$$h_{\mathrm{A}} = \frac{k_{\mathrm{L}}}{D} Re_{\mathrm{L}}^{0.9} Pr_{\mathrm{L}} F(Re_{\mathrm{L}}, x), \tag{6.27}$$

where

$$F = \frac{F_1(x)}{F_2(Re_L, Pr_L)}$$

and

$$F_1(x) = 0.15[\chi + 2.0\chi^{0.32}] \quad (6.28)$$

in which

$$\chi = \left(\frac{\rho_L}{\rho_V}\right)^{0.5}\left(\frac{\mu_V}{\mu_L}\right)^{0.1}\left(\frac{x}{1-x}\right)^{0.9} \quad (6.29)$$

and $F_2(Re_L, Pr_L)$ is given as follows:

$$\left.\begin{array}{ll} Re_L > 1125: & F_2 = 5\,Pr_L + 5\ln(1 + 5\,Pr_L) + 2.5\ln(3.1 \times 10^{-3} Re_L^{0.81}) \\ 60 < Re_L < 1125: & F_2 = 5\,Pr_L + 5\ln[1 + Pr_L(9.6 \times 10^{-2} Re_L^{0.58} - 1)] \\ Re_L < 60: & F_2 = 0.707\,Pr_L Re_L^{0.5}. \end{array}\right\} \quad (6.30)$$

The Prandtl number is defined by $Pr_L = \nu_L/\kappa_L$ and the Reynolds number by

$$Re_L = \frac{GD}{\mu_L}(1 - x).$$

Numerical illustrations of annular flow boiling are provided by problems Q6.12–6.14 in Chapter 8.

6.4.5 Performance limits

It has been noted above that changing either the heat flux or the mass flux creates substantial alterations in the régimes encountered in an evaporator tube. In the illustrations given, it was assumed that evaporation was carried to completion, pure vapour emerging further downstream of the dry-out point. Dry-out thus defined the performance limit, beyond which tube wall temperatures become unacceptably high. The paths A → B on Figs 6.17 and 6.19 were chosen with this failure limit at B.

Consider now the effect of increasing the heat flux to a tube in which the mass flux is fixed. The path A → B, the initial path, is also drawn as a horizontal line in Fig. 6.22, which is a plot of heat flux density $\dot{q}$ versus quality x at various locations along the tube. For this particular level of heat flux, the path is seen to traverse the nucleate boiling and annular evaporation regions. Most important is the location of the failure limit B. It is observed that an increased heat flux density shortens the annular régime, moving the dry out point further upstream, i.e. towards A. This is reflected on Fig. 6.22 by a shift of the limit at B to a new point higher

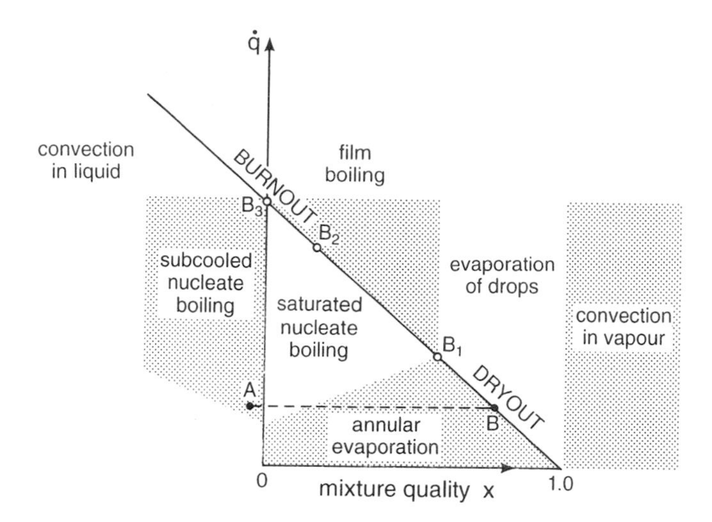

Fig. 6.22 Hypothetical régime map for flow boiling in a tube (after Collier 1981). A → B corresponds to the paths shown in Figs. 6.17 and 6.19.

and further left. With additional increases in $\dot{q}$, the failure limit continues to shift on a diagonal in this manner until the annular régime barely exists.

Eventually, the failure point moves into the saturated nucleate boiling régime at the condition B_1 on Figs 6.17 and 6.22. This marks a significant transition along the diagonal limit line through B and B_1 on Fig. 6.22. Above B_1 on this diagonal, failure is attributed to the attainment of the Leidenfrost temperature at the wall; for simplicity, this will be called *burn-out* because it is associated with film boiling. As the heat flux is further increased to very high values, the limit point moves further up along the edge of the saturated nucleate boiling régime, eventually reaching the subcooled régime at B_3. Similar comments may be made on the failure limits for horizontal tubes, as Fig. 6.19 suggests. However, it is important to recall that the annular régime in a horizontal tube is usually more complex. With a rivulet still running along the bottom surface of the tube, the upper surface is intermittently depleted of liquid; burnout is likely in this location.

Failure in the nucleate boiling régime occurs at higher heat transfer rates in flow boiling than in pool boiling. This is indicated in both Fig. 6.20 and Fig. 6.22. Under saturated conditions, the fully developed region of the flow boiling curve in Fig. 6.20 extends well above the critical heat flux obtained in pool boiling. This may be attributed to the role of bulk liquid velocity: firstly, in delaying the onset of nucleation, and thus transferring the effect of nucleate boiling to a higher range of substrate temperatures; secondly, in hindering the subsequent development of the large vapour pockets found in pretransitional pool boiling. With increasing heat flux, these changes shift the emergence, development and failure of the macro-layer to the liquid films surrounding vapour plugs or slugs. However, as the failure point moves towards the subcooled régime it is the crowding and coalescence of smaller bubbles at the tube wall that becomes the cause of transition to film boiling; the liquid is believed to undergo a classical separation from the wall. Subcooling itself may alter this failure mechanism by permitting individual bubbles, anchored at specific sites, to spread over the tube wall, evaporating at their heated peripheries while condensing onto their subcooled liquid domes. In any event, subcooled flow boiling is capable of raising the critical heat flux density to extremely high values, e.g. $10\,\mathrm{MW\,m^{-2}}$.

6.4.6 Boiling in tube bundles

The principal difference between two-phase flow inside a single tube and that within the intertube spaces formed by a tube bundle is caused by alterations in the confining effect of the tube wall. The space between adjacent tubes is generally open to surrounding spaces, thus reducing the confinement. Even so, it is to be expected that boiling in the interstices of a vertical tube bundle would be similar to that inside a vertical tube.

The geometry of the tube layout presents three or four *convex* tube surfaces to the boiling flow, thus altering the shape of any vapour plug or annular film. Symmetry about a tube axis no longer exists because the smallest gap between adjacent tubes is less than the effective diameter of the intertube space. Conditions in this narrow gap are thus further beyond the bubbly flow régime than at the geometric centre of the intertube space. When the spacing–diameter ratio of the tubes is 1.0 or greater, however, this circumferential variation in conditions may be unimportant, especially in the plug and annular régimes.

As a rough approximation, a tube bundle may be regarded as a porous medium in which bulk flow is driven by Archimedean buoyancy combined with a pressure gradient. In a vertical tube bundle, for example, liquid will be drawn in through the vertical boundary of the bundle; the upper and lower boundaries are usually blanked off. This entraining effect, especially in the lower reaches, creates a lateral flow superimposed on the main vertical flow. Bubbles may therefore be forced to move laterally, a tendency resisted by plug bubbles which must neck down to pass between adjacent tubes.

Much the same situation occurs in a horizontal tube bundle where the vertical motion of a bubble is more restricted. For either orientation, the distributed bubbles produce a void fraction which increases with height. Bubbly flow in the lower reaches of the bundle may thus be accompanied by plug flow or annular flow in the upper reaches. Even in the bubbly régime, boiling in a horizontal bundle may exhibit plug-like behaviour because any bubbles intercepted by a tube above are forced to negotiate around it. The result is a 'sliding' bubble which, as its name suggests, moves over the tube surface on a liquid film formed when the bubble first butts again the tube. Heat transfer and evaporation into sliding bubbles are governed by considerations similar to those discussed earlier for bubbles in falling films. Most of the evaporation takes place across a thin basal macrolayer which, in moving relative to the sliding bubble, continually replenishes the liquid evaporated. Evaporation thus alters the thickness of the macrolayer as it travels under the bubble. The macrolayer may evaporate completely near the trailing edge of the bubble if the heat transfer rate is high enough.

Selected bibliography

Bar-Cohen, A. (1992). Hysteresis phenomena at the onset of nucleate boiling. In *Pool boiling and external flow boiling* (eds V. K. Dhir and A. E. Bergles). ASME, New York.

Bergles, A. E., Collier, J. G., Delhaye, J. M., Hewitt, G. F., and Mayinger, F. (eds). (1981). *Two phase flow and heat transfer in the power and process industries*. Hemisphere, Washington.

Bergles, A. E. and Rohsenow, W. M. (1964). The determination of forced convection surface boiling heat transfer. *Journal of Heat Transfer*, **86**, 365–82.

Bjorge, R. W. and Rohsenow, W. M. (1982). Correlation of forced convection boiling heat transfer data. *International Journal of Heat Mass Transfer*, **25**(26), 753–57.

Carey, V. P. (1992). *Liquid-vapour phase-change phenomena*. Hemisphere, Washington.

Cerza, M. and Sernas, V. (1985). A bubble growth model for nucleate boiling in thin, falling, superheated, laminar, water films. *International Journal of Heat Mass Transfer*, **28**(7), 1307–16.

Chun, K. R. and Seban, R. A. (1971). Heat transfer to evaporating liquid films. *Journal of Heat Transfer*, **93**, 391–96.

Collier, J. G. (1972). *Convective boiling and condensation*. McGraw-Hill, London.

Collier, J. G. (1981). Forced convective boiling. In *Two phase flow and heat transfer* (eds A. E. Bergles, J. G. Collier, J. M. Delhaye, G. F. Hewitt and F. Mayinger) Chapter 8. Hemisphere, Washington.

Dhir, V. K. and Bergles, A. E. (eds) (1992). *Pool boiling and external flow boiling*. American Society of Mechanical Engineers, New York.

Farber, E. A. and Scorah, R. L. (1948). Heat transfer to water boiling under pressure. *Transactions of the ASME*, **70**, 369–84.

Forster, H. K. and Zuber, N. (1955). Bubble dynamics and boiling heat transfer. *AIChE Journal* **1**, 532.

Gaertner, R. F. and Westwater, J. W. (1960). Population of active sites in nucleate boiling heat transfer. *Chem. Eng. Prog. Symp. Ser #30*, **56**, 39–48.

Hahne, E. and Grigull, U. (1977). *Heat transfer in boiling*. Hemisphere, Washington.

Hewitt, G. W. (1981). Burnout. In *Handbook of multiphase systems*. (ed.) G. Hetsroni) Hemisphere, Washington.

Katto, Y. (1992). Critical heat flux in pool boiling. In *Pool boiling and external flow boiling*. (eds V. K. Dhir and A. E. Bergles). ASME, New York.
Kim, H.-K., Fakeeha, A., and Mesler, R. (1963). Nucleate boiling in flowing and horizontal liquid films. *ASME National Heat Transfer Conference*, 61–5.
Kopchikov, I. A., Voronin, G. I., Kolach, T. A., Labuntsov, D. A., and Lebedev, P. D. (1969). Liquid boiling in a thin film. *International Journal of Heat Mass Transfer*, **12**, 791–6.
Kutaleladze, S. S. (1961). Boiling heat transfer. *International Journal of Heat Mass Transfer*, **4**, 31.
Lock, G. S. H. (1992). *The tubular thermosyphon*. Oxford University Press, Oxford.
Mesler, R. B. and Mailen, G. (1977). Nucleate boiling in thin liquid films. *AIChE Journal* **23**(6), 954–57.
Mikic, B. B. and Rohsenow, W. M. (1969). New correlation of pool boiling data including the effect of heating surface characteristics. *Journal of Heat Transfer*, **91**, 245–50.
Palen, J. W. (1986). *Heat exchanger sourcebook*. Hemisphere, Washington.
Rohsenow, W. M. (1952). A method of correlating heat transfer data for surface boiling of liquids. *Transactions of the ASME*, **74**, 1969.
Rohsenow, W. M. (1973). Boiling. In *Handbook of heat transfer*. McGraw-Hill, New York.
Rohsenow, W. M. (1981). Forced convection boiling. In *Handbook of multiphase systems*, (ed. G. Hetsroni) Hemisphere, Washington.
Stephan, K. (1992). *Heat transfer in condensation and boiling*. Springer, Berlin.
Traviss, D. P., Rohsenow, W. M., and Baron, A. B. (1973). Forced convection condensation inside tubes: a heat transfer equation for condenser design. *ASHRAE Transactions*, Pt. 1, 157–65.
Van Stralen, S. J. D. and Cole, R. (1979). *Boiling phenomena*, Vols. 1 and 2. Hemisphere, Washington.
Whalley, P. B. (1987). *Boiling, condensation and gas-liquid flow*. Oxford University Press, Oxford.
Yu, C.-L. and Mesler, R. B. (1977). A study of nucleate boiling near the peak heat flux through measurement of transient surface temperature. *International Journal of Heat Mass Transfer*, **20**, 827–40.
Zuber, N. (1958). On the stability of boiling heat transfer. *Transactions of the ASME*, **80**, 711–20.
Zuber, N. (1963). Nucleate boiling. The region of isolated bubbles and similarity with natural convection. *International Journal of Heat Mass Transfer*, **6**, 53–78.
Zuber, N. and Tribus, M. (1958). Further remarks on the stability of boiling heat transfer. *AEC Report* No. 3631, AEC Technical Information Service, Oak Ridge, Tennessee.

Exercises

1. Show that the time for complete evaporation of a disc-like drop on a substrate at temperature T_0 is given by

$$t_{\mathrm{f}} = \frac{\rho_{\mathrm{L}} \lambda'_{\mathrm{LV}} W_{\mathrm{i}}^2}{2k_{\mathrm{L}}(T_0 - T_{\mathrm{I}})}$$

if the initial drop thickness is W_i, the vapour superheat is $T_\infty - T_I$ and $\lambda'_{LV} = \lambda_{LV}(1 - Ja_V)$.

2. Estimate the time interval during which transient conduction occurs in a 18.9 μm diameter hemispherical substrate drop if $\kappa_L = 1.68 \times 10^{-7}\,m^2\,s^{-1}$.

   (Ans: 0.53 ms.)

3. Consider a laminar, shear-driven liquid film on which the interfacial shear stress τ_I is constant. Show that film completely evaporates into saturated vapour in a substrate length

$$X_f = \frac{\lambda_{LV}\tau_I\Delta_0^3}{3k_L(T_0 - T_I)\nu_L}$$

   if the substrate temperature is $T_0 > T_I$ and the initial (upstream) film thickness is Δ_0.

4. Show that laminar film evaporation on the inside of a cylinder, diameter D, length $L \gg D$, and rotating at a high speed Ω about its own axis, is described by

$$Nu = C(Ja, Pr)\left(\frac{\lambda_{LV}\Omega^2 DL^3\Delta\rho}{\nu_L k_L \theta}\right)^{1/5}$$

   where the remaining symbols have their usual meaning.

5. Demonstrate that $Bo_b = O(1)$ for a spherical bubble pinching off on a vertical surface:

   (a) if the bubble is small,

   (b) if the bubble is large.

6. Derive eqn (6.7) from first principles.

7. Derive eqn (6.17) from $I = O(B) = O(T)$.

8. Following the example of Q6.6 in Chapter 8, construct the boiling curves for the same tube when the water pressure is (a) 0.1 bar, (b) 10 bar.

   (Ans: see table.)

Pressure (bar)	T_{sat} (K)	Onset			Maximum			Minimum		
		$\dot{q}$ (W m^{-2})	ΔT (K)	T_b (K)	$\dot{q}_{max}$ (W m^{-2})	ΔT (K)	T_d (K)	$\dot{q}_{min}$ (W m^{-2})	ΔT (K)	T_e (K)
0.1	319	4.0×10^6	53.4	372.4	4.70×10^5	26.1	345	3.93×10^3	13	332
1.0	373	1.2×10^4	6.68	380	1.27×10^6	31.6	405	3.13×10^4	201	574
10.0	453	37.3	0.8	454	3.00×10^6	34.9	488	2.29×10^5	1164	1617

The results given above reveal the increased difficulty in initiating boiling as the pressure is lowered, but they also indicate the limitations of extrapolation: at low pressures they predict that $T_b > T_d > T_e$. Also evident is the flattening of the transitional régime with increasing pressure as the minimum heat flux and Leidenfrost temperature T_e both increase.

9. Consider a water drop of radius r supported by vapour above a horizontal substrate whose temperature T_0 exceeds the Leidenfrost temperature. Show that the thickness Δ of the vapour film supporting the drop is given by

$$\frac{\Delta}{r} = \left(\frac{9Ja_V}{4Ar_V Pr_V}\right)^{1/4}$$

where $Ar_V = gr^3\Delta\rho/\rho_V\nu_V^2$. Hint: balance drop weight against supporting pressure.

10. Cold water with a bulk temperature T_L flows at a rate $\dot{m}$ in a vertical pipe having a diameter D and a uniform wall temperature T_w. If h is the internal heat transfer coefficient, show that

$$T_w - T_L = (T_w - T_{Li}) \exp\left(\frac{-h\pi DX}{\dot{M}c_{pL}}\right),$$

where T_{Li} is the inlet temperature, describes the water temperature as a function of X, the distance from the inlet.

11. Derive eqn (6.22).

12. For the flow boiling conditions considered in Q6.10 in Chapter 8, calculate the heat transfer coefficient where $x = x_t$, the transitional quality.

(Ans: 29.6 kW m^{-2} K^{-1}.)

Evaporation and boiling projects

1. Review and assess evaporation from binary liquid films.
2. Review and assess inverted flow boiling.
3. Develop a physical and mathematical model of the wet bulb thermometer.
4. Survey enhancement techniques for flow boiling.
5. Review and assess pool boiling in liquid metals.
6. Review and assess boiling in horizontal tube bundles.
7. Develop a physical and mathematical model for reflux boiling in a tubular thermosyphon.

8. Review and assess evaporation from ocean surfaces.
9. Review and assess the quenching of metal components.
10. Develop a physical and mathematical model of low-gravity pool boiling.

7
DIRECT CONTACT PROCESSES

It will be recalled from Chapter 1 that there are many situations in which latent heat transfer may occur in the absence of a separating wall. Spray condensers, combustion chambers, and crystallizers are representative industrial examples. In an environmental context, pollution and water conservation have stimulated interest in such devices as the cooling tower and the cooling pond. Even wider is the traditional interest in latent heat found in the environmental disciplines of glaciology, meteorology, and oceanography. In any of these contexts the situation is essentially the same: two media, one hotter than the other, lie in intimate contact with each other. Phase change at the boundary between the media is then no longer influenced by an intervening wall with its characteristic ability to promote heterogeneous nucleation. In general, the media are not pure components, as when sea water freezes or water vapour condenses from the atmosphere. In certain circumstances, therefore, the exchange of latent heat produces concentration gradients at the common interface. The absence of an impermeable wall then implies diffusive mass transfer through and between the media.

The first three sections of this chapter extend the previous discussion of heat and mass transfer at the interface to bipartitioned and dispersed fluid media which are essentially independent of bounding walls. The remaining sections provide an introduction to crystal and reactive systems. In keeping with the general philosophy of the book each section concentrates more on fundamentals than on applications, which are mentioned only to remind the reader of the significance of the problem being discussed.

7.1 Condensation and evaporation domains

The various circumstances under which condensation and evaporation may occur are illustrated in Fig. 7.1 using a *Ts* diagram. Four separate domains, based on the state of the vapour, are shown grouped around the interface state at I on the saturation curve. The isotherm through T_I divides the condition for vapour heat gain at the interface ($T_V < T_I$) from the condition of vapour heat loss ($T_V > T_I$). During phase change of a pure vapour, the pressure is essentially the saturation pressure P_I and hence for practical purposes the vapour state lies on this isobar: superheated, saturated, or supersaturated. Heat transfer to or from the interface may then be determined from expressions such as eqn (5.43); condensation or evaporation rates are very high. This situation is illustrated using the examples of a steam desuperheater in problems Q7.1–7.4 of Chapter 8.

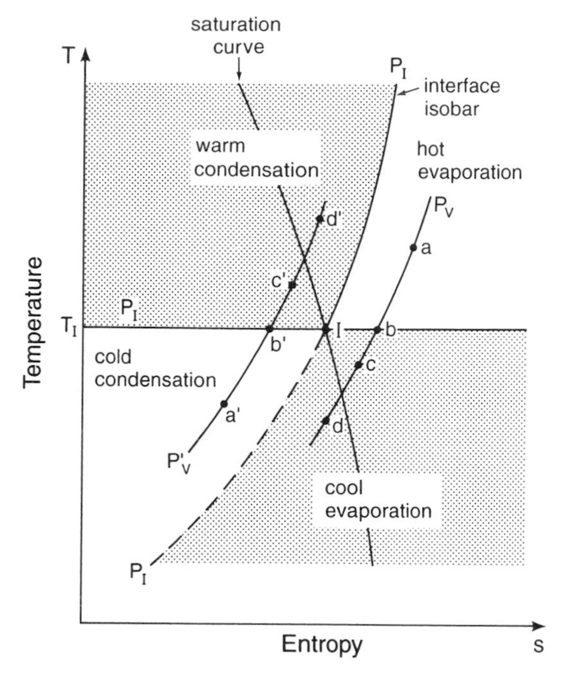

Fig. 7.1 Condensation and evaporation domains for a vapour–gas mixture.

In the presence of an incondensable gas, the vapour pressure may lie above or below P_I. Under neutral conditions, when neither evaporation nor condensation takes place, $\dot{m}_{VI} = 0$. More precisely, the mass flux *density* is given by

$$\mathbf{j}_{VI} = -\mathbf{G}_{VI} = 0. \tag{7.1}$$

Thus, neutrality is defined by

$$\mathbf{j}_{VI} = -\rho\mathfrak{D}(\nabla m_V)_I \simeq -\frac{\mathfrak{D}}{R_V T}(\nabla P_V)_I = 0 \tag{7.2}$$

in the common situation when $m_V \ll 1$. The isobar P_I then marks the neutral boundary.

Consider now the four domains of a vapour-gas mixture shown in Fig. 7.1. Evaporation processes are indicated by the vapour isobar $P_V < P_I$. The point a is representative of *hot evaporation*; $T_V > T_I$, and heat transfer from the superheated vapour–gas mixture drives the process. At b, $T_V = T_I$ and this drive ceases. When $T_V \leqslant T_I$, evaporation may proceed only if the parent liquid is superheated and thus supplies the necessary heat. The point c is representative of this situation, described as *cool evaporation*; it may extend to supersaturated vapours, as indicated by the point d, provided there is a corresponding increase in liquid superheat. In general, cool evaporation is limited by the liquid superheat, i.e. the liquid Jakob member.

Condensation processes are indicated by the vapour isobar $P'_V > P_I$. *Cold condensation* is defined by $T_V < T_I$ and is typified by the state point a′. Heat transfer to the supersaturated vapour–gas mixture then drives the process. At b′, this drive ceases. When $T_V \geqslant T_I$, condensation may only proceed if the liquid is subcooled and thus removes the necessary heat. The point c′ is representative of this situation, described as *warm condensation*; it may extend to superheated vapours, as indicated by the point d′, provided there is a corresponding increase in liquid subcooling. In general, warm condensation is limited by the liquid subcooling.

In many direct contact situations, the condensed phase is present in discrete amounts, e.g. as drops or jets, and hence the amount of liquid sensible heat associated with superheating or subcooling is finite. In warm condensation or cool evaporation, the total amount of phase change is limited by the liquid Jakob number. Under steady conditions, the mass of vapour m_V at temperature T_∞ formed or removed by conversion of the liquid mass m_L, initially at T_0, is given by the expression

$$m_V\{c_{pV}|T_\infty - T_I| + \lambda_{LV}\} \leqslant m_L c_{pL}|T_I - T_0|. \tag{7.3}$$

Hence

$$m_V \leqslant m_L\left(\frac{Ja_L}{1 + Ja_V}\right), \tag{7.4}$$

where $Ja_L = c_{pL}|T_I - T_0|/\lambda_{LV}$ and $Ja_V = c_{pV}|T_\infty - T_I|/\lambda_{LV}$ are the liquid and vapour Jakob numbers, respectively. In typical examples of cool evaporation or warm condensation under direct contact conditions, $T_\infty \simeq T_I$, and hence $Ja_V \ll 1$. Relation (7.4) then reduces to

$$m_V \leqslant m_L Ja_L. \tag{7.5}$$

It is often found that $Ja_L \ll 1$ also, thus revealing that large quantities of liquid are usually required to condense or produce small quantities of vapour under these particular conditions.

7.2 Direct condensation

In a typical industrial situation, we are concerned with the exchange of heat between a slightly superheated vapour at temperature T_∞ and a highly subcooled liquid, initially at a temperature T_0; this is an example of warm condensation. We thus face a Jakob number limited problem governed by relation (7.5) in which the inequality further restricts the amount of condensation, partly because heat transfer rates are finite and partly because of heating from the vapour ($Ja_V \neq 0$). A common strategy to improve heat transfer is to increase the surface area of the liquid. This has led to the development of spray and jet condensers.

7.2.1 Condensation on drops

In Chapter 3 it was noted that liquid sprays, natural or artificial, are seldom monodisperse. For our purpose, however, it is enough to treat condensation on drops of a given diameter. We will focus on industrial processes where, typically, the drop diameter is of the order of 1 mm. With reference to Fig. 7.2, the basic problem is posed as warm condensation of a pure, saturated vapour on a drop of liquid coolant not necessarily having the same composition as the vapour. Assuming a positive spreading coefficient and immiscibility, the condensate will form a liquid shell, thickness Δ, around the coolant drop, diameter d.

Using the relation (7.5), the final thickness of the condensate shell Δ_f is found to be

$$\frac{\Delta_f}{d} = \frac{\rho_c \, Ja_c}{6\rho_L}. \tag{7.6}$$

Since the liquid coolant density ρ_c is typically of the same order as the condensate density ρ_L, it follows that $\Delta_f \ll d$ when $Ja_c \ll 1$. This suggests a condensate shell in which a purely conductive thermal resistance will often be negligible. Under these conditions, the interface equation reduces to

$$\rho_L \lambda_{LV} \frac{d\Delta}{dt} = h_L(T_I - T_s) \simeq \frac{k_L}{\Delta}(T_I - T_s), \tag{7.7}$$

where h_L is the liquid shell heat transfer coefficient and $T_s(t)$ is the unknown surface temperature of the drop. Heat transfer within the coolant

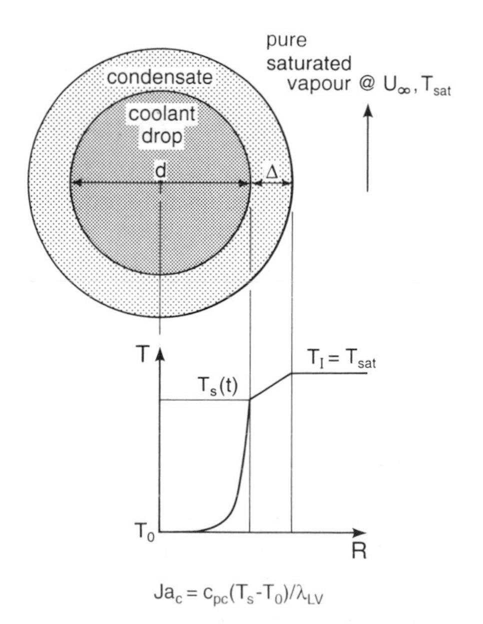

Fig. 7.2 Early condensation of saturated vapour on a small moving coolant drop.

drop depends upon the magnitude of the external Biot number which may be determined from eqns (7.6) and (7.7) as $Bi_d = h_L d/k_L \simeq 6\rho_L k_L/\rho_c k_c Ja_c$. With the exception of liquid metal drops, the ratio $\rho_L k_L/\rho_c k_c$ is typically $O(10)^{-1}$ and hence $Bi_d \geqslant O(1)$ for most condensates and coolants. This implies significant radial temperature gradients within the coolant.

Consider heat transfer within the drop as it moves with velocity U_∞ through the vapour: for example, during terminal flight. The relative motion of the vapour tends to induce a circulation within the drop but this convective effect may often be ignored. Viscosity and density are normally much greater in the coolant than in the surrounding vapour; this ensures a negligible circulation within the confines of a small diameter drop. Heat is transferred radially inward from the drop surface by conduction.

The coolant energy equation under these conditions is given by

$$\frac{\partial\theta}{\partial t} = \frac{\kappa_c}{R^2}\frac{\partial}{\partial R}\left(R^2\frac{\partial\theta}{\partial R}\right) = \frac{\kappa_c}{R^2}\frac{\partial}{\partial Y}\left(R^2\frac{\partial\theta}{\partial Y}\right), \tag{7.8}$$

where $\theta = T - T_0$ and $Y = d/2 - R$. Once the flight of the drop begins, $T_s \simeq T_I$ is essentially fixed, and a thermal boundary layer of thickness δ^T begins to penetrate beneath the drop surface. Normalization of eqn (7.8) using the Y coordinate reveals that

$$\delta^T = 0(\kappa_c t)^{1/2}. \tag{7.9}$$

Providing $\delta^T \ll d/2$, the temperature distribution within this boundary layer is given by

$$\theta \simeq (T_I - T_0)\,\mathrm{erfc}\,[Y/2(\kappa_c t)^{1/2}] \tag{7.10}$$

and hence the heat flux density at the coolant drop surface ($Y = 0$) is

$$\dot{q}_0 = -k_c\left(\frac{\partial\theta}{\partial y}\right)_0 = k_c(T_I - T_0)/(\pi\kappa_c t)^{1/2}. \tag{7.11}$$

This enables us to write the average heat transfer coefficient over the period t as

$$\bar{h}_c = 2h_c = 2k_c/(\pi\kappa_c t)^{1/2}. \tag{7.12}$$

Alternatively, we may introduce the travel distance $X = U_\infty t$, and write

$$\bar{h}_c = \frac{2k_c}{\pi^{1/2}}\left(\frac{U_\infty}{\kappa_c X}\right)^{1/2}$$

which permits us to define the *travel* Nusselt number by

$$\overline{Nu_X} = \frac{\bar{h}_c X}{k_c} \simeq \frac{2}{\pi^{1/2}} Pe_X^{1/2}, \tag{7.13}$$

where $Pe_X = U_\infty X/\kappa_c$ is the travel Peclet number.

The rate of condensate growth may now be determined from the interface eqn (7.7) rewritten

$$\rho_L \lambda_{LV} \frac{d\Delta}{dt} = \dot{q}_0 = k_c(T_I - T_0)/(\pi \kappa_c t)^{1/2}$$

which integrates to yield

$$\Delta(t) = \frac{2k_c(T_I - T_o)t^{1/2}}{\rho_L \lambda_{LV}(\pi \kappa_c)^{1/2}} = \frac{2Ja_c\rho_c}{\rho_L}\left(\frac{\kappa_c t}{\pi}\right)^{1/2}, \tag{7.14}$$

if $\Delta(0) = 0$. This predicts condensate growth when the thermal boundary layer is thin. Since $\Delta(t) \leqslant \Delta_f$, eqns (7.6) and (7.14) may be combined to reveal that

$$\frac{\kappa_c t^e}{d^2} = Fo_d \leqslant \frac{\pi}{144} \tag{7.15}$$

in which t^e is the equilibrium period defined by $\Delta(t^e) = \Delta_f$. The ratio

$$\frac{\Delta(t)}{\Delta_f} = \frac{12}{\pi^{1/2}} Fo_d^{1/2} \tag{7.16}$$

for $t < t^e$, is shown plotted as a full line in Fig. 7.3 along with the asymptote $\Delta(t) = \Delta_f$, for $t > t^e$.

A complete prediction, suggested by the dashed curve, requires the solution of the radial conduction problem for a greater time period. While this theoretical solution is available, at least for a constant $T_s \simeq T_I$, i.e. a thin condensate shell, it may not be as useful or as flexible as a numerical result based on discretization of the energy equation under more general

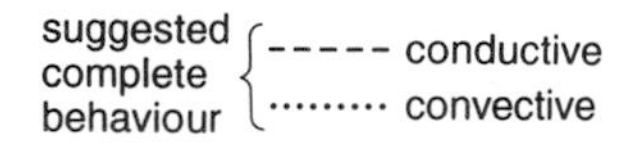

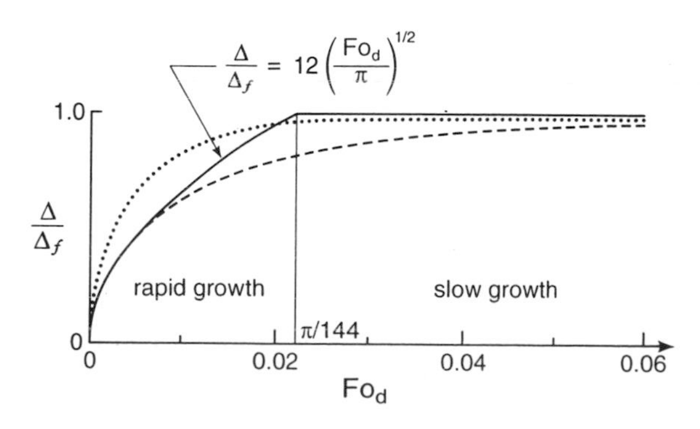

Fig. 7.3 Saturated growth characteristics of condensate on a small subcooled drop.

boundary conditions, including the effect of internal circulation. Here we adopt a simple and direct approach by normalizing eqn (7.8). This reveals that the time scale for the bulk of the subcooling to be removed is $t^e = 0\ (d^2/4\kappa_c)$. Thus the designer of a spray condenser, for example, is also presented with a much longer time and travel scale during which the subcooling in the drop has a diminishing effect on the growth of the condensate shell; we are then in the slow growth régime. With a 1 mm water drop travelling at $1\ \mathrm{m\,s^{-1}}$, the rapid growth result, relation (7.15), indicates that $t^e \simeq 0.14$ s or the travel $X \simeq 14$ cm; the slow growth result, on the other hand, yields $t^e \simeq 1.7$ s and $X \simeq 1.7$ m.

While these estimates are an order of magnitude apart, their difference is not much greater than the uncertainty introduced through other design features such the characteristics of the spray, the movement of the vapour bulk, and the presence of an incondensable gas. Further details of flow conditions are required for more accurate predictions. With larger drops, for example, the conduction models used above may overestimate the time and length scales. When the convective effect of drag-induced circulation within the drop is included the loss of subcooling is more rapid. It has also been observed that the flight of some larger drops is accompanied by oscillations in their shape. These evidently induce internal mixing resembling turbulence and are capable of increasing the radial heat transfer rate substantially. The effect of mixing and circumferential convection on condensate growth is suggested by the dotted curve sketched on Fig. 7.3; the corresponding length and time scales may be reduced significantly.

7.2.2 Condensation on jets

Turning now to bipartitioned systems, let us consider direct condensation of a warm, pure, saturated vapour ($T_V = T_I$) on the smooth surface of a steady, cylindrical jet of liquid coolant. This is another Jakob number limited problem governed by heat transfer within the subcooled jet. For a vertical jet of constant diameter d, outlet temperature T_0, and velocity U_0, the situation is described by Fig. 7.4 if the spreading coefficient is positive. The figure suggests the film of condensate growing progressively on the jet surface beneath which there exists a thermal boundary layer of thickness δ^T. In the X, R coordinate system shown, the problem is steady. From the point of view of an observer travelling with the jet, however, the situation is similar to that treated in the previous subsection; instead of a moving sphere, a moving disc of coolant is gradually heated as condensate collects on its rim.

Using previous arguments, we may neglect the thermal resistance of the condensate film and assume that the jet surface temperature $T_s(t)$ is essentially fixed at the interface temperature T_I; vapour superheat may be neglected providing $Ja_V \ll 1$. Similarly, if it is assumed that viscous drag from the surrounding vapour has a negligible effect on the jet velocity profile, which retains a block shape, the problem again reduces to radial

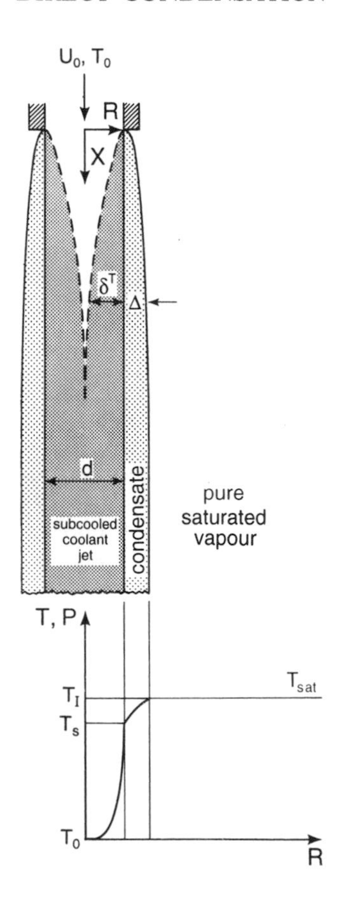

Fig. 7.4 Condensation of vapour on a circular liquid coolant jet of diameter d.

heat conduction within the liquid coolant. Once more we divide our enquiry into two time periods: an early period of rapid heat transfer rate when $\delta^{\mathrm{T}} \ll d/2$; and a longer time period during which $\delta^{\mathrm{T}} = O(d/2)$ and the heat transfer rate is lower.

For the rapid growth period, the solution is again given by eqn (7.10) and hence the heat flux density is provided by eqn (7.11). The average coolant heat transfer coefficient over a jet length $X = U_0 t$ is therefore given by

$$\bar{h}_{\mathrm{c}} = \frac{2k_{\mathrm{c}}}{\pi^{1/2}}\left(\frac{U_0}{\kappa_{\mathrm{c}} X}\right)^{1/2}, \tag{7.17}$$

and the average, diameter-based Nusselt number by

$$\bar{N}u_{\mathrm{d}} = \frac{\bar{h}_{\mathrm{c}} d}{k_{\mathrm{c}}} = \frac{2}{\pi^{1/2}} Gz_{\mathrm{d}}^{1/2}, \tag{7.18}$$

where the Graetz number $Gz_{\mathrm{d}} = U_0 d^2/\kappa_{\mathrm{c}} X$.

The rate of condensate growth follows automatically from the interface equation (7.7) which, since $d\Delta/dt = U_0(d\Delta/dX)$, integrates to yield

$$\frac{\Delta(X)}{d} = \frac{2}{\pi^{1/2}} Ja_c \frac{\rho_c}{\rho_L} \Big/ Gz_d^{1/2}. \tag{7.19}$$

Comparing this with the final thickness of the condensate film given by

$$\frac{\Delta_f}{d} = \frac{Ja_c \rho_c}{4\rho_L} \tag{7.20}$$

we find that

$$\frac{\Delta(X)}{\Delta_f} = \frac{8}{\pi^{1/2}} \Big/ Gz_d^{1/2}. \tag{7.21}$$

This is not unlike the rapid growth result seen in eqn (7.16). It is plotted in Fig. 7.5 along with a sketch of the complete solution shown dashed. Equation (7.21) indicates an equilibrium length given by

$$X^e = \frac{\pi U_0 d^2}{64\kappa_c}, \tag{7.22}$$

whereas the slow growth estimate is $X^e = U_0 t^e = O(U_0 d^2/4\kappa_c)$. For a 1 mm water jet issuing at 1 m/s^{-1}, the equilibrium estimates are $X^e = 0.3$ m and 1.7 m for rapid and slow growth, respectively. More accurate predictions again require further details of flow conditions.

For larger diameter jets, the above estimates are conservative. Drag induced convection will enhance boundary layer growth and condensate growth, as will turbulence. The dotted curve in Fig. 7.5 suggests the resulting growth pattern of the condensate. Capillary waves will also improve the heat transfer rate, and may ultimately lead to the formation of drops. In a high-speed jet, small drops may be torn from the jet surface but upstream of this the heat transfer rate may be estimated from the empirical relation

$$Nu_d = 3.2\, Re_d^{0.8} Pr^{0.3} Fr^{0.18}/Su^{0.19}(L/d)^{0.57}, \tag{7.23}$$

where $Fr = U_0^2/dg$ is a Froude number, and $Su = \sigma d/\rho\nu^2$ is the Suratman number (Kim and Mills 1989). At low speeds, the entire jet may break into drops of about the same diameter as the jet itself. The presence of an incondensable gas, even in small amounts, significantly reduces performance (Jacobs and Nadig 1987).

7.2.3 Other condensing systems

While the drop and the cylindrical jet constitute two common and practical configurations, there are several others worth mentioning briefly. Some of these are variations on the jet theme. The plane jet, for example, is the two-dimensional equivalent of the cylindrical jet. The fan jet and cone jet, on

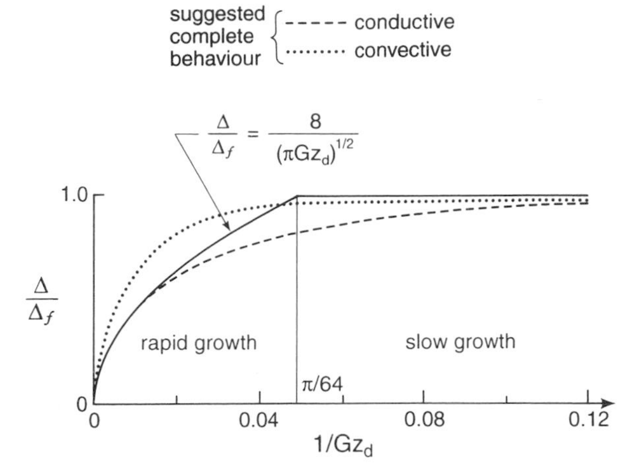

Fig. 7.5 Saturated growth characteristics of condensate on a subcooled cylindrical jet.

the other hand, are attempts to reduce the jet thickness as the distance from the orifice increases; the length and time scales for deep thermal penetration are thus reduced. The fan jet is simply a planar jet in which the discharge from the orifice fans out more or less radially within a well-defined sector; an increasing span thus creates a diminishing thickness. Heat transfer comparisons are given in Fig. 7.6 using the same general conditions discussed earlier. The figure reveals that the equilibrium Graetz number $Gz_d^e = 64/\pi$ estimated from eqn (7.22) adequately marks the overall loss of subcooling.

Sometimes the vapour itself forms the jet, as in the steam heating of water. These jets are invariably turbulent and, as noted in Chapter 3 and elsewhere, are subject to interfacial instabilities. Surface wave growth may create an internal spray of drops, further enhancing condensation. At low velocities the jet will break up into a chain of bubbles. Bubble collapse is yet another example of direct condensation, one in which the liquid surrounds the vapour. This technique is exemplified in the bubble condenser. When the bubbles are large, they first accelerate above the discharge nozzle and then decelerate as they shrink. Transient circumferential convection at the bubble surface is then important. Here we limit our enquiry to smaller bubbles for which the effect of translation is less important. A description of the situation is shown in Fig. 7.7.

Collapse of a stagnant vapour bubble in a subcooled liquid induces a radially-inward velocity V and thus creates radial convection. When this velocity is very high the collapse is controlled by inertial forces. For lower velocities, the energy and interface equations together are sufficient to describe heat transfer during which a point symmetric thermal boundary

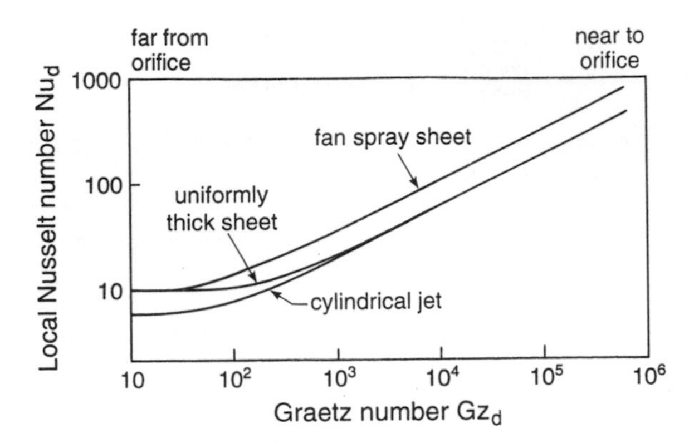

Fig. 7.6 Nusselt number–Graetz number relation for conduction heat transfer in jet and sheets (after Hasson *et al.* 1964).

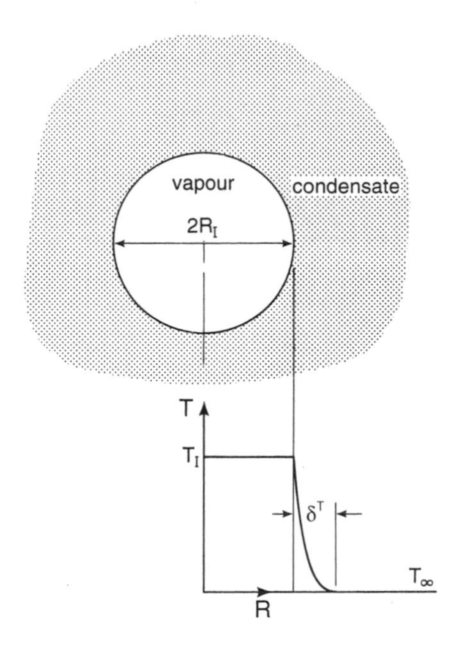

Fig. 7.7 Collapse model for a pure vapour bubble in its own condensate.

layer of thickness δ^{T} surrounds the bubble. Under these circumstances, the energy equation in the liquid may be stated as

$$\frac{\partial\theta}{\partial t}+V\frac{\partial\theta}{\partial R}=\frac{\kappa_{\mathrm{L}}}{R^{2}}\frac{\partial}{\partial R}\left(R^{2}\frac{\partial\theta}{\partial R}\right), \tag{7.24}$$

where $\theta = T - T_{\infty}$, while the interface equation reduces to

$$\rho_V \lambda_{LV} \frac{dR_I}{dt} = k_L \left(\frac{\partial \theta}{\partial R} \right)_I.$$

The subcooling $\theta^c = T_I - T_\infty$ therefore represents the temperature scale. Normalizing the interface equation yields

$$\frac{\rho_V \lambda_{LV} R^c d_i}{2 k_L t^c \theta^c} = O(1) \tag{7.25}$$

by taking $R_I = O(d_i/2)$ where d_i is the initial diameter of the bubble: $R^c = \delta^T$. Similarly, the energy eqn (7.24) yields

$$\frac{V^c R^c}{\kappa_L} = \frac{(R^c)^2}{\kappa_L t^c} = O(1) \tag{7.26}$$

provided radial convection is important in these transient conditions.

The three unknown scales V^c, R^c, and t^c may now be determined from eqns (7.25) and (7.26). In particular, we find that

$$R^c = \delta^T = \frac{d_i}{2} \left(\frac{\rho_V}{\rho_L Ja_L} \right) \tag{7.27}$$

and

$$t^c = t_f = \frac{d_i^2}{4 \kappa_L} \left(\frac{\rho_V}{\rho_L Ja_L} \right)^2. \tag{7.28}$$

The first of these reveals that the thermal boundary layer thickness δ^T is proportional to the bubble diameter. Typically, $\rho_L Ja_L / \rho_V \gg 1$, except near the critical point, and hence $\delta^T \ll d_i$; in Fig. 7.7, δ^T is shown enlarged for clarity. In a water–steam system, for example, $\rho_L Ja_L / \rho_V \simeq 3\theta^c$ near atmospheric pressure. The collapse time varies as the square of the bubble diameter according to eqn (7.28). Again considering water and steam, $t_f = O(d_i/\theta^c)^2 \times 10^5$ s. A 1 mm diameter bubble in water subcooled by 10 K collapses completely in about 1 ms. In the final stages of collapse, however, bubble behaviour is more complex: inertial forces come into play; residual incondensable gases become important; and, ultimately, the continuum model breaks down. Even so, the above description provides us with a useful overall estimate of the bubble collapse rate when the relative velocity of the bubble and surrounding liquid is small. Problems Q7.9–7.12 in Chapter 8 apply the ideas to a bubble tray condenser.

7.3 Direct evaporation

Just as condensation produces a heating effect, evaporation creates a cooling effect. In some industrial applications, such as cooling towers, this effect is sought explicitly but other applications have different, though related, purposes. In ocean thermal energy conversion (OTEC), for example, the main aim is to generate an energetic vapour for power production, while in a solar still it is the process of distillation which is facilitated. As with condensation, direct evaporation processes are found in both bipartitioned and dispersed systems.

In some situations, the amount of latent heat transferred during direct evaporation is limited to the excess sensible heat of the liquid; this poses a Jakob number limited problem of cool evaporation, as noted earlier. In others, the limit is imposed only by the mass of liquid being evaporated. Our enquiry thus divides naturally into problems of two types, each characterized by the heat transfer process controlling the evaporation. For the Jakob number limited problem, attention is directed to heat transfer within the liquid; otherwise, attention is focused on heat transfer in the vapour.

7.3.1 Evaporation from a drop surrounded by vapour

It has been noted previously that a spray of subcooled liquid drops can provide a very efficient cooling source if the drops are made to evaporate. Broadly speaking, spray cooling may be divided into two categories: surface cooling and bulk cooling. The former was treated in Section 6.1. Here we consider the spray cooling of a hot gas, i.e. hot evaporation. We will concentrate on evaporation from a single drop but in contrast to the pure vapour considered previously, we will analyse the more general situation in which the vapour is mixed with an incondensable gas. The problem is described in Fig. 7.8 in which it is assumed that the gas temperature T_∞ greatly exceeds the initial drop temperature T_0; the interface temperature T_I is unknown. Heat transferred from the gas is balanced by evaporation at the drop surface together with conduction into the drop. The interface equation is thus written

$$\rho_\mathrm{L}\lambda_\mathrm{LV}\frac{\mathrm{d}R_\mathrm{I}}{\mathrm{d}t} = -k_\mathrm{m}\left(\frac{\partial T_\mathrm{m}}{\partial R}\right)_\mathrm{I} + k_\mathrm{L}\left(\frac{\partial T_\mathrm{L}}{\partial R}\right)_\mathrm{I}, \tag{7.29}$$

where the subscript m refers to the vapour–gas mixture.

This general description also fits cold condensation on a drop surrounded by *supersaturated* vapour, except that the vapour temperature and pressure profiles are inverted. Here we are interested in evaporation in the presence of *superheated* vapour. Figure 7.9 describes the vapour states for both situations. Drop growth under supersaturated conditions is caused by heat loss between the temperatures T_I and T'_∞; sensible heat in the liquid drop is usually negligible. Similarly, drop shrinkage under superheated conditions is attributable to the temperature differenc $T_\infty - T_\mathrm{I}$, while liquid sensible heat again plays only a minor role. Cold condensation is explored in problems Q7.5–7.8 in Chapter 8.

Neglecting liquid sensible heat, eqn (7.29) reduces to

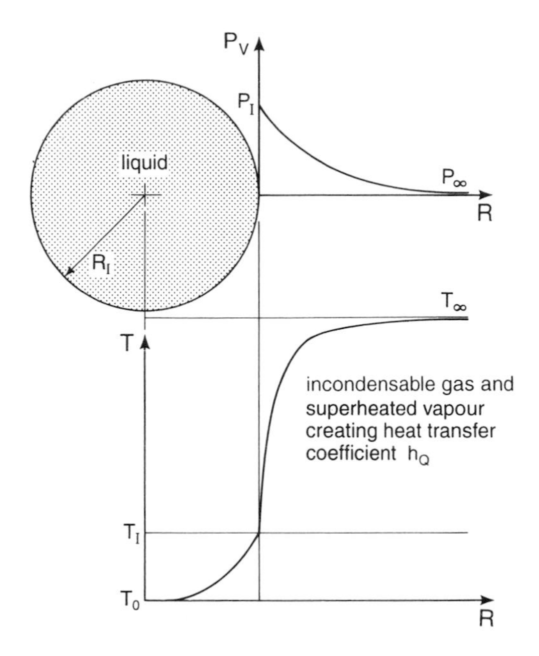

Fig. 7.8 Evaporation from a spherical drop, showing temperature and vapour pressure profiles.

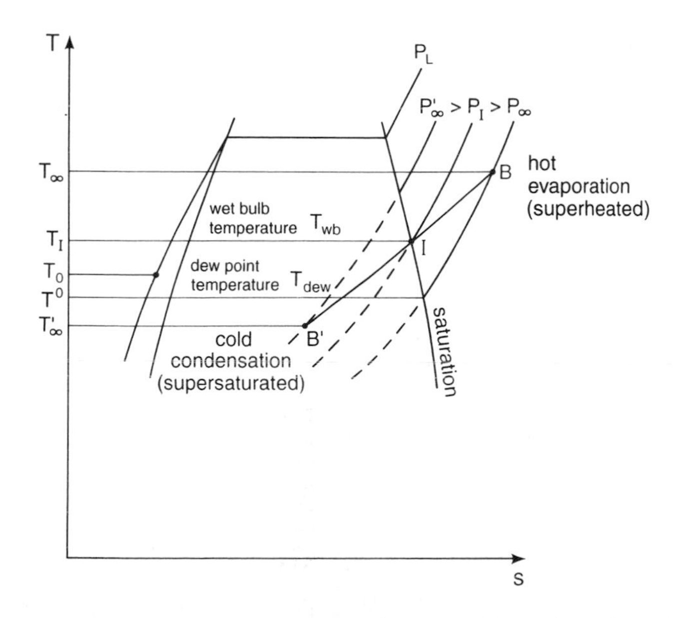

Fig. 7.9 Vapour states during hot evaporation (or cold condensation) showing the wet bulb and dew point temperatures.

$$\rho_L \lambda_{LV} \frac{dR_I}{dt} = -k_m \left(\frac{\partial T_m}{\partial R} \right)_I = -h_Q(T_\infty - T_I). \tag{7.30}$$

After normalizing, we obtain

$$\left[\frac{\rho_L \lambda_{LV} d_i}{2h_Q(T_\infty - T_I)t^c} \right] \frac{dr_I}{d\tau} = -1,$$

thus permitting the time scale to be written

$$t^c = \frac{\rho_L \lambda_{LV} d_i}{2h_Q(T_\infty - T_I)} \tag{7.31}$$

in which d_i is the initial drop diameter. Similarly, the interfacial mass balance is given by

$$\rho_L \frac{dR_I}{dt} = -h_M(m_I - m_\infty) \tag{7.32}$$

which, after normalization, gives

$$t^c = \frac{\rho_L d_i}{2h_M(m_I - m_\infty)}. \tag{7.33}$$

Since these two time scales are identical, eqns (7.31) and (7.33) may be combined to yield

$$m_I - m_\infty = \frac{h_Q}{h_M} \frac{Ja_m}{c_{pm}}. \tag{7.34}$$

We now follow the ideas advanced in Section 5.4.3. To begin with, eqn (5.61) enables us to write

$$\frac{h_Q}{h_M} = \frac{h_Q B}{h_M^0 \ln(1 + B)}$$

where $B = (m_I - m_\infty)/(1 - m_I)$, and hence

$$\frac{h_Q}{h_M} = \frac{k_m B}{\rho_m \mathfrak{D} \ln(1 + B)}, \tag{7.35}$$

using the analogy between heat and mass transfer. Equation (7.34) may therefore be written

$$Ja_m = (1 - m_I) \ln(1 + B), \tag{7.36}$$

if the vapour-gas Lewis number $Le_m = \kappa_m / \mathfrak{D}$ assumes the representative value of 1. For low vapour concentrations, $m_V \ll 1$ and $B \ll 1$, in which case eqn (7.36) reduces to $Ja_m \simeq B$, and the temperature depression is given by

$$T_\infty - T_\mathrm{I} \simeq \frac{\lambda_{\mathrm{LV}}}{c_{\mathrm{pm}}} \;(m_\mathrm{I} - m_\infty). \tag{7.37}$$

Equations (7.36) and (7.37) provide explicit relations between the two unknowns m_I and T_I. The second of these is particularly important because it is the *wet bulb temperature* T_{wb} commonly used in psychrometry. It is the temperature at any adiabatic interface where evaporation (or condensation) is balanced by sensible heat transfer in the vapour.

To solve eqn (7.31) or eqn (7.33) it is first necessary to incorporate another relation between m_I and T_I. This is provided by the Clausius–Clapeyron equation integrated from the reference valaues of P_∞ and T^0, the saturation temperature corresponding to a vapour pressure P_∞. The latter is often called the *dew point temperature* T_{dew} and may be found from steam tables, for example. After integration, we obtain the vapour molar fraction at the interface as

$$x_\mathrm{I} = \frac{P_\mathrm{I}}{P_\mathrm{m}} = \frac{P_\infty}{P_\mathrm{m}} \exp \lambda_{\mathrm{LV}} \left(\frac{T_\mathrm{I} - T^0}{R_\mathrm{V} T_\mathrm{I} T^0} \right). \tag{7.38}$$

Bearing in mind that $m_\mathrm{I} = x_\mathrm{I} R_\mathrm{m}/R_\mathrm{V}$, this enables us to solve eqn (7.36) iteratively: guessing T_I to determine x_I, and hence determining R_m, m_I, and B. Equation (7.36) or (7.37) closes the loop. Problem Q7.13 in Chapter 8 provides a numerical example using the results of Q7.7 and 7.8.

The drop evaporation time may now be estimated for prescribed conditions in the vapour: specifically, a given mixture pressure P_m, bulk vapour temperature T_∞, bulk vapour pressure P_∞ (and saturation temperature T^0), and heat (or mass) transfer coefficient. Using eqn (7.31), for example, the heat transfer coefficient may be substituted from a general relation of the form

$$Nu_\mathrm{d} = 2 + f(Pr, Re). \tag{7.39}$$

For drops with sufficient inertia, and laminar flow in the adjacent vapour, the second term $f(Pr, Re) = O(Pe_\mathrm{i}^{1/2})$ dominates, and hence

$$t^\mathrm{c} = O\left[\frac{d_i^2}{2\kappa_\mathrm{m}} \left(\frac{\rho_\mathrm{L}}{\rho_\mathrm{m} Ja_\mathrm{m}} \right) \Big/ Pe_i^{1/2} \right], \tag{7.40}$$

where $Pe_\mathrm{i} = U_\infty d_\mathrm{i}/\kappa_\mathrm{m}$ is the initial Peclet number. For smaller drops, especially those which are fast approaching the vanishing point, the first (conduction) term in eqn (7.39) dominates, in which case

$$t^\mathrm{c}_{\mathrm{cond}} = O\left[\frac{d_i^2}{4\kappa_\mathrm{m}} \left(\frac{\rho_\mathrm{L}}{\rho_\mathrm{m} Ja_\mathrm{m}} \right) \right]. \tag{7.41}$$

By comparison,

$$t^\mathrm{c}_{\mathrm{cond}}/\mathrm{t}^\mathrm{c} = O(Pe_i^{1/2}), \tag{7.42}$$

thus demonstrating that the spray characteristics U_∞ and d must be carefully specified if the life (or travel) of the drop, and its cooling capacity,

$$\dot{Q}_d = \pi d_i^3 \rho_L \lambda_{LV}/6t^c$$

are to be estimated accurately.

7.3.2 *Evaporation from a jet surrounded by vapour*

Conditions for evaporation from a cylindrical jet are described in Fig. 7.10. For a pure vapour, this differs from Fig. 7.4, describing condensation on a jet, in only two significant ways. Firstly, the temperature and density profiles are inverted; secondly, the condensate film is replaced by a region of liquid depletion. In both situations, the magnitude of $|T_0 - T_I|$ determines the maximum amount of latent heat which may be transferred; both systems are Jakob number limited. The Graetz problem thus defined has the solution given by eqn (7.18). Similarly, the *reduction* of jet width caused by evaporation is given by eqn (7.21), derived for condensate *growth*. The

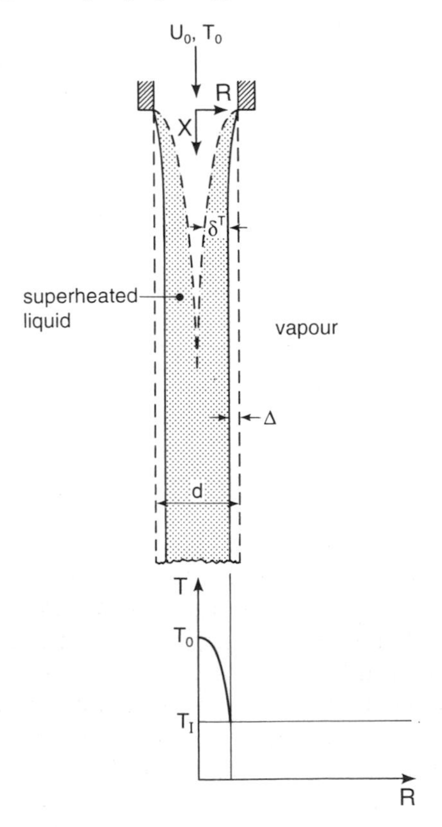

Fig. 7.10 Evaporation from a cylindrical jet of initial diameter d.

magnitude of $Gz_d = O(64/\pi)$ indicated in Fig. 7.5 again separates the rapid growth régime from the slower growth régime. If the jet is turbulent, mixing within the liquid enhances the rate of heat flow towards the interface and thus increases the evaporation rate. The equilibration length will then be shortened but at the cost of increased pumping power. The hydrodynamic problem is very similar to that discussed in Section 7.2.2, and the heat transfer rate for a coherent jet is again provided by eqn (7.23). In general, the effects of surface tension and viscosity tend to subdue turbulent fluctuations and thus re-create laminar flow. Evaporation reduces the jet width and enhances this tendency. At low velocities, the jet may disintegrate into drops, as noted previously.

Vapour and liquid states in the presence of an incondensable gas are shown in Fig. 7.11 which indicates that evaporation depends principally on the liquid sensible heat excess $c_{pL}(T_0 - T_I)$ which consists of two parts: $c_{pL}(T_0 - T_{LS})$ attributable to the initial superheat, if any, above the liquid saturation temperature T_{LS}; and $c_{pL}(T_{LS} - T_I)$ attributable to the subcooling which results from the incondensable gas lowering the vapour interfacial pressure beneath the total pressure P_L. For given overall temperature limits, T_0(at A) and T_∞(at B), it is possible to shift the division between superheat and subcooling by reducing the jet (and gas) pressure while maintaining the vapour pressure at a fixed value. The result is shown in Fig. 7.12. Lowering P_L to P_{L1} shifts the curve AC to A_1C_1 which has no effect on the overall sensible heat but increases the fraction attributable to superheat. In the limit, as $P_{L1} \rightarrow P_I$, all of the sensible heat becomes superheat. In certain circumstances, such increases in superheat may be sufficient to promote nucleation around any particles or gas bubbles entrained in the liquid. The resultant boiling action further facilitates the evaporation process but may be sufficiently disruptive to shatter the jet in a process not unlike that occurring during quenching (see Section 6.3.3). Behaviour is then very different from that in a stable, cohesive jet and heat transfer rates are increased substantially.

7.3.3 *Evaporation from a drop surrounded by liquid*

In Section 7.3.1 we considered evaporation from a dispersed system of liquid drops surrounded by a hot vapour. Here we extend the discussion to include hot liquid surroundings having a different composition. The immediate problem that arises in this type of direct evaporation is that the vapour released cannot diffuse very far from the drop. When the vapour is soluble in the surrounding liquid, a certain amount of diffusion may take place but the speed of this diffusion, as indicated by the reciprocal of the Lewis number $\kappa/\mathfrak{D}$, is usually not very great. To a first approximation, the loss of vapour mass into the surrounding liquid may be ignored. Even so, internal evaporation may create substantial alterations in the geometry and hydrodynamics of the drop.

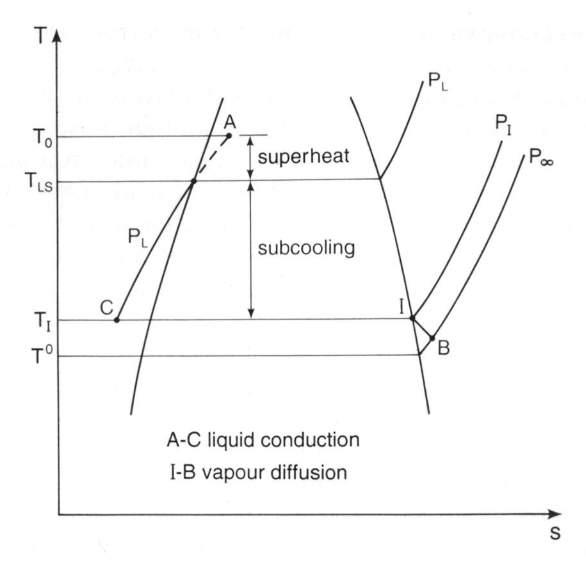

Fig. 7.11 Vapour and liquid states during cool evaporation from a warm liquid into a cooler vapour–gas mixture.

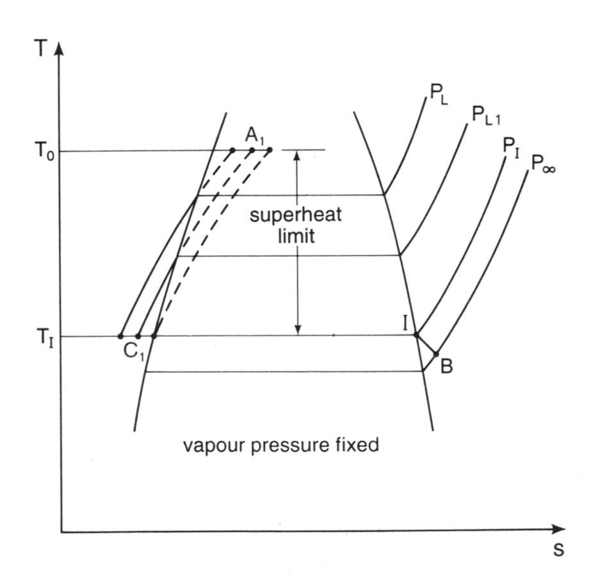

Fig. 7.12 Effect of lowering liquid and gas pressure on cool evaporation.

Experimental observations of a buoyant drop surrounded by a different liquid with a temperature T_∞ above the drop saturation temperature indicate that nucleation begins at, or near, the forward stagnation point where the heat transfer rate is likely to be greatest. Vapour released and collected in this region is bound to the drop by surface tension forces and thus adds to the buoyancy force. Figure 7.13 suggests the evolution of the resulting dual sphere composite. For simplicity, the bubble-drop interface A_I is shown planar but will generally be curved, depending on the three surface tensions acting on the common ring interface and on any drag-induced circulation.

The low thermal conductivity of the vapour greatly reduces heat transfer at the front of the composite leaving the vapour close to the saturation temperature T_I. Heat transfer into the drop is thus largely restricted to its common interface A_S (@ T_S) with the surrounding liquid and is strongly influenced by wake behaviour. This heat is then conducted towards the bubble-drop interface A_I where evaporation continues to occur. The dual sphere model enables estimates to be made of both A_S and A_I at any stage of evaporation.

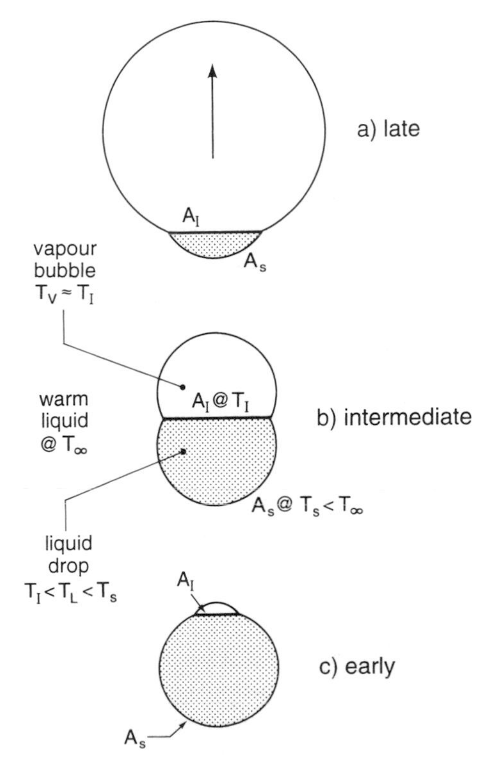

Fig. 7.13 Schematic representation of evaporation from a buoyant liquid drop rising in a different liquid at temperature T_∞.

In practice, this simple model must be modified. It will be recalled from Chapter 3 that if a vapour bubble expands as it accelerates upward the geometry of its surface may change substantially. Since $\rho_L \gg \rho_V$, a small spherical drop may balloon into a bubble which is skirted or capped, even assuming quasi-steady conditions. If the 'balloon' is large enough it may break into two; its behaviour will also be complicated by the role of the unevaporated liquid 'undercarriage'. In addition, large numbers of bubbles many swarm or coalesce.

7.4 Direct crystallization

The formation of crystalline solids without the intervention of foreign surfaces was mentioned briefly in Chapters 2 and 4. It will be recalled that the absence of any substrate both reduces the opportunity for heterogeneous nucleation and limits subsequent growth to an amount set by supercooling or supersaturation. This has profound implications for the growth of crystals in both industrial and environmental settings. In this section, it is the latter which will receive the greater emphasis.

7.4.1 Growth from the vapour phase

Spontaneous nucleation throughout a vapour bulk invariably calls for supersaturated conditions and, in the production of crystals, for temperatures below the triple point value. The presence of foreign particles or molecules, however, may reduce the degree of supersaturation necessary for the formation of critical clusters. Dispersed particles in the atmosphere, for example, are often as large as the critical ice cluster radius R^* and may thus promote heterogeneous nucleation if their molecular characteristics are suitable: that is, if hydrogen bonding at their surface is similar to that of the H_2O molecule, and their crystallographic structure is compatible with the structure of ice. While the particles found in the atmosphere are numerous, e.g. 10^{11} per cubic metre, only a small number are efficient nucleation sites, e.g. 10^3 per cubic metre; the latter value therefore determines the number of ice crystals which may appear.

Natural ice crystals have the hexagonal prismatic shape illustrated in Fig. 7.14. This reflects the geometry of the H_2O molecule and its tendency to create tetrahedral lattice structures. The result is a crystal with faces of two types: the prism face, lying normal to the *a* axis, is rectangular, while the basal face, lying normal to the *c* axis, is hexagonal. A striking feature of these faces is that they propagate outwards at speeds which depend upon the prevailing temperature. The relationship between growth velocity and temperature is not monotonic, however, and the relationship is not the same for the prism and basal faces. For certain temperature ranges, the basal velocity is higher and the crystal tends to elongate into a needle form; for other temperature ranges, the prism face velocity is higher thereby creating

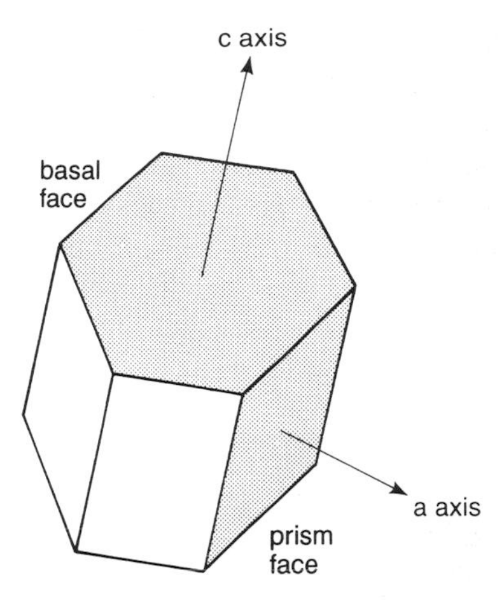

Fig. 7.14 A hexagonal prism of ice showing basal and prism faces.

a hexagonal plate form. Examples of these fundamental crystal forms are shown in Fig. 7.15. The degree of supersaturation also has an effect on crystal shape, principally by altering growth near the edges and corners. A needle may thus be converted into a hollow prismatic column, while a hexagonal plate may develop a branched dendritic structure.

The growth of an atmospheric ice crystal may also be enhanced by accretion. Falling through a cloud, the crystal collects drops and smaller crystals in its path and may eventually become heavily rimed while still retaining the outline of its original shape. If the crystals are needle-like, air turbulence may cause them to fracture upon collision. An efficient *secondary nucleation* process is thus established. On the other hand, if they are plate-like, and sufficiently numerous, their mutual proximity may lead to aggregation: partly through sintering after a collision, and partly through the intertwining of their dendritic extensions. The resulting structure is a snowflake.

In other circumstances, crystals falling through a dense cloud of supercooled drops will grow extensively through accretion. Their crystalline geometric identity is then lost as they tumble and rotate, creating a roughly spherical (though occasionally conical) lump of ice. Hailstone growth, for example, is a special case of accretion on a quasi-spherical substrate; namely, ice itself. Typically, the hailstone temperature lies close to the equilibrium freezing temperature T_I; the hailstone surface is therefore an adiabatic. During the fall, the ambient temperature and water content may both vary, sometimes exposing the hailstone to periodic surface conditions. Examination of hailstones frequently reveals a multilayered structure reflecting the complex history of their descent.

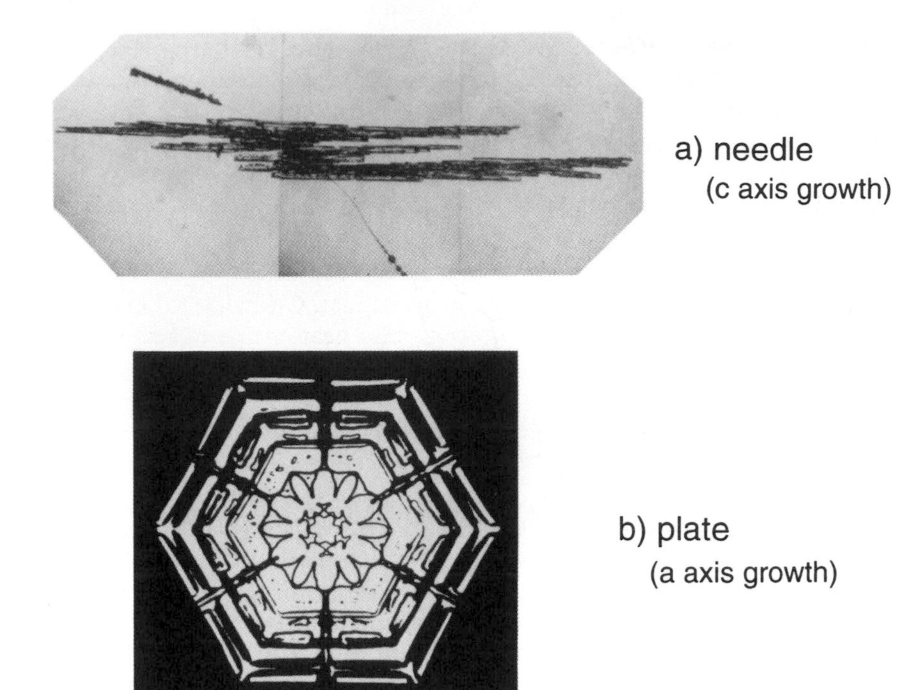

Fig. 7.15 Fundamental ice crystal forms (after Bentley and Humphreys, 1962).

7.4.2 Growth from the liquid phase

Hailstone growth is typically a wet process in which the impact of collected drops establishes a thin film of supercooled water. Ice in the body of the hailstone then grows dendritically into this supercooled film. Under these low Stefan number conditions, some of the water may remain unfrozen in the interdendritic spaces near the hailstone surface. Heat is not removed fast enough to freeze all of this water and the result is a porous structure often described as spongy ice. Dendritic growth also occurs at the surface of the polar and subpolar seas where it has a valuable end result: namely, desalination.

In an industrial setting, freeze desalination often consists of the production of distributed ice crystals in a supercooled aqueous solution. The initiation of these crystals may occur on suspended particulates or as the result of vigorous mixing near the liquid free surface if it is highly supercooled. We thus generate the frazil crystals described in Section 4.4.1. To increase the rate of ice production, new particles may be created through mechanical action: for example, with a propeller. This causes the initial population of ice crystals to multiply through collision-induced fracture which is especially effective when the frazil discoids have developed dendritic arms that break easily. This *secondary nucleation* process is facilitated by turbulence, which also enhances the heat and mass transfer rates at the crystal surfaces and thus increases the ice growth and salt rejection rates. The same general technique is used in other industrial crystallizers, although secondary nucleation may be explicitly avoided if uniformity of the final crystals is paramount.

In desalination, ice is the primary product but there are other direct contact applications where it is merely a by-product. Consider, for example, the production of mechanical power in a polar heat engine which extracts heat from water beneath an ice cover and discharges it into colder air above the ice cover. This is not unlike the OTEC system mentioned earlier. One way of extracting heat from the water is to bubble drops of a suitable engine working fluid through it, e.g. a hydrocarbon injected near its saturated liquid condition. The water bulk then acts as an evaporator in which a drop of working fluid gradually vaporizes as it travels upward under the influence of a buoyancy force. This situation was discussed earlier in Section 7.3.3. The essential difference here arises from the fact that the water is usually close to its equilibrium freezing temperature; heat removed from the water is largely latent heat, thus requiring the production of ice. We therefore have a three-phase system: solid, liquid, and vapour.

It appears that no study of this system has yet been undertaken but Fig. 7.16 suggests a model based on the earlier discussion. Relative to the dual sphere composite, water flowing downwards would be supercooled by the cooler working liquid in the drop; $T_L < T_w$. Depending upon the drop Reynolds number, this cooling effect would be distributed over the flanks

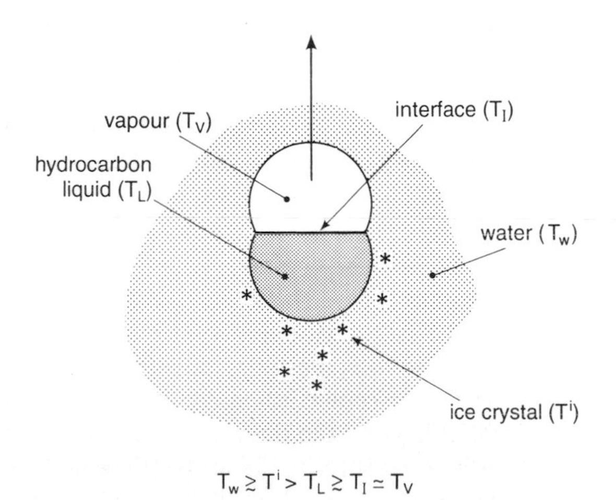

Fig. 7.16 Ice crystal formation in cold water near a colder hydrocarbon drop rising and evaporating.

of the drop and into the wake where heat transfer is most effective. Nucleation may be delayed initially but as the supercooling of the water adjacent to the drop grows with time and elevation, nucleation is most likely to occur in the coolest regions, as suggested. The ice crystals are then left to grow behind the passing drop and its successors.

7.4.3 Re-crystallization

When a polycrystalline solid is deformed e.g. during manufacturing, internal grains and grain boundaries may undergo severe distortion. The original crystals are then forced to change shape and are thus heavily strained. This is well illustrated by the cold working of metallic components. Raising the temperature of components thus worked helps relieve the internal strain in the process known as *annealing*. This is a re-crystallization process in which new grains appear at crystal discontinuities and grow at the expense of the old grains. The molecules thus re-distribute themselves into a new granular structure. Re-crystallization is a subtle, but complex, problem in latent heat transfer and illustrates the importance of molecular mobility, even in the solid phase. It also emphasizes the significance of abrupt alterations in the intermolecular force field over a crystal surface. Strain-induced discontinuities, like the geometrical discontinuities which occur in the natural atmospheric ice crystals discussed earlier, facilitate molecular capture and thus enable new grains to form and grow. The vigour of this process increases more or less monotonically with the temperature. Thermal history thus becomes a very important aspect of recrystallization in metals.

The temperature–time relationship is also important in certain re-crystallization phenomena found in porous media. When the fluid filling the pores is a concentrated liquid solution, a rapid and substantial drop in temperature may create *vitrification*, a condition in which nucleation and crystal growth cease; kinetic and diffusion rates are then very low and the 'frozen' liquid behaves like a glass. Upon warming, nucleation and crystal growth may begin again. The resulting process tends to be a continuation of previous events unless, of course, the frozen material has been deformed in the interim and the original crystals have been highly strained.

A good example of this type of re-crystallization is found in the cryogenic treatment of biological tissue where the adverse effect of ice crystals is to be minimized. If the tissue is cooled rapidly, it may be stored below the vitrification temperature for long periods, during which nucleation and crystal growth are inhibited, if not eliminated. It is the reheating process which introduces the difficulty. As the thermal wave penetrates through the tissue from the surface it raises the local temperature above the de-vitrification point long enough for nucleation, followed by new or renewed crystal growth, to occur. The tissue is then vulnerable to the damaging effects of ice. In small bodies of tissue the warming may proceed rapidly and more or less uniformly, but temperature gradients in cryopreserved human organs, for example, may allow crystal growth to become lethal.

Another important example of re-crystallization in porous media is observed when the pores are filled with water vapour, not water, and the matrix of the medium is ice itself. In other words, we are dealing with snow of one type or another. Under steady winter conditions, a freshly-fallen snow cover tends to reach a condition where the ice matrix is in thermodynamic equilibrium with the pore vapour. However, naturally occurring temperature gradients in the snow create corresponding vapour pressure gradients and thus give rise to a vapour flux. Hence, as the air temperature falls during the early stages of winter, sublimation may occur at depth in the (warmer) snow while deposition takes place nearer the (cooler) surface. The angularity of the snow crystals again facilitates the redistribution process which has the overall effect of producing a very fragile, almost skeletal, ice structure well beneath the surface. Providing the snow cover is not too thick this process may help small mammals in their winter tunnelling practices. On the other hand, it can create a strategic weakness beneath the surface of a snow slope and thus give rise to conditions leading to a catastrophic avalanche.

7.5 Combustion of condensed phases

Combustion is the exothermic reaction in which a fuel and an oxidant form gaseous products having a high temperature. In an industrial setting, this temperature is sought explicitly for the raising of steam, the production of mechanical power, the melting of metals, etc.; the combustion process is carefully controlled. In many situations, the fuel is a liquid or a solid which must somehow change phase in the generation of hot gaseous products.

This section serves as an introduction to combustion-induced phase change processes. At the outset it should be noted that most emphasis will be placed on liquid fuels and only simple chemical reactions will be considered. Despite these restrictions, however, the nature of latent heat transfer processes during combustion will become clear, thus paving the way for more advanced studies.

7.5.1 Combustion near the surface of condensed phases

Many solids and most liquids do not actually burn. They vaporize to create a gaseous fuel which diffuses away from their surface; at the same time, a gaseous oxidant diffuses towards the surface. Under suitable circumstances the two fluxes meet and chemically recombine in an envelope usually known as the flame. The flame is seen as a thin, luminous sheath often following the contours of the solid or liquid surface being consumed. Figure 7.17 provides a schematic representation of the situation showing temperature and mass fraction profiles.

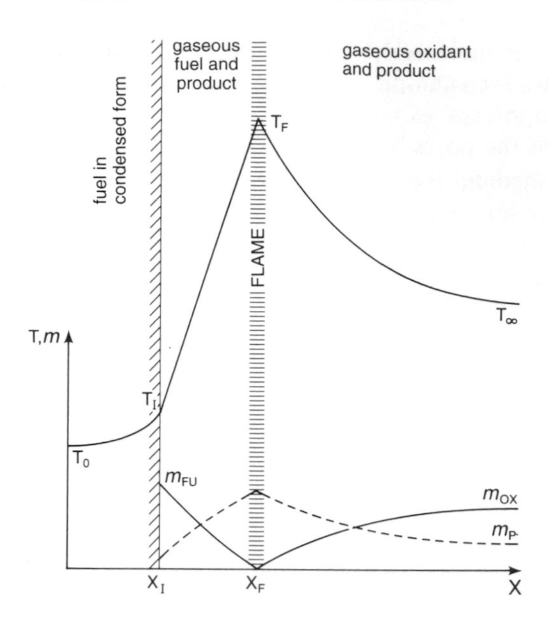

Fig. 7.17 Model of combustion near the surface of a condensed phase, showing temperature and mass fraction profiles.

Vaporization from the condensed phase surface at T_{I}, the saturation temperature of the fuel vapour, is caused by heating from the flame which also heats the surrounding gas containing the oxidant: $T_{\mathrm{F}} > T_{\infty} > T_{\mathrm{I}}$. The flame is nourished by the supply of fuel vapour, from one side, and gaseous oxidant from the other; this is suggested by the mass fraction profiles m_{FU} and m_{OX} for the fuel and oxidant, respectively. A surface of discontinuity at X_{F} represents the flame itself. As indicated, the reaction products decrease in mass fraction m_{P} on both sides of the flame where they are created.

Movement of the gaseous reactants and products, along with any diluent gases such as nitrogen, is governed mainly by the diffusion eqn (2.22). Thus, under steady state conditions, the flow of component k is given by

$$\nabla \cdot (\mathbf{G}_{\mathrm{k}}) + \nabla \cdot \mathbf{j}_{\mathrm{Mk}} = -\dot{s}_{\mathrm{k}}, \tag{7.43}$$

where $\dot{s}_{\mathrm{k}}$ is the volumetric *consumption* rate. Now $\mathbf{G}_{\mathrm{k}} = m_{\mathrm{k}}\mathbf{G}$, where $\mathbf{G}$ is the mixture mass flux density and $\mathbf{j}_{\mathrm{Mk}} = -\Gamma_{\mathrm{k}} \nabla m_{\mathrm{k}}$, where $\Gamma_{\mathrm{k}} = \rho_{\mathrm{k}} \mathfrak{D}_{\mathrm{k}}$ is an exchange coefficient. Hence eqn (7.43) may be re-written

$$\nabla \cdot (m_{\mathrm{k}}\mathbf{G}) - \nabla \cdot (\Gamma_{\mathrm{k}} \nabla m_{\mathrm{k}}) = -\dot{s}_{\mathrm{k}}. \tag{7.44}$$

For the oxidant in particular,

$$\nabla \cdot (m_{\mathrm{OX}}\mathbf{G}) - \nabla \cdot (\Gamma_{\mathrm{OX}} \nabla m_{\mathrm{OX}}) = -\dot{s}_{\mathrm{OX}}. \tag{7.45}$$

The flow of energy under the same conditions is governed by a similar equation. Using eqn (2.10), for example, we may write

$$\nabla \cdot (\mathrm{h}\mathbf{G}) - \nabla \cdot (\mathrm{k}\nabla T) = 0 \tag{7.46}$$

in which the mixture specific enthalpy $\mathrm{h} = \Sigma_{\mathrm{k}} m_{\mathrm{k}} \mathrm{h}_{\mathrm{k}}$, and

$$\mathrm{h}_{\mathrm{k}} = \mathrm{h}_{\mathrm{k}}^{0} + \mathrm{h}_{\mathrm{k}}^{\mathrm{T}}, \tag{7.47}$$

where $\mathrm{h}_{\mathrm{k}}^{0}$ is the enthalpy of formation for component k, while

$$\mathrm{h}_{\mathrm{k}}^{\mathrm{T}} = \int_{T^0}^{T} c_{\mathrm{pk}}\,\mathrm{d}T \tag{7.48}$$

represents the sensible heat at T. Equation (7.47) may thus be written

$$\nabla \cdot (\mathrm{h}^{\mathrm{T}}\mathbf{G}) - \nabla \cdot (\mathrm{k}\nabla T) = -\sum_{\mathrm{k}} \mathrm{h}_{\mathrm{k}}^{0} \nabla \cdot \mathbf{G}_{\mathrm{k}}. \tag{7.49}$$

Now the heat of combustion $\mathcal{H}$ per unit mass of fuel may be defined by

$$\mathcal{H}\dot{s}_{\mathrm{FU}} = -\sum_{\mathrm{k}} \mathrm{h}_{\mathrm{k}} \nabla \cdot \mathbf{G}_{\mathrm{k}},$$

where $\dot{s}_{\mathrm{FU}}$ is the fuel consumption rate per unit volume. Therefore, since $\mathcal{H} \gg \mathrm{h}_{\mathrm{k}}^{\mathrm{T}}$, eqn (7.49) may be approximated by

$$\nabla \cdot (\mathrm{h}^{\mathrm{T}}\mathbf{G}) - \nabla \cdot (\mathrm{k}\nabla T) = \mathcal{H}\dot{s}_{\mathrm{FU}}. \tag{7.50}$$

Once combustion begins, the fuel temperature may rise rapidly during a period of Newtonian heating (see Section (1.6.2).) In dispersed fuel, a low drop Biot number ensures a fairly uniform liquid temperature which soon reaches and remains at the wet bulb temperature, i.e. $T_{\mathrm{I}} = T_{\mathrm{wb}}$. The interface equation may then be written

$$\lambda \mathbf{G}_{\mathrm{FU}} = (\mathrm{k}\nabla T)_{\mathrm{I}}, \tag{7.51}$$

where λ is the appropriate latent heat (vaporization or sublimation) and $\mathbf{G}_{\mathrm{FU}}$ is the mass flux density of the fuel leaving the interface where it is essentially the sole gaseous component. Equations (7.50), (7.51), and the component diffusion eqns (7.44), collectively describe simple combustion processes near the surfaces of pure, condensed phases under steady, isobaric conditions.

7.5.2 Combustion of liquid fuels

Figure 7.17 may be used to model combustion above a pool of liquid fuel. Since the liquid surface is nominally horizontal, the temperature below the flame, increasing with height, tends to stabilize the vaporization process, but the adverse temperature gradient above the flame destabilizes the nearby air and thus creates a turbulent mixture of air and reaction products. Equations (7.44), (7.50), and (7.51) continue to apply providing the exchange coefficients can accommodate the effects of turbulent fluctuations. Commonly, the flame will create a chimney effect, thus giving rise to strong air currents parallel to the pool surface.

The recession of a liquid fuel surface during combustion may be offset by a wick which continually replenishes the liquid vaporized. The flow of liquid is then maintained by a suction beneath the menisci formed at the wick surface, as shown schematically in Fig. 7.18. The capillary pressure in the liquid is defined by

$$\Delta P_{\mathrm{cap}} = \frac{2\sigma}{R_{\mathrm{I}}}, \tag{7.52}$$

where R_{I} is the effective radius of curvature at the interface. Suction in the liquid is created by evaporation at the interface, thus reducing R_{I} beneath the value R_{I}^{0} observed in the absence of evaporation. For low combustion rates, as for a candle (where vaporization is preceded by melting), a laminar, three-dimensional flame may form. For the much higher rates sought in industry, the surface area of the liquid must be increased considerably. A bipartitioned system must be converted into a dispersed system, usually a spray.

Liquid fuel sprays exhibit the behaviour discussed briefly in Chapter 3. For our present purpose it is sufficient to note that the breakup of a liquid fuel jet may be facilitated in two main ways: by its shape (e.g. a fan disintegrating under a Weberian instability) or by collision with another jet (usually a gas, but sometimes a fraction of the same liquid jet previously divided). While these mechanisms do not produce a monodisperse spray, we will consider only drops of a given initial diameter. Consider the quasi-steady combustion of a single drop falling through air, as shown schematically in Fig. 7.19. In general, this poses a problem in at least two dimensions because of translation-induced convection on each side of the flame. Here we concentrate on conditions where circumferential convection may be neglected. A point symmetric model of this situation is suggested in the temperature and mass fraction profiles sketched in Fig. 7.20, which also contains an additional simplification with respect to the drop temperature distribution. It is assumed not only that the drop has a uniform wet bulb temperature $T_{\mathrm{wb}} = T_{\mathrm{I}}$, but that T_{I} has risen to the boiling point T_{B} corresponding to the given system pressure, i.e. $m_{\mathrm{FU}} \simeq 1$.

For this model of drop combustion, the energy eqn (7.50) reduces to

$$\frac{1}{R^2}\frac{\mathrm{d}}{\mathrm{d}R}(R^2 \mathrm{h}^{\mathrm{T}} G) - \frac{1}{R^2}\frac{\mathrm{d}}{\mathrm{d}R}\left(R^2 \mathrm{k}\frac{\mathrm{d}T}{\mathrm{d}R}\right) = \mathcal{H}\dot{s}_{\mathrm{FU}}$$

or, since $\mathrm{dh}^{\mathrm{T}} = c_{\mathrm{p}}\mathrm{d}T$ under isobaric conditions and $\dot{s}_{\mathrm{OX}} = r\dot{s}_{\mathrm{FU}}$, where r is the oxygen/fuel ratio,

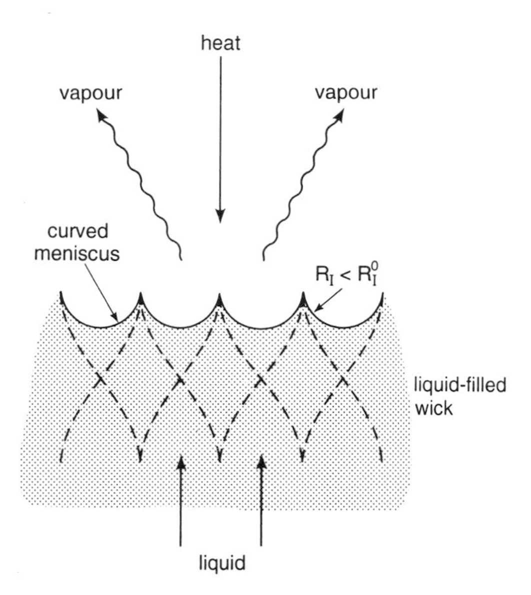

Fig. 7.18 Evaporation at a wick surface. The same model also describes 'sweat' cooling at a porous surface.

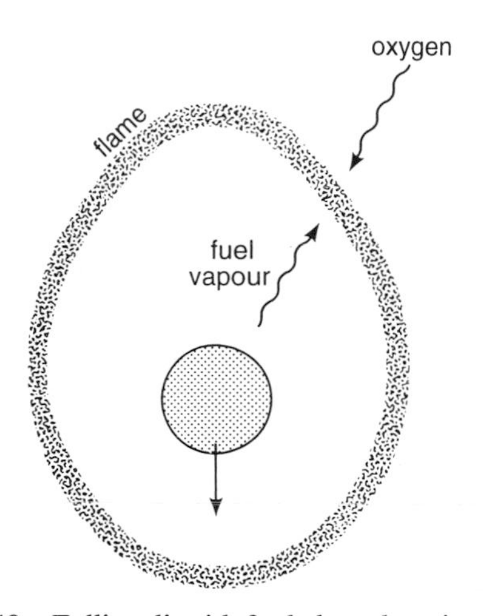

Fig. 7.19 Falling liquid fuel drop burning in air.

$$\frac{1}{R^2}\frac{\mathrm{d}}{\mathrm{d}R}(R^2\mathrm{h}^{\mathrm{T}}G) - \frac{1}{R^2}\frac{\mathrm{d}}{\mathrm{d}R}\left(R^2\Gamma\frac{\mathrm{dh}^{\mathrm{T}}}{\mathrm{d}R}\right) = \frac{\mathcal{H}\dot{s}_{\mathrm{OX}}}{r}, \tag{7.53}$$

where $\Gamma = \mathrm{k}/c_{\mathrm{p}}$. Similarly, eqn (7.45) becomes

$$\frac{1}{R^2}\frac{\mathrm{d}}{\mathrm{d}R}(R^2 m_{\mathrm{OX}}G) - \frac{1}{R^2}\frac{\mathrm{d}}{\mathrm{d}R}\left(R^2\Gamma_{\mathrm{OX}}\frac{\mathrm{d}m_{\mathrm{OX}}}{\mathrm{d}R}\right) = -\dot{s}_{\mathrm{OX}}.$$

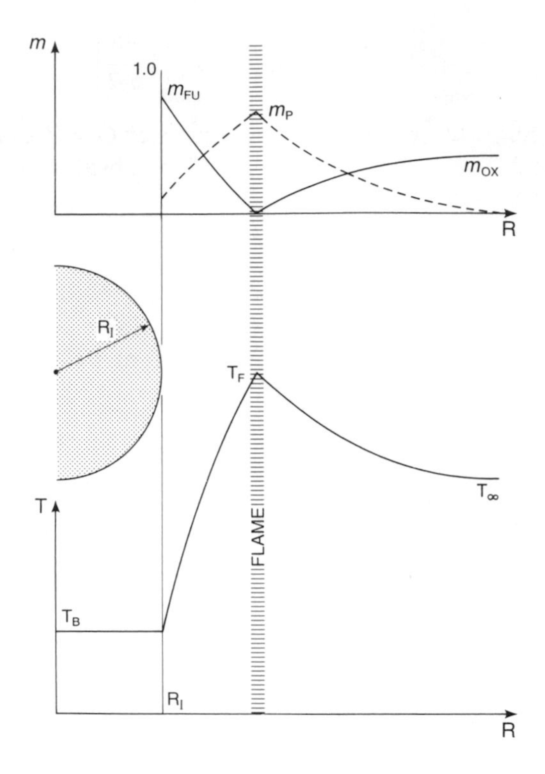

Fig. 7.20 Temperature and mass fraction profiles near a burning drop.

Multiplying this by the constant $\mathcal{H}/r$ and adding to eqn (7.53) yields

$$\frac{d}{dR} R^2 G \left(h^T + m_{OX} \frac{\mathcal{H}}{r} \right) - \frac{d}{dR} R^2 \Gamma \frac{d}{dR} \left(h^T + m_{OX} \frac{\mathcal{H}}{r} \right) = 0, \qquad (7.54)$$

providing $\Gamma/\Gamma_{OX} = k/\rho_{OX} c_p \mathcal{D}_{OX} \simeq 1$, a reasonable assumption in gases where the Lewis number $\simeq 1$. The new dependent variable $\tilde{h} = h^T + m_{OX}\mathcal{H}/r$ is seen to behave conservatively; there is no source term. Other properties are also found to be conservative; problem Q7.14 in Chapter 8 provides another example.

Equation (7.54) may be integrated as follows. After one integration we obtain

$$R^2 \left[G\tilde{h} - \Gamma \frac{d\tilde{h}}{dR} \right] = R_I^2 \left[G\tilde{h} - \Gamma \frac{d\tilde{h}}{dR} \right]_I \qquad (7.55)$$

by satisfying the condition at the interface where $R = R_I$. If we now define the excess $\theta = \tilde{h} - \tilde{h}_I$ where $\tilde{h} = h^T(T_I)$ is taken as a datum at the interface where $m_{OX} = 0$, this result may be rewritten as

$$R^2\Gamma\frac{\mathrm{d}\theta}{\mathrm{d}R} = R_{\mathrm{I}}^2 G_{\mathrm{FU}}\theta + R_{\mathrm{I}}^2\left(\Gamma\frac{\mathrm{d}\theta}{\mathrm{d}R}\right)_{\mathrm{I}}$$

after substituting the continuity requirement $R^2G = R_{\mathrm{I}}^2 G_{\mathrm{FU}}$. Integrating once more yields the *burning rate* G_{FU}. After substituting $\phi = \theta + (\Gamma\,\mathrm{d}\theta/\mathrm{d}R)_{\mathrm{I}}/G_{\mathrm{FU}}$, we obtain

$$\frac{G_{\mathrm{FU}}R_{\mathrm{I}}^2}{\Gamma R} = \ln\frac{\phi_\infty}{\phi} \tag{7.56}$$

by satisfying the boundary condition $\phi \to \phi_\infty$ as $R \to \infty$.

This result may be simplified by noting that at the drop surface

$$G_{\mathrm{FU}} = \frac{(\mathrm{k}\nabla T)_{\mathrm{I}}}{\lambda} = \left(\frac{\Gamma}{\lambda}\frac{\mathrm{d}\theta}{\mathrm{d}R}\right)_{\mathrm{I}}$$

using eqn (7.51) and the definitions of Γ, h^{T}, $\tilde{\mathrm{h}}$, and θ. Hence $(\Gamma\,\mathrm{d}\theta/\mathrm{d}R)_{\mathrm{I}}/G_{\mathrm{FU}} = \lambda$, the latent heat of vaporization of the fuel. Equation (7.56) therefore simplies to

$$\frac{G_{\mathrm{FU}}R_{\mathrm{I}}}{\Gamma} = \ln\left(1 + \frac{\theta_\infty}{\lambda}\right) \tag{7.57}$$

at the drop surface where $R = R_{\mathrm{I}}$ and $\theta = 0$. Problem Q7.16 in Chapter 8 provides an illustration of its use.

It is worth noting that

$$\theta_\infty = \tilde{\mathrm{h}}_\infty - \mathrm{h}_{\mathrm{I}}^{\mathrm{T}} = c_{\mathrm{p}}(T_\infty - T_{\mathrm{B}}) + m_{\mathrm{OX},\infty}\mathcal{H}/r,$$

and hence

$$\theta_\infty/\lambda = Ja_m + m_{\mathrm{OX},\infty}\mathcal{H}/r\lambda \tag{7.58}$$

which contains two enthalpy ratios: a mixture Jakob number and $\mathcal{H}/\lambda$. For most liquid hydrocarbons burning in air (for which $m_{\mathrm{OX},\infty} = 0.232$), the quantity $\theta_\infty/\lambda = O(10)$. The initial burning rate for a 50 μm radius drop of liquid octane burning in air ($\theta_\infty/\lambda \simeq 9.7$) is readily calculated from eqn (7.57). Assuming that $\mathrm{k} = 7.3\times10^{-2}$ W m^{-1} K^{-1} and $c_{\mathrm{p}} = 2.1\times10^3$ J kg^{-1} K^{-1}, we obtain $G_{\mathrm{FU}} = 1.65$ kg m^{-2} s^{-1}. Some fuels, e.g. diesel oil, require a substantial addition of sensible heat to bring them to the boiling point. To account for this, the latent heat may be written in the modified form $\lambda' = \lambda_{\mathrm{LV}} + c_{\mathrm{pc}}(T_{\mathrm{B}} - T_{\mathrm{R}})$, where T_{R} is the injection temperature and c_{pc} is the fuel specific heat.

To obtain the drop radius R_{I} as a function of time we note that $G_{\mathrm{FU}} = -\rho_{\mathrm{c}}\mathrm{d}R_{\mathrm{I}}/\mathrm{d}t$ and rewrite eqn (7.57) as

$$-\frac{\rho_{\mathrm{c}}R_{\mathrm{I}}}{\Gamma}\frac{\mathrm{d}R_{\mathrm{I}}}{\mathrm{d}t} = \ln\left(1 + \frac{\theta_\infty}{\lambda}\right)$$

which integrates to give

$$R_{\mathrm{I}}^2 = R_0^2 - \frac{2\Gamma t}{\rho_{\mathrm{c}}} \ln\left[1 + \frac{\theta_\infty}{\lambda}\right], \tag{7.59}$$

where R_0 is the drop radius when burning begins at $t = 0$. The time to completely burn the drop quasi-steadily is thus

$$t_{\mathrm{f}} = \frac{\rho_{\mathrm{c}} R_0{}^2}{2\Gamma \ln\left[1 + \dfrac{\theta_\infty}{\lambda}\right]}. \tag{7.60}$$

For the same liquid octane drop considered above ($\rho_{\mathrm{c}} = 700\,\mathrm{kg\,m^{-3}}$), we obtain $t_{\mathrm{f}} = 10.6\,\mathrm{ms}$, or 13.1 ms if we take drop sensible heat into account. A time scale of this order is usually acceptable in a furnace but may prevent complete combustion in a high-speed, internal combustion engine. Problem Q7.16 in Chapter 8 provides another illustration.

More generally, the effect of drop translation may be incorporated by replacing simple heat conduction with forced convection. Under laminar flow conditions, the drop Nusselt number is given by $Nu_{\mathrm{d}} \propto Re_{\mathrm{d}}^{1/2}$, instead of a constant. The time for the drop to disappear completely is then given by $t_{\mathrm{f}} \propto R_0^{3/2}$, as compared with eqn (7.60). This result also applies to the drop evaporation problem discussed earlier in Section (7.3.1); see for example, eqn (7.40).

Selected bibliography

Hasson, D., Luss, D. and Peck, R. (1964). Theoretical analysis of vapour condensation on laminar jets. *International Journal of Heat Mass Transfer*, **7**, 969–81.

Haydock, J. L. (1985). Electrical energy production from water/air temperature differences in the arctic. *Report for Canadian Electrical Association*, Acres International, Toronto.

Jacobs, H. R. (1988). Direct contact heat transfer for process technologies. *Journal of Heat Transfer*, **110**, 1259–70.

Kim, S. and Mills, A. F. (1989). Condensation on coherent turbulent liquid jets: Part I—Experimental study. *Journal of Heat Transfer*, **111**, 1068–74.

Kleinstreuer, C., Wang, T.-Y. and Chiang, H. (1989). Interfacial heat and mass transfer of single and multiple drops in a hot gas stream. In *Multiphase Flow, Heat and Mass Transfer*, pp. 93–102. American Society of Mechanical Engineers, New York.

Kreith, F. and Boehm, R. F. (1988). *Direct contact heat transfer*. Hemisphere, Washington.

Lerner, Y., Kalman, H. and Letan, R. (1984). Condensation of an accelerating-decelerating bubble: experimental and phenomenological studies. In *Basic Aspects of Two Phase Flow and Heat Transfer*, pp. 1–10. American Society of Mechanical Engineers, New York.

Lock, G. S. H. (1989). A benign, small-scale power unit for the Arctic: the Carnot cycle concept. *Arctic*, **42**(3), 253–64.

Lock, G. S. H. (1990). *The growth and decay of ice*. Cambridge University Press, Cambridge.

Pegg, D. E. and Karow, A. M. (eds.) (1987). *The biophysics of organ cryopreservation*. NATO ASI Series A: Life Sciences Vol. 147. Plenum, New York.
Pruppacher, H. R. and Klett, J. D. (1978). *Microphysics of clouds and precipitation*. Reidel, Dordrecht.
Rogers, B. A. (1965). *The nature of metals*. MIT Press, Cambridge, Massachusetts.
Rollason, E. C. (1955). *Metallurgy for engineers*. Edward Arnold, London, 2nd edn.
Sideman, S. and Moalem-Maron, D. (1982). Direct contact condensation. In *Advances in Heat Transfer*, Vol. 15, pp. 227–81. Academic Press, New York.
Spalding, D. B. (1979). *Combustion and mass transfer*. Pergamon, Oxford.
Williams, A. (1990). *Combustion of liquid fuel sprays*. Butterworth, London.

Exercises

1. Derive eqns (7.6) and (7.20).
2. Show that the thermal resistance of the condensate shell on a spherical drop surrounded by saturated vapour may be ignored through the slow growth period when

$$\frac{T_\mathrm{I} - T_\mathrm{s}}{T_\mathrm{s} - T_0} = \frac{Ja_\mathrm{L}}{6Fo_\mathrm{d}^{1/2}} \ll 1$$

   if $(k\rho c_\mathrm{p})_\mathrm{c} \ll (k\rho c_\mathrm{p})_L$. Thus demonstrate that it may also be neglected over most of the rapid growth period if $Ja_\mathrm{L} \ll 1$.
3. Using normalization, prove that condensation on a drop may be treated quasi-steadily if $\rho_\mathrm{V} \ll \rho_\mathrm{L}$.
4. Show that the quasi-steady solution for the above situation is given by

$$\rho_\mathrm{V} = \rho_{\mathrm{V}\infty} - (\rho_{\mathrm{V}\infty} - \rho_\mathrm{VI}) R_\mathrm{I}/R.$$

5. The wet and dry bulb thermometers of a psychrometer read 20°C and 22°C, respectively. Find the dew point temperature and moisture content of the air. Locate your results on a psychrometric chart. (Ans: $T_\mathrm{dew} = 17.6$°C, $m_\infty = 0.0124$ kg kg^{-1}.)
6. Show that the equation for the condensate film thickness $\Delta(X)$ on a laminar coolant jet as a function of distance from the orifice (X) is

$$\frac{\Delta(X)}{d} = \frac{2\rho_\mathrm{c}}{\rho_\mathrm{L}} \left(\frac{X}{\pi d}\right)^{1/2} Ja_\mathrm{c}/Pe_\mathrm{d}^{1/2}.$$

7. A laminar, evaporating jet breaks up into spherical drops. Determine the change in evaporation rate, based on liquid surface area, if
   (a) the drops form from cylindrical jet slugs having a length of one diameter;
   (b) the drop diameters equal the original jet diameter.
   (Ans: (a) 31 per cent increase, (b) 50 per cent increase.)
8. Pure, saturated steam at a pressure of 2.339 kPa surrounds an 8 cm

long, 4 mm diameter water jet moving at $12\,\mathrm{m\,s^{-1}}$. If the jet injection temperature is 10°C, find the condensation rate ($\dot{M}_V$) as a fraction of the water flow rate ($\dot{M}_L$) and compare the result with the water Jakob number.
(Ans: $\dot{M}_V/\dot{M}_L = 6.79 \times 10^{-3}$, $Ja_L = 1.68 \times 10^{-2}$.)

9. Derive eqns (7.27) and (7.28) from first principles.
10. Derive eqn (7.36) from eqns (7.31) and (7.33).
11. Consider a small control volume at the surface of which the mass flux density of component k is $\mathbf{G}_k$, where k covers the reactants and products of combustion within the volume. If $\mathcal{H}$ is the heat (enthalpy) of combustion per unit mass of fuel, show that

$$\mathcal{H}\dot{s}_{FU} = -\sum_{k} h_k \nabla \cdot \mathbf{G}_k$$

where $\dot{s}_{FU}$ is the fuel consumption rate per unit volume.

12. For an intensive property ψ, we may define a flux density

$$\dot{m}_\psi = \psi \mathbf{G} - \Gamma_\psi \nabla \psi$$

consisting of bulk and diffusive components. If this satisfies the conservative requirement $\nabla \cdot \dot{m}_\psi = 0$, show that

$$\frac{G_I R_I^2}{\Gamma_\psi}\left[\frac{1}{R_I} - \frac{1}{R}\right] = \ln\left[1 + \frac{\psi - \psi_I}{\left[\Gamma_\psi \dfrac{d\psi}{dR}\right]_I \Big/ G_I}\right]$$

for point symmetric flow with $R_I < R < \infty$.

Direct contact projects

1. Review and assess combustion in high-speed, reciprocating, internal combustion engines using liquid fuel injection.
2. Develop a physical and mathematical model of 'sweat cooling' for gas turbine blades.
3. Review and assess the growth of ice crystals in the cryopreservation of human organs.
4. Compare theory and practice in the design of spray condensers.
5. Review and assess fluidized combustion.
6. Compare theory and practice in the design of small cooling towers.
7. Develop a physical and mathematical model of comet ablation.
8. Review and assess OTEC direct contact condensers.
9. Compare theory and practice in the design of grain dryers.
10. Review and assess methods for the desalination of water.

8
WORKED EXAMPLES

Fundamentals are the core of this book. Even so, it is often helpful to illustrate their use through application. The exercises at the end of each chapter provide the student with the opportunity to test his or her understanding and at the same time gain a sense of the physical magnitudes likely to be encountered in practice. This chapter is an extension of the same learning process. It is designed as a series of fully worked examples in which theoretical ideas are developed and applied to real situations. In some circumstances, a precise computational scheme is necessary to provide an accurate prediction of temperature, heat flux, etc; in others, an order of magnitude estimate suffices. This provides a mix of situations, some quantitative but ideal and some qualitative but real.

Each chapter is represented by a series of problems coded QX.Y in which X designates an appropriate chapter number. Frequently, this series is broken into several groups, each of which corresponds to a more substantial problem: for example, Q5.1–5.9 are all specific aspects of condenser analysis. An attempt has thus been made to lead the student through a series of smaller steps, at the end of which a much larger problem will have been solved. In keeping with the priorities of the book, there are more examples for the main chapters (4–7) than for the preparatory chapters (2 and 3).

In many of the situations considered, expressions from the appropriate chapter(s) may be used directly, but on occasion additional information is needed and has been provided. Thermophysical properties are given without comment, and are taken to be constant. Although properties may not be constant in practice, this assumption reduces the complexity of the calculations and thus permits the limited space to be devoted entirely to the essential features of the problems. Physical and mathematical models play a central role.

Latent heat transfer problems are so numerous and varied in scope that they cannot be treated exhaustively in one book. The worked examples below are therefore only representative selections from a much wider field. In a one-term course, however, they should provide enough variety to meet the particular needs of particular students, especially when combined with the end-of-chapter exercises. Before working with either, the student is advised to review the appropriate textual material carefully.

Q2.1 Quiescent water fills a domestic water pipe which has a diameter D and wall thickness t. Find expressions for the circumferential stress and strain in the pipe resulting from the freeze-induced excess pressure P represented by the following approximation to the Clausius equation:

$$\frac{dP}{dT} = -A$$

in which $P = 0$ at $T = T_T$, the triple point.

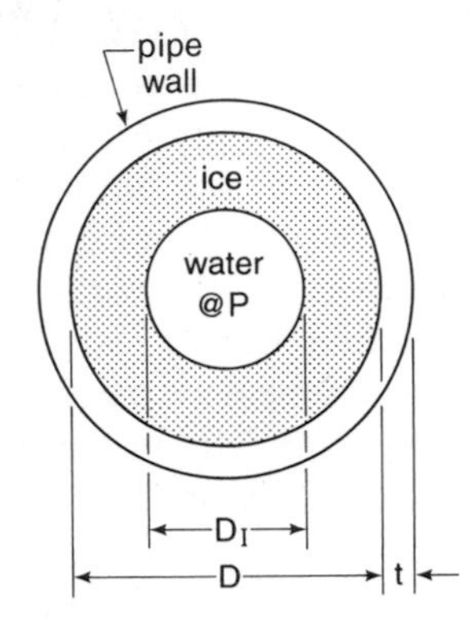

Solution: From static equilibrium, a change in circumferential (hoop) stress σ_h in the pipe is given by

$$d\sigma_h = \frac{D}{2t} dP$$

if we assume that the ice and water sustain only hydrostatic stress. Hence, if $\theta = T_T - T$,

$$d\sigma_h = \frac{D}{2t} A d\theta.$$

Integrating,

$$\sigma_h = \frac{DA\theta}{2t}$$

if changes in D, t, and A are neglected.

From Hooke's law,

$$d\sigma_h = \frac{EdD}{D}$$

where E is the modulus of elasticity. Hence

$$\frac{\Delta D}{D} = \frac{DA\theta}{2Et}$$

gives the circumferential strain corresponding to a diameter enlargement of ΔD if the effect of axial stress is ignored.

Q2.2 Using the above results, find an expression for the diameter of the ice–water interface D_I when $\theta > 0$. Hence find the temperature θ_f required to freeze the pipe completely shut.

Solution: Continuity requires that the mass of the ice and water at any temperature must be equal to the initial mass of water when $D = D_i$. Thus

$$\rho_L \frac{\pi}{4} D_i^2 = \rho_L \frac{\pi}{4} D_I^2 + \rho_S \frac{\pi}{4}(D^2 - D_I^2),$$

and hence

$$D_I = \left(\frac{\rho_L D_i^2 - \rho_S D^2}{\rho_L - \rho_S}\right)^{1/2}$$

in which

$$D^2 = (D_i + \Delta D)^2$$

$$\simeq D_i^2\left(1 + \frac{A\theta D_i}{Et}\right).$$

The pipe is completely frozen shut when $D_I = 0$, i.e.

$$D^2 = \frac{\rho_L}{\rho_S} D_i^2.$$

This occurs when

$$\theta_f = \frac{Et}{AD_i}\left(\frac{\rho_L - \rho_S}{\rho_S}\right)$$

which corresponds to a strain of

$$\frac{\Delta D}{D} = \frac{1}{2}\frac{\rho_L - \rho_S}{\rho_S}.$$

For water and ice, $\rho_L = 10^3\,\text{kg m}^{-3}$ and $\rho_S = 917\,\text{kg m}^{-3}$, and this strain is about 4.5 per cent.

Q2.3 The bursting stress of a copper domestic water pipe is $\sigma_{cr} = 360 \times 10^6\,\text{Pa}$. If the modulus of elasticity of copper is $120 \times 10^9\,\text{Pa}$, determine whether or not the pipe would freeze shut before bursting.

Solution: Using the results of Q2.1,

$$\sigma_h = \frac{E\Delta D}{D} = \frac{DA\theta}{2t}.$$

Hence the temperature which will produce the bursting stress is

$$\theta_{cr} = \frac{2t\sigma_{cr}}{D_i A}.$$

Using the result from Q2.2, this temperature may be expressed as a ratio with the temperature for shut off. Thus

$$\frac{\theta_{cr}}{\theta_f} = \frac{2\sigma_{cr}\rho_S}{E(\rho_L - \rho_S)} = 0.075.$$

Therefore the pipe will burst before freezing shut.

Q2.4 Find an expression for the elevation of boiling temperature of a dilute solution of A (a volatile solvent) and B (a non-volatile solute) under constant pressure.

Solution: Before or after the addition of the solute, chemical equilibrium requires that

$$\mu_{AV} = \mu_{AL}.$$

The addition of solute therefore creates the isobaric changes

$$(\Delta\mu_{AV})_p = (\Delta\mu_{AL})_p.$$

For the (pure) vapour A,

$$(d\mu_{AV})_p = \left(\frac{\partial \mu_{AV}}{\partial T}\right)_{p,x} dT = -s_{AV}\, dT,$$

whereas for liquid A,

$$(d\mu_{AL})_p = \left(\frac{\partial \mu_{AL}}{\partial T}\right)_{p,x} dT + \left(\frac{\partial \mu_{AL}}{\partial x_A}\right)_{p,T} dx_A$$

$$= -s_{AL}\, dT + R_A T d(\ln x_A).$$

using Raoult's law. Hence

$$(s_{AL} - s_{AV})\, dT = -\frac{\lambda_A dT}{T} = R_A T d(\ln x_A),$$

where λ_A is the latent heat of evaporation of the solvent. Integrating, we obtain

$$\frac{\lambda_A(T - T^0)}{R_A T T^0} = -\ln x_A,$$

where T^0 is the boiling point of the pure solvent ($x_A = 1$).

Now $x_A = 1 - x_B$, and therefore if $x_B \ll 1$,

$$\ln x_A \simeq -x_B,$$

and hence

$$T - T^0 \simeq \frac{R_A (T^0)^2}{\lambda_A} x_B$$

if $T - T^0 \ll T^0$.

Q2.5 Calculate the solute mole fraction required to raise the boiling temperature of an aqueous solution by 3 K above the atmospheric boiling point of pure water.

Solution: From the result above

$$x_B = \frac{\lambda \Delta T}{R_w (T^0)^2} = 0.105$$

if $\lambda = 2.25 \times 10^6$ J kg^{-1} and $R_w = 461.9$ J kg^{-1} K^{-1}.

This gives an approximate value which is only applicable to an ideal solution. It does not account for dissociation of the solute.

Q3.1 Water and water vapour in thermal equilibrium flow cocurrently upward in a long, vertical 10 cm diameter tube. If the water flow rate is 1.6 kg s^{-1} and the vapour flow rate is 0.12 kg s^{-1}, demonstrate that annular flow exists.

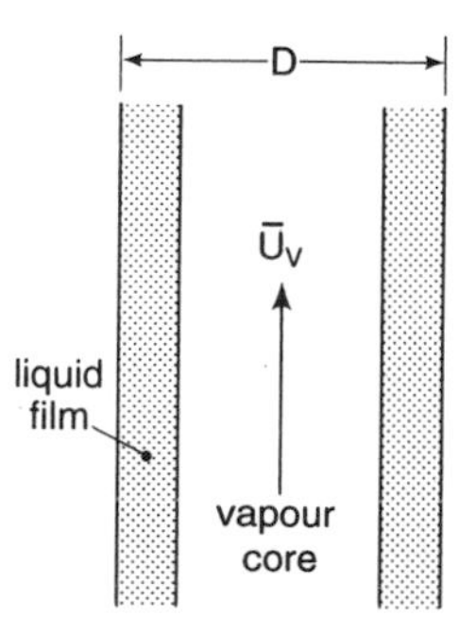

Solution: For the vapour:

$$G_V = 0.12 \times 4/\pi D^2 = 15.3 \text{ kg m}^{-2}\text{s}^{-1}.$$

Hence

$$U_V^s = G_V/\rho_V = 25.9 \text{ m s}^{-1}$$

if $\rho_V = 0.59$ kg m^{-3}.

For the liquid:

$$G_L = 1.6 \times 4/\pi D^2 = 204 \text{ kg m}^{-2}\text{ s}^{-1}.$$

Hence

$$U_L^s = G_L/\rho_L = 0.21 \text{ m s}^{-1}$$

if $\rho_L = 960$ kg m^{-3}.

Using these superficial velocity coordinates in Fig. 3.13, we locate a point in the annular régime. Alternatively, if the film is thin, the superficial vapour velocity is equal to the mean vapour velocity $\bar{U}_V$ and may be used to calculate the Kutateladze number

$$Ku = \left(\frac{\rho_V \bar{U}_V^4}{\sigma g \Delta\rho}\right)^{1/4} = 4.$$

if $\sigma = 0.059\,\mathrm{N\,m^{-1}}$. Thus $Ku > 3$, confirming annular flow.

Q3.2 Determine if the vapour and liquid flows in the above system are laminar or turbulent.

Solution: For the vapour:

$$Re_V = \frac{\bar{U}_V D}{\nu_V} = 12 \times 10^4$$

if $\nu_V = 2.16 \times 10^{-5}\,\mathrm{m^{-2}\,s^{-1}}$.

Since $Re_V > 2500$ the vapour core is turbulent.

For the liquid:

$$\Gamma = \frac{G_L D}{4} = 5.09\,\mathrm{kg\,ms^{-1}}$$

and

$$Re_L = \frac{\Gamma}{\mu_L} = 18.2 \times 10^3$$

if $\mu_L = 2.8 \times 10^{-4}\,\mathrm{kg\,ms^{-1}}$.

Using the critical value of $Re_L \simeq 400$ for a falling film, the liquid is turbulent.

Q3.3 Assuming the surface of the above film to be smooth, estimate the wall shear stress and the pressure gradient.

Solution: We begin by noting that the film is driven by the vapour and assume that the shear stress is constant across the film, i.e. $\tau_w \simeq \tau_I$.

Now

$$\tau_I = f \rho_V \bar{U}_V^2 / 2,$$

where

$$f = \frac{0.079}{Re_V^{1/4}} = 4.24 \times 10^{-3}$$

is the frictional coefficient for turbulent vapour flow in a smooth pipe. Hence

$$\tau_w \simeq \tau_I = 0.839\,\mathrm{N\,m^{-2}}.$$

The pressure gradient is given by

$$\left|\frac{\partial P}{\partial X}\right| = \tau_W \Big/ \left(\frac{D}{4} + \bar{\Delta}\right)$$

in which we will assume that the mean film thickness $\bar{\Delta} \ll D/4$. Thus

$$\left|\frac{\partial P}{\partial X}\right| \simeq \frac{4\tau_w}{D} = 33.6 \text{ Pa m}^{-1}.$$

The above figures must be regarded as very rough estimates until the thickness and surface condition of the film are known.

Q3.4 Estimate the mean thickness of the above water film, and determine its surface condition.

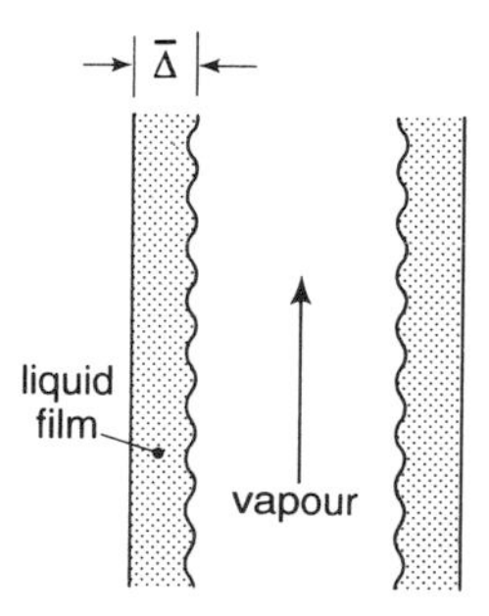

Solution: The film thickness may be estimated from the triangular relation

$$\Gamma = \Gamma(\bar{\Delta}, \tau_w)$$

in which

$$\Gamma = \rho_L \int_0^{\Delta} U_L \, dY,$$

and

$$\Gamma^+ = \int_0^{\delta^+} u^+ \, dy^+.$$

Integrating the velocity profiles in eqn (3.28), we obtain

$$\Gamma^+ = 0.5(y^+)^2 \qquad \text{for } y^+ \leqslant 5,$$

$$\Gamma^+ = 12.5 - 8.05\, y + 5\, y^+ \ln y^+, \qquad \text{for } 5 \leqslant y^+ \leqslant 30,$$

and

$$\Gamma^+ = -64 + 3\, y^+ + 2.5\, y^+ \ln y^+, \qquad \text{for } y^+ \geqslant 30.$$

By inspection, the third of these relations satisfies the requirement (from Q3.2)

$$\frac{\Gamma}{\mu_L} = \Gamma^+ = -64 + 3\delta^+ + 2.5\delta^+ \ln \delta^+,$$

yielding

$$\delta^+ = 911,$$

and

$$\bar{\Delta} = \frac{\mu_L}{(\rho_L \tau_w)^{1/2}} \delta^+ = 9\,\text{mm}$$

if the value of the τ_w calculated in Q3.3 is used. This is a very thick film for which

$$We_D = \frac{\rho_V \bar{U}_V^2 \bar{\Delta}}{\sigma} = 60.3 \gg 1$$

indicating a wavy surface.

These preliminary estimates reveal that the smooth film assumption made in Q3.3 is suspect. Since the system is turbulent throughout we must take a relation such as

$$f = f_s\left(1 + 360\frac{\bar{\Delta}}{D}\right),$$

where $f_s = 4.24 \times 10^{-3}$, calculated in Q3.3 for a smooth interface.

The solution of the triangular relation must now be obtained by combining the expression for f immediately above with the expression for $\bar{\Delta}$ above and the expression for τ_w (i.e. τ_I) in Q3.3. Hence we obtain

$$\tau_w = 9.1\,\text{N m}^{-2},$$

$$\left|\frac{\partial P}{\partial X}\right| = 364\,\text{Pa m}^{-1}$$

and

$$\bar{\Delta} = 2.72\,\text{mm}.$$

Note how these values differ considerably from the earlier estimates.

Finally, we may validate our latest assumptions by noting that

$$\bar{\Delta} \ll D/4$$

and

$$We_\Delta = \frac{\rho_V \bar{U}_V^2 \bar{\Delta}}{\sigma} = 18.2 \gg 1.$$

Q3.5 A natural cloud droplet size distribution is given by

$$n_d = ad^{\alpha} \exp(-bd)$$

where n_d is the number of droplets per m^3 having a diameter d in μm. If $\alpha = 6$, $b = 0.6\,\mu\text{m}^{-1}$, and $a = 62 \times 10^3\,\text{m}^{-3}\,\mu\text{m}^{-6}$, find the total number of droplets n_T per unit volume.

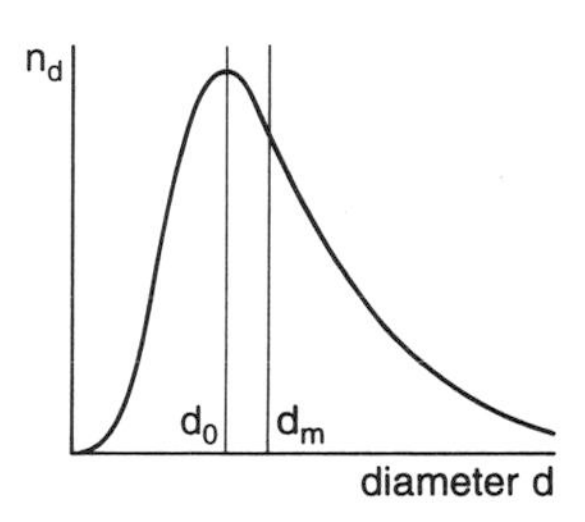

Solution:

$$n_T = \int_0^{\infty} n_d\, d(d) = a \int_0^{\infty} d^{\alpha} \exp(-bd) d(d).$$

Therefore,

$$n_T = \frac{a\Gamma(\alpha + 1)}{b^{\alpha+1}}$$

or, since $\Gamma(\alpha + 1) = \alpha!$, $n_T = 1590 \times 10^6\,\text{m}^{-3}$.

Q3.6 Find the modal diameter d_0 of the cloud described in Q3.5.

Solution: This is the maximal diameter and therefore occurs when $\mathrm{d}n_d/\mathrm{d}(d) = 0$. Differentiating the distribution function,

$$\frac{\mathrm{d}n_d}{\mathrm{d}(d)} = a\alpha d^{\alpha-1} \exp(-bd) - ad^{\alpha} b \exp(-bd) = 0.$$

Hence

$$d_0 = \frac{\alpha}{b} = 10\,\mu\text{m}.$$

Q3.7 Find the volumetric mean diameter d_m of the cloud described in Q3.5.

Solution: This is defined in terms of the liquid volume $\mathcal{V}_L$ expressed as a fraction of the total volume $\mathcal{V}_T$ occupied by the cloud. Thus

$$\mathcal{V}_L/\mathcal{V}_T = n_T \frac{\pi}{6} d_m^3 = a \int_0^\infty d^\alpha \exp(-bd) \frac{\pi}{6} d^3 d(d).$$

Hence

$$d_m^3 = \frac{a}{n_T} \int_0^\infty d^{\alpha+3} \exp(-bd) d(d)$$

$$= \frac{a}{n_T} \frac{\Gamma(\alpha+4)}{b^{\alpha+4}}$$

$$= \frac{b^{\alpha+1}}{\Gamma(\alpha+1)} \cdot \frac{\Gamma(\alpha+4)}{b^{\alpha+4}},$$

using the result in Q3.5. Therefore, since

$$\Gamma(\alpha+4) = (\alpha+3)!,$$

$$d_m^3 = 2338\ \mu\text{m}^3$$

and

$$d_m = 13.3\ \mu m,$$

slightly greater than d_0.

Q3.8 Find the liquid water content LWC of the cloud described in Q3.5, and hence determine the water flux density $\dot{m}_w$, if the cloud moves with a velocity of $\bar{U} = 50\,\text{m s}^{-1}$.

Solution: The liquid water content is defined as the mass of water per unit cloud volume. Hence

$$LWC = n_T \frac{\pi}{6} d_m^3 \rho_L = 1.95\ \text{g m}^{-3}$$

if $\rho_L = 10^3\,\text{kg m}^{-3}$. The water flux density is therefore

$$\dot{m}_w = \bar{U}(LWC) = 97\ \text{g m}^{-2}\,\text{s}^{-1}.$$

Q3.9 Would you expect much of the above water flux to collide with a small aircraft wing, thickness 10 cm, if it travels with a relative velocity of $50\,\text{m s}^{-1}$?

Solution: Equation (3.44) states

$$\frac{F_i}{F_v} = O\left(\frac{Re_D \rho_L d^2}{18 \rho_V D^2}\right),$$

where d is the droplet diameter and D is the body diameter. This provides a measure of the collision efficiency η.

The body Reynolds number is

$$Re_D = \frac{\bar{U}D}{\nu_V} = 5 \times 10^5$$

if $\nu_V = 10^{-5}\,\mathrm{m^2\,s^{-1}}$. This suggests laminar flow over the leading face. Taking $\rho_L = 10^3\,\mathrm{kg\,m^{-3}}$ and $\rho_V = 1.4\,\mathrm{kg\,m^{-3}}$, we find that

$$\frac{F_i}{F_v} = 0.2 \ll 1,$$

for $d = d_o = 10\,\mu\mathrm{m}$. This indicates that most of the droplets smaller than the modal diameter would avoid collision. For $d = d_m = 13.3\,\mu\mathrm{m}$, $F_i/F_v = 0.35$, while for $d = 30\,\mu\mathrm{m}$, $F_i/F_v = 1.8$. This indicates that collisions would be mainly attributable to the larger and rarer droplets.

Q4.1 A steady, $-10°\mathrm{C}$ (T_a) wind blowing continually over a river near its freezing point creates an ice cover 1.0 m thick. If the heat transfer coefficient h_a attributable to the wind is $40\,\mathrm{W\,m^{-2}\,K^{-1}}$, calculate the time required to form the cover and determine the final temperature of the ice upper surface (T_0).

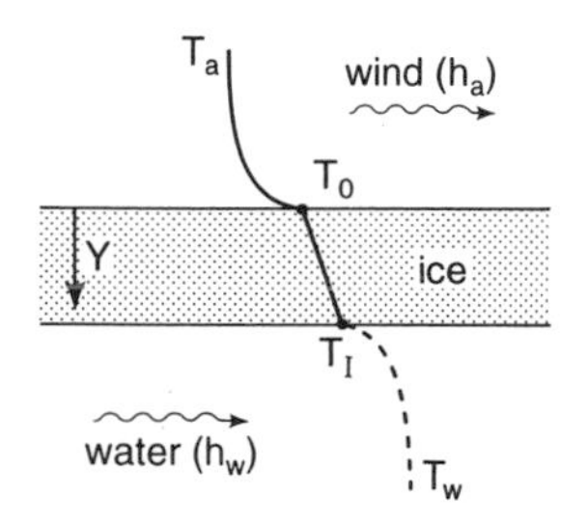

Solution: The Stefan number of the ice is

$$Ste_S = (T_I - T_0)c_{pS}/\lambda < (T_I - T_a)c_{pS}/\lambda = 0.063$$

if $c_{pS} = 2.1 \times 10^3\,\mathrm{J\,kg^{-1}\,K^{-1}}$ and $\lambda = 334 \times 10^3\,\mathrm{J\,kg^{-1}}$. Since $Ste_S \ll 1$, the problem takes a quasi-steady form.

Adapting eqn (4.18), the time corresponding to a cover Y_I thick is

$$t = \frac{\lambda\rho_S}{2k_S(T_I - T_a)}\left[Y_I^2 + \frac{2k_S Y_I}{h_a}\right] = 8.06 \times 10^6\,s$$

if $k_S = 2.1\,\mathrm{W\,m^{-1}\,K^{-1}}$ and $\rho_S = 917\,\mathrm{kg\,m^{-3}}$. Hence $t = 93.3$ days.

A heat balance at the ice upper surface is

$$h_a(T_0 - T_a) = \frac{k_S(T_I - T_0)}{Y_I}$$

from which we obtain

$$T_0 = \frac{T_\mathrm{I} + BiT_\mathrm{a}}{1 + Bi},$$

where $Bi = h_\mathrm{a} Y_\mathrm{I}/k_\mathrm{S} = 19$. Hence $T_0 = -9.5°\mathrm{C}$. This suggests that $T_0 \simeq T_\mathrm{a}$ is a useful approximation in these circumstances.

Q4.2 Warm discharge at 20°C (T_w) leaves a power plant and suddenly floods the under side of the above ice cover. Demonstrate that this would melt the ice completely if it generates a heat transfer coefficient h_w of $100\,\mathrm{W\,m^{-2}\,K^{-1}}$, and estimate the time before the ice cover disappears.

Solution: For the ice to melt completely, $T_0 \geqslant T_\mathrm{I}$. If and when this happens, the final heat balance at the interface requires

$$h_\mathrm{a}(T_0 - T_\mathrm{a}) = h_\mathrm{w}(T_\mathrm{w} - T_0),$$

in which case

$$T_0 = \frac{h_\mathrm{w} T_\mathrm{w} + h_\mathrm{a} T_\mathrm{a}}{h_\mathrm{a} + h_\mathrm{w}} = 11.4°\mathrm{C}.$$

Hence all the ice melts.

During the melting process, the interface equation is

$$\rho_\mathrm{S}\lambda \frac{\mathrm{d}Y_\mathrm{I}}{\mathrm{d}t} = \frac{k_\mathrm{S}(T_\mathrm{I} - T_0)}{Y_\mathrm{I}} - h_\mathrm{w}(T_\mathrm{w} - T_\mathrm{I}).$$

Normalizing,

$$\left[\frac{\rho_\mathrm{S}\lambda Y^\mathrm{c}}{h_\mathrm{w}(T_\mathrm{w} - T_\mathrm{I})t^\mathrm{c}}\right]\frac{\mathrm{d}y_\mathrm{I}}{\mathrm{d}\tau} = \left[\frac{k_\mathrm{S}(T_\mathrm{w} - T_\mathrm{a})}{Y^\mathrm{c}h_\mathrm{w}(T_\mathrm{w} - T_\mathrm{I})}\right]\frac{\phi_\mathrm{S}}{y_\mathrm{I}} - 1$$

in which $\theta^\mathrm{c} = T_\mathrm{w} - T_\mathrm{a} = 30\,\mathrm{K}$ is the overall temperature difference.

The coefficient of $\phi_\mathrm{S}/y_\mathrm{I}$ is

$$\frac{k_\mathrm{S}(T_\mathrm{w} - T_\mathrm{a})}{Y^\mathrm{c}h_\mathrm{w}(T_\mathrm{w} - T_\mathrm{I})} = 0.03,$$

taking $Y^\mathrm{c} = 1.0\,\mathrm{m}$, the initial ice thickness. This confirms that the water heat supply rate dominates and thus enables us to estimate

$$t^\mathrm{c} = O\left[\frac{\rho_\mathrm{S}\lambda Y^\mathrm{c}}{h_\mathrm{w}(T_\mathrm{w} - T_\mathrm{I})}\right] = 1.53 \times 10^5\,\mathrm{s}.$$

Hence the ice will disappear in about 2 days.

Q4.3 Determine the final equilibrium thickness Y_f of a 1.0 m thick ice cover in a region, distant from the aforementioned power plant discharge, where $T_w = 1°C$ and $h_w = 20\ W\ m^{-2}\ K^{-1}$. Assume $T_a = -10°C$.

Solution: A final heat balance at the interface requires

$$\frac{k_S(T_I - T_0)}{Y_f} = h_w(T_w - T_I)$$

or, since $T_0 \simeq T_a$,

$$Y_f \simeq \frac{k_S(T_I - T_a)}{h_w(T_w - T_I)} = 1\ m.$$

Since $Y_f \simeq Y^i$, the initial thickness, the ice would neither grow nor decay in this region.

Q4.4 A heat exchanger tube is cast from pure aluminium poured at its melting temperature $T_I = 660°C$ into a mould having an initial temperature $T_\infty = 20°C$. Find expressions for the temperature distributions in the mould and the aluminium.

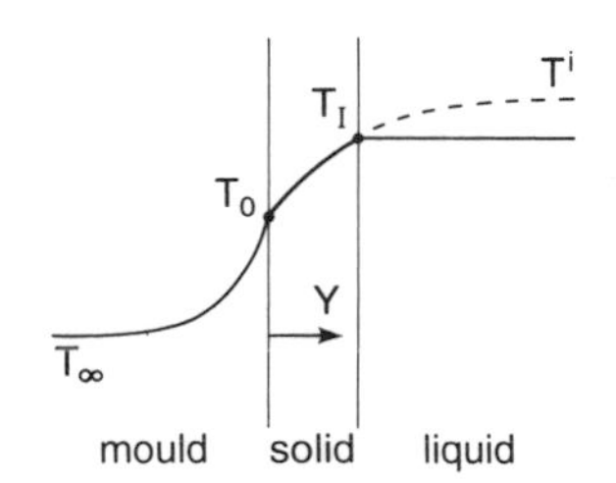

Solution: The Stefan number of the aluminium is

$$Ste_S = \frac{c_{pS}(T_I - T_0)}{\lambda} < \frac{c_{pS}(T_I - T_\infty)}{\lambda} = 1.44$$

if $c_{pS} = 896\ J\ kg^{-1}\ K^{-1}$ and $\lambda = 400 \times 10^3\ J\ kg^{-1}$. Since Ste_S may not be small, the problem must be treated in transient form.

Using the Neumann method with the origin at the mould surface, we seek solutions of the form

$$T_{mo} - T_\infty = \theta_{mo} = A_{mo} + B_{mo}\ \mathrm{erf}\ Y/2(\kappa_{mo}t)^{1/2}$$

and

$$T_S - T_\infty = \theta_S = A_S + B_S\ \mathrm{erf}\ Y/2(\kappa_S t)^{1/2}$$

for the mould and aluminium, respectively. Initially, $T_{mo} = T_\infty = 20°C$ and $T_S = T_I = 660°C$.

Appropriate boundary conditions for the mould are:

$$\theta_{mo}(-\infty) = 0$$

and

$$k_{mo}\left(\frac{\partial\theta_{mo}}{\partial Y}\right)_0 = k_S\left(\frac{\partial\theta_S}{\partial Y}\right)_0$$

at the mould suface. For the aluminium:

$$\theta_{mo}(0) = \theta_S(0)$$

and

$$\theta_S(Y_I) = \theta_I, \text{ at the interface } Y = Y_I.$$

Satisfying these conditions we obtain

$$\theta_{mo} = \frac{\theta_I}{(1 + K\operatorname{erf}\beta)}\left[1 + \operatorname{erf} Y/2(\kappa_{mo}t)^{1/2}\right]$$

and

$$\theta_S = \frac{\theta_I}{\left(\operatorname{erf}\beta + \dfrac{1}{K}\right)}\left[\frac{1}{K} + \operatorname{erf} Y/2(\kappa_S t)^{1/2}\right],$$

where

$$\beta = Y_I/2(\kappa_S t)^{1/2} \quad \text{and} \quad K = \frac{k_{mo}}{k_S}\left(\frac{\kappa_S}{\kappa_{mo}}\right)^{1/2} = \frac{(k\rho c_p)^{1/2}_{mo}}{(k\rho c_p)^{1/2}_S}.$$

Q4.5 Calculate the time required to solidify the heat exchanger tube during the above conditions if its thickest section is $2W = 1.0$ cm across.

Solution: The interface equation is

$$\rho_S\lambda\frac{dY_I}{dt} = k_S\left(\frac{\partial\theta_S}{\partial Y}\right)_I.$$

Using the distribution for θ_S given above, the interface equation reduces to

$$\beta\exp\beta^2\left[\operatorname{erf}\beta + \frac{1}{K}\right) = \frac{c_{pS}(T_I - T_\infty)}{\lambda\pi^{1/2}}$$

in which $c_{pS}(T_I - T_\infty)/\lambda = 1.44$, as noted above. Taking

$$k_{mo} = 1.6\ \text{W m}^{-1}\,\text{K}^{-1} \qquad k_S = 220\ \text{W m}^{-1}\,\text{K}^{-1}$$
$$\rho_{mo} = 3.2\times10^3\ \text{kg m}^{-3} \qquad \rho_S = 2.73\times10^3\ \text{kg m}^{-3}$$
$$c_{pmo} = 10^3\ \text{J kg}^{-1}\,\text{K}^{-1} \qquad c_{pS} = 896\ \text{J kg}^{-1}\,\text{K}^{-1}$$

we find that $1/K = 10.27$, and hence $\beta = 0.078$.

The time required to freeze half of the 1.0 cm section is thus given by

$$t_f = \frac{W^2}{4\beta^2\kappa_S} = 11.3\ \text{s}$$

which is scarcely enough time to fill the mould.

Q4.6 Estimate the effect of pouring the aluminium at $T^i = 700°\text{C}$ in the above casting process.

Solution: The Stefan number of the liquid metal is

$$\frac{c_{pS}(T^i - T_I)}{\lambda} = 0.0896.$$

Since this is much less than 1, the effect of the liquid sensible heat is small. Likewise the effect of sensible heat in the solid is also small: $(T_I - T_0)c_{pS}/\lambda = 1.25 \times 10^{-3}$, much less than the Stefan number estimated in Q4.4. This suggests that a Stefan solution

$$t_f = \frac{\rho_S \lambda Y_I^2}{2k_S(T_I - T_0)}$$

would be adequate. Substituting, we find that $t_f = 11.1\text{s}$, very close to the accurate result in Q4.5.

The time period during which liquid sensible heat is influential divides into two parts: the initial period t_i before the thermal wave penetrates to the centre of the tube cross section; and the final period t_f during which the remaining sensible heat is removed.

The initial period may be estimated from

$$t_i = \frac{W^2}{\kappa_L}.$$

If the liquid and solid properties are the same, this gives $t_i = 0.278\ \text{s}$.

The final period is characterized by the heat balance

$$k_L \frac{(\bar{T} - T_I)}{W/2} \simeq -\rho_L c_{pL} W \frac{d\bar{T}}{dt},$$

where $\bar{T}$ is the (mean) liquid temperature at $W/2$, the mid-point of half of the tube cross section. Hence

$$\bar{\theta} \simeq \bar{\theta}_i \exp\left(-\frac{2\kappa_L t}{W^2}\right),$$

where $\bar{\theta} = \bar{T} - T_I$ and $\bar{\theta}_i$ is its initial value when $t \simeq t_i$. The additional time period may thus be estimated from

$$\frac{2\kappa_L t_f}{W^2} = O(1),$$

or

$$t_f \simeq \frac{t_i}{2}.$$

Hence the freezing time is increased by about 0.417 s or 3.7 per cent, a negligible change.

Q4.7 A liquid metal at temperature T^i is poured into a mould having a temperature T_∞. Find an expression for the mould surface temperature T_0.

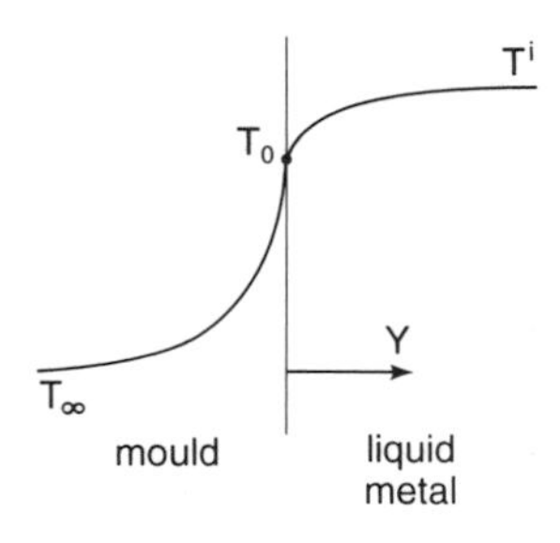

Solution: Following the Neumann method, we assume

$$T_L = A + B \operatorname{erf} Y/2(\kappa_L t)^{1/2}$$

and

$$T_{mo} = C + D \operatorname{erfc} Y/2(\kappa_{mo} t)^{1/2}$$

for the metal and mould, respectively. Taking $\theta = T - T_\infty$, the boundary conditions may be stated as

$$\theta_{mo}(-\infty) = 0$$

and

$$\theta_{mo}(\infty) = \theta_0$$

for the mould, if $Y = 0$ represents the mould surface where

$$k_{mo}\left(\frac{\partial \theta_{mo}}{\partial Y}\right)_0 = k_L\left(\frac{\partial \theta_L}{\partial Y}\right)_0.$$

For the metal,

$$\theta_L(0) = \theta_0$$

and

$$\theta_L(\infty) = \theta^i.$$

Satisfying these conditions we obtain

$$\theta_L = \theta^i\,[1 + K\,\mathrm{erf}\; Y/2(\kappa_L t)^{1/2}]/(1+K)$$

and

$$\theta_{mo} = \frac{\theta^i}{(1+K)}\,\mathrm{erfc}\; Y/2(\kappa_{mo} t)^{1/2},$$

where

$$K = (k\rho c_p)^{1/2}_{mo}/(k\rho c_p)^{1/2}_L.$$

When $Y = 0$, we find that

$$\theta_0 = \frac{\theta^i}{(1+K)}.$$

Q4.8 A valve flange 2.2 cm think is formed from an alpha brass alloy poured into a mould having an initial temperature $T_\infty = 25°C$. If freezing takes place over the temperature range $T_2 = 1055°C$ to $T_1 = 1045°C$, find the minimum pour temperature to prevent instantaneous freezing and estimate the time period before freezing begins.

Solution: No freezing takes place initially if $T_0 > T_I$,

or

$$\theta_0 = \frac{\theta^i}{1+K} > \theta_I.$$

Taking

$$k_{mo} = 1.6\ \mathrm{W\,m^{-1}\,K^{-1}} \qquad k_L = 109\ \mathrm{W\,m^{-1}\,K^{-1}}$$

$$\rho_{mo} = 3.2 \times 10^3\ \mathrm{kg\,m^{-3}} \qquad \rho_L = 8.52 \times 10^3\ \mathrm{kg\,m^{-3}}$$

$$c_{pmo} = 10^3\ \mathrm{J\,kg^{-1}\,K^{-1}} \qquad c_{pL} = 385\ \mathrm{J\,kg^{-1}\,K^{-1}}$$

we find that $K = 0.12$. Hence initial freezing is prevented if

$$T^i - T_\infty > 1.12(T_I - T_\infty)$$

in which T_I corresponds to the liquidus point T_2. The minimum pour temperature is therefore

$$T^i = 1.12(T_2 - T_\infty) + T_\infty = 1179°C.$$

The time interval before the effect of the mould surface temperature reaches the mid section of the flange may be estimated from

$$t = O\left(\frac{W^2}{\kappa_L}\right),$$

where $W = 1.1$ cm is the flange half width. Hence $t = 0(3.6\text{ s})$. After this brief period, the mould surface temperature falls rapidly, as noted in Q4.6, until $T_0 = T_2 = 1055°\text{C}$ and freezing begins.

Q4.9 The alpha brass alloy discussed above is to be used in the production of 10 cm diameter ornate cylinders using the slush casting process in which the slushy core is poured back out of the mould once a thin solid cover has been produced inside the entire mould surface. Find the time necessary to create a cover with an average thickness of 2 mm if the mould temperature is initially $T_\infty = 25°\text{C}$ and the alloy is poured at $T_2 = 1055°\text{C}$.

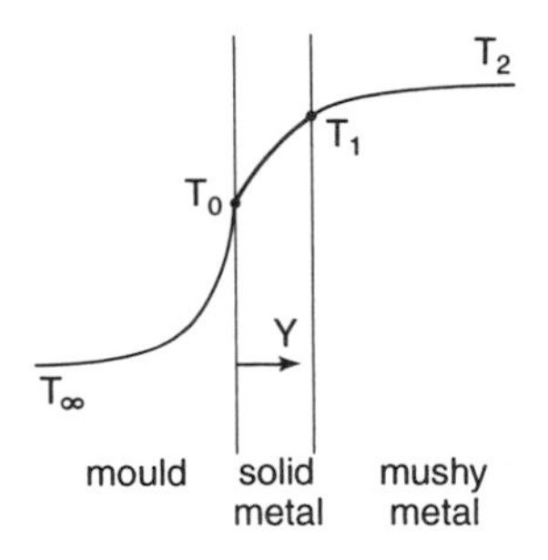

Solution: Freezing begins immediately. During the brief period required to produce a thin surface coating, the slushy core temperature is assumed to remain close to 1055°C, thus making it possible to apply Neumann's method. The thickness of the surface coating is therefore given by

$$Y_\text{I} = 2\beta(\kappa_\text{S} t)^{1/2}$$

in which β is found from eqn (4.42):

$$\frac{\exp[(\kappa_\text{S} - \kappa_\text{m})\beta^2/\kappa_\text{m}]\,\text{erfc}\,\beta(\kappa_\text{S}/\kappa_\text{m})^{1/2}}{\text{erf}\,\beta + \dfrac{(k\rho c_\text{p})_\text{S}^{1/2}}{(k\rho c_\text{p})_\text{mo}^{1/2}}} = \frac{(T_2 - T_1)(k\rho c_\text{p})_\text{m}^{1/2}}{(T_1 - T_\infty)(k\rho c_\text{p})_\text{S}^{1/2}}.$$

Taking the same thermophysical property data used in Q4.8 and assuming

$$k_\text{S} = k_\text{m}, \quad \rho_\text{S} = \rho_\text{m} \quad \text{and} \quad \lambda = 345 \times 10^3\ \text{J kg}^{-1},$$

we obtain

$$\kappa_\text{S} = \frac{k_\text{S}}{\rho_\text{S} c_\text{pS}} = 3.32 \times 10^{-5}\ \text{m}^2\,\text{s}^{-1}, \quad \kappa_\text{m} = \frac{k_\text{S}}{\rho_\text{S} c'_\text{pm}} = 3.66 \times 10^{-7}\ \text{m}^2\,\text{s}^{-1}$$

in which

$$c'_{pm} = c_{pS} + \frac{\lambda}{T_2 - T_1} = 3.49 \times 10^4 \text{ J kg}^{-1}\text{ K}^{-1}.$$

Hence

$$\frac{\kappa_S - \kappa_m}{\kappa_m} = 89.7, \left(\frac{\kappa_S}{\kappa_m}\right)^{1/2} = 9.52,$$

$$\frac{(k\rho c_p)_S^{1/2}}{(k\rho c_p)_{mo}^{1/2}} = 8.36, \frac{(k\rho c_p)_m^{1/2}}{(k\rho c_p)_S^{1/2}} = 9.51$$

and

$$\frac{T_2 - T_1}{T_1 - T_\infty} = 9.8 \times 10^{-3}.$$

Using eqn (4.42) we obtain

$$\frac{\exp(89.7\beta^2)\operatorname{erfc}(9.52\beta)}{\operatorname{erf}\beta + 8.36} = 0.0932$$

from which we find that $\beta = 0.025$. The time required to produce a surface coating Y_I thick is therefore

$$t = \frac{Y_I^2}{4\beta^2\kappa_S} = 48.2\,\text{s}.$$

Comparing this with the time scale for penetration of the surface temperature to the centre of the casting, i.e.

$$t^c = \frac{D^2}{4\kappa_S} = 75\,\text{s},$$

the use of the Neumann method is validated.

Q4.10 A supercooled fog with the same droplet size distribution used in Q3.5 moves steadily across a 2.0 cm diameter overhead power cable. If the fog speed U_∞ is 16.2 kilometres per hour, determine whether or not most of the droplets will collide with the cable.

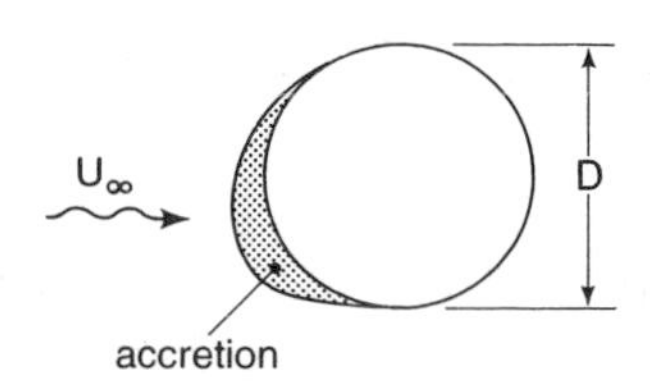

Solution: Using eqn (3.44)

$$\frac{F_v}{F_i} = \frac{18\rho_a D^2}{Re_D \rho_w d^2},$$

where d is the droplet diameter and D is the cable diameter. The cable Reynolds number

$$Re_D = \frac{U_\infty D}{\nu_a} = 7.5 \times 10^3$$

if $\nu_a = 1.2 \times 10^{-5}\,m^2\,s^{-1}$. This implies laminar flow.

From Q3.7, the volumetric mean diameter is 13.3 μm. Hence, taking $\rho_a = 1.4\,kg\,m^{-3}$, and $\rho_w = 10^3\,kg\,m^{-3}$, $F_v/F_i = 7.59$. Since this is much greater than 1.0, most of the droplets will not collide with the cable. However, the collision rate is not zero.

Q4.11 For the above situation, calculate the mass flux density $\dot{m}_w$ of the fog, and determine the collision efficiency η given by

$$\eta = 0.5[\log_{10}(8K')]^{1.6},$$

where

$$K' = K/[0.087\ Re_d^{0.76/Re_d^{0.027}} + 1],$$

$$K = 2F_i/F_v$$

and Re_d is the droplet Reynolds number.

Solution: From Q3.8, the liquid water content (LWC) was found to be $1.95\,g\,m^{-3}$ Hence

$$\dot{m}_w = U_\infty LWC = 8.78 \times 10^{-3}\,kg\,m^{-2}\,s^{-1}.$$

The droplet Reynolds number

$$Re_d = \frac{U_\infty d}{\nu_a} = 5,$$

while $F_v/F_i = 7.59$, from Q4.10. Hence $K = 0.263$, $K' = 0.205$, and $\eta = 0.043$. This value of η confirms the prediction given in Q4.10 above.

Q4.12 Estimate the average radius of the glaze ice produced under the above conditions if they are maintained for a full day.

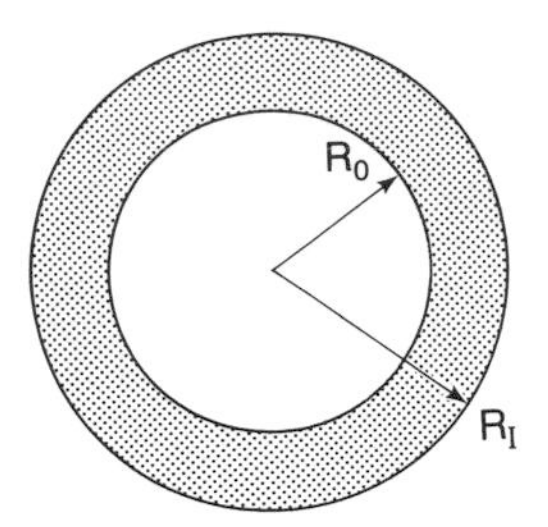

Solution: In 24 hours, the icing rate $\dot{m}_i = \eta \dot{m}_w D$ yields an addition of $0.652\ \mathrm{kg\,m^{-1}}$. This is collected near the forward stagnation point and thus creates a moment about the cable axis. The resulting torque tends to twist the cable, which is usually very long, and thus exposes a different surface on which the accretion continues. This gradual twisting of the cable creates a fairly uniform ice coating.

The average coating thickness $R_I - R_0$ is given by

$$\pi(R_I^2 - R_0^2)\rho_S = \dot{m}_i t$$

over the time interval t. Taking $\rho_S = 917\ \mathrm{kg\,m^{-3}}$, and recalling that $R_o = 1$ cm, we find that $R_I = 1.8$ cm after one day.

The average coating thickness is therefore 0.8 cm.

Q4.13 Estimate the effect of the cumulative accretion on the icing rate in the above circumstances.

Solution: An increase in the effective diameter of the accreting surface alters the flow conditions. After 24 hours

$$Re_D = \frac{2U_\infty R_I}{\nu_a} = 13.5 \times 10^3$$

thus increasing F_v/F_i from 7.59 to 13.7. This decreases the collision efficiency. Thus we find that $\eta \to 0$, and the icing rate is reduced to a very low value.

The definition of F_v/F_i reveals that the icing rate is highest initially and may, in some circumstances, be reduced to a negligible level.

Q5.1 Using mass and heat balances on a longitudinal element dX of a condensate film on a plane substrate inclined at α to the horizontal, show that

$$\frac{3\mu_L k_L(T_I - T_0)}{\rho_L g\lambda(\rho_L - \rho_V)} = \Delta \frac{d}{dX}(\Delta^3 \sin\alpha)$$

where T_0 is the substrate temperature.

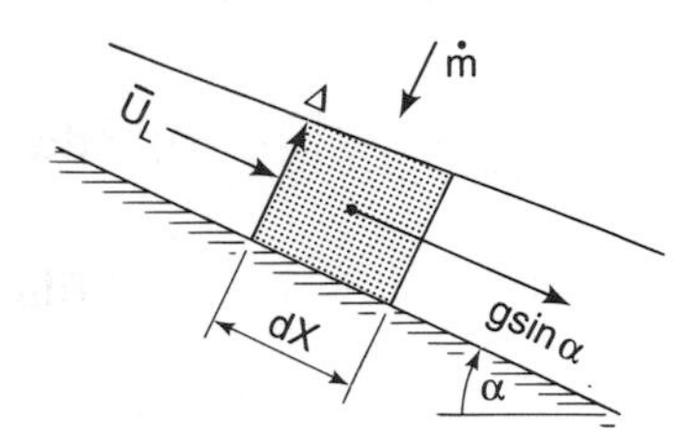

Solution: The Jakob number of the film $Ja = c_{pL}(T_I - T_0)/\lambda$ is assumed to be much less than 1.0.

Mass balance:

$$\dot{m}\mathrm{d}X = \rho_L d(\bar{U}\Delta),$$

where $\dot{m}$ is the condensation rate at the interface.

Heat balance:

$$\mathrm{d}\dot{q} = \frac{k_L(T_I - T_0)\,\mathrm{d}X}{\Delta} = \lambda\dot{m}\,\mathrm{d}X.$$

Hence, since $\bar{U}_L = (\rho_L - \rho_V)\Delta^2 g \sin\alpha/3\mu_L$,

$$\frac{\mathrm{d}\dot{q}}{\mathrm{d}X} = \frac{k_L(T_I - T_0)}{\Delta} = \lambda\rho_L\frac{\mathrm{d}}{\mathrm{d}X}\left\{\frac{g\sin\alpha\Delta^3(\rho_L - \rho_V)}{3\mu_L}\right\}.$$

Therefore

$$\frac{3\mu_L k_L(T_I - T_0)}{\rho_L\lambda g(\rho_L - \rho_V)} = \Delta\frac{\mathrm{d}}{\mathrm{d}X}(\Delta^3\sin\alpha).$$

Q5.2 Use the above result to show that the average condensate film thickness on a horizontal tube of diameter D is given by

$$\bar{\Delta} = 1.38\left[\frac{\mu_L k_L D(T_I - T_0)}{\rho_L g\lambda(\rho_L - \rho_V)}\right]^{1/4}.$$

Solution: Taking $\beta = 3\,\mu_L k_L(T_I - T_0)D/2\rho_L g\lambda(\rho_L - \rho_V)$, the expression derived in Q5.1 may be restated as

$$\beta = \Delta\frac{\mathrm{d}}{\mathrm{d}\alpha}(\Delta^3\sin\alpha)$$

for the arc $\mathrm{d}\alpha = 2\,\mathrm{d}X/D$. Substituting $b = \Delta^4/\beta$, this may be rewritten as

$$\frac{3}{4}\sin\alpha\frac{\mathrm{d}b}{\mathrm{d}\alpha} + b\cos\alpha = 1$$

which has the solution

$$b(\alpha) = \frac{4}{3(\sin\alpha)^{4/3}} \int_0^\alpha (\sin\alpha)^{1/3}\, d\alpha.$$

The local heat transfer coefficient and the local film thickness are related through the expression

$$\frac{h}{k_L} = \frac{1}{\Delta} = \frac{1}{(\beta b)^{1/4}}.$$

Hence the averages $\bar{h}$ and $\bar{\Delta}$ may be determined from $1/b^{1/4}$ obtained graphically or numerically.

We find that

$$\bar{\Delta} = 1.38 \left[\frac{\mu_L k_L D (T_I - T_0)}{\rho_L g \lambda (\rho_L - \rho_V)} \right]^{1/4}.$$

Q5.3 Calculate the average thickness and average heat transfer coefficient of a water film on a 2 cm horizontal tube when the tube wall is subcooled by 10K.

Solution: The Jakob number $Ja = c_{pL}(T_I - T_0)/\lambda = 1.69 \times 10^{-2}$ confirms that the thin film analysis is valid. Using the result from Q5.2 above, $\bar{\Delta} = 76\ \mu\text{m}$ if we take $k_L = 0.659\ \text{W m}^{-1}\text{K}^{-1}$, $c_{pL} = 4.22 \times 10^3\ \text{J kg}^{-1}\ \text{K}^{-1}$, $\mu_L = 1.75 \times 10^{-3}\ \text{kg m}^{-1}\text{s}^{-1}$, $\lambda = 2.5 \times 10^6\ \text{J kg}^{-1}\text{K}^{-1}$ and assume $(\rho_L - \rho_V) \simeq \rho_L$.

The average heat transfer coefficient is then given by

$$\bar{h} = \frac{k_L}{\bar{\Delta}} = 8.7 \times 10^3\ \text{W m}^{-2}\text{K}^{-1}.$$

Q5.4 A water-cooled steam condenser consists of 2.5 m long horizontal copper tubes with an inside diameter $D = 2$ cm and a wall thickness of 2 mm. Calculate the thermal resistance of the coolant and the tube wall if the water flow rate is $\dot{M}_w = 0.31\ \text{kg s}^{-1}$ inside each tube.

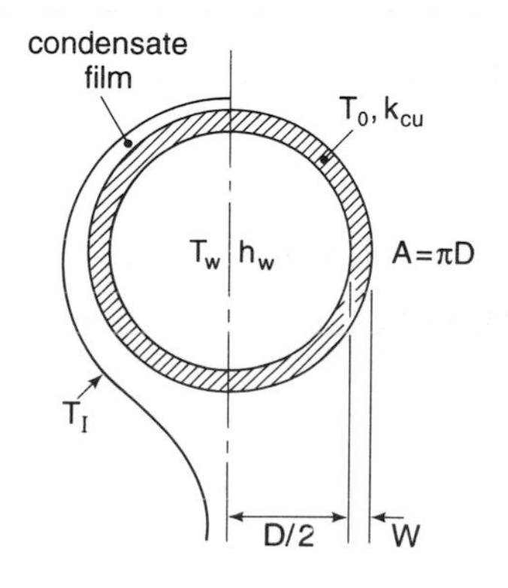

Solution: Since the water is heated by the external condensation, we take the (Dittus–Boelter) internal heat transfer correlation as

$$Nu_D = 0.023 Re_D^{0.8} Pr^{0.4}.$$

The Reynolds number of the coolant is

$$Re_D = \frac{4\dot{M}_w}{\rho_w \pi D^2} \cdot \frac{D}{\nu_w} = 1.07 \times 10^4$$

if $\rho_w = 0.99 \times 10^3 \text{kg m}^{-3}$ and $\mu_w = 1.85 \times 10^{-3} \text{kg m}^{-1} \text{s}^{-1}$. Also, if $c_{pw} = 4.2 \times 10^3 \text{J kg}^{-1} \text{K}^{-1}$ and $k_w = 0.57 \text{W m}^{-1} \text{K}^{-1}$, $Pr_w = \mu_w c_{pw}/k_w = 13.6$. Hence $Nu_D = 109.3$.

The heat transfer coefficient in the water is

$$h_w = \frac{Nu_D k_w}{D} = 3114 \text{W m}^{-2} \text{K}^{-1}.$$

The water thermal resistance is therefore

$$R_w = 1/h_w A = 2.04 \times 10^{-3} \text{K W}^{-1},$$

where A is the pipe surface area. The thermal resistance of the pipe wall

$$R = W/Ak_{cu} = 3.18 \times 10^{-5} \text{K W}^{-1},$$

where W is the wall thickness, and $k_{cu} = 400 \text{W m}^{-1} \text{K}^{-1}$. By comparison, the wall thermal resistance may be neglected.

Q5.5 If the above condenser operates at a steam pressure of 10 kPa estimate the tube wall outer temperature when the cooling water is admitted with a temperature of 20°C.

Solution: A heat balance based on a tube wall temperature T_0 which varies neither radially (see above) nor longitudinally, yields

$$h_w(T_0 - T_w) = \bar{h}(T_I - T_0)$$

in which h_w has been calculated in Q5.4, $\bar{h}$ may be determined from the expression found in Q5.2, and $T_I = 45.6$°C from steam tables. Hence

$$h_w(T_0 - T_w) = 0.725(T_I - T_0)^{3/4} \left(\frac{\rho_w g \lambda k_w^3 (\rho_L - \rho_V)}{\mu_w D} \right)^{1/4}.$$

Assuming T_w is uniform throughout the tube length, we obtain $T_o = 39.9$°C if $\lambda = 2.395 \times 10^6 \text{J kg}^{-1}$.

However, this tube wall temperature implies a rise in the water temperature given by

$$\Delta T_w = \frac{h_w(T_0 - T_w) A}{\dot{M}_w c_{pw}} = 7.47 \text{K},$$

so that the average water temperature $\bar{T}_w$ would be closer to $20 + 3.7 = 23.7°C$. Using this value raises the corresponding value of T_0 by approximately 1 K to 40.9°C but reduces $T_0 - T_w$ from 19.9 K to 17.2 K and hence lowers $\bar{T}_w$ to 23.2°C. Thus we estimate $T_0 \simeq 40.8°C$.

Q5.6 Calculate the overall heat transfer coefficient and hence find the overall condensation rate for a bundle of the tubes discussed above if they are stacked

(a) 10 rows deep,

(b) 20 rows deep,

and each row contains 30 tubes.

Solution: From Q5.5,

$$\bar{h}_1 = \frac{h_w(T_0 - \bar{T}_w)}{T_I - T_0} = 1.35 \times 10^4 \text{ W m}^{-2}\text{ K}^{-1}.$$

Equation (5.33) states that

$$\bar{h}_n = \frac{\bar{h}_1}{n^{1/6}},$$

where n is the number of rows and $\bar{h}_1$ is the average heat transfer coefficient in the first row. Thus

$$\bar{h}_n = 9193 \text{ W m}^{-2}\text{ K}^{-1} \text{ for 10 rows,}$$

and

$$\bar{h}_n = 8193 \text{ W m}^{-2}\text{ K}^{-1} \text{ for 20 rows.}$$

An energy balance on the bundle yields the overall condensate rate $\dot{M}$ as

$$\dot{M} = 30n\pi DL(T_I - T_0)\bar{h}_n/\lambda,$$

where L is the tube length. Hence

$$\dot{M} = 0.863 \text{ kg s}^{-1} \text{ for 10 rows,}$$

and

$$\dot{M} = 1.54 \text{ kg s}^{-1} \text{ for 20 rows.}$$

Q5.7 If the steam in the above condenser is admitted with a superheat of 40 K, estimate the change in condensation rate.

Solution: The effect of superheat in a pure vapour is incorporated in the modified latent heat

$$\begin{aligned} \lambda' &= \lambda + c_{pV}(T_V - T_I) \\ &= \lambda[1 + Ja_V], \end{aligned}$$

in which the vapour Jakob number $Ja_V = 0.033$ if $c_{pV} = 2 \times 10^3\ \mathrm{J\,kg^{-1}\,K^{-1}}$. Since

$$\bar{h} = 0.725\left[\frac{\rho_L g \lambda k_L^3(\rho_L - \rho_V)}{\mu_L D(T_I - T_0)}\right]^{1/4},$$

the effect on heat transfer rate is given by $(1 + Ja_V)^{1/4}$. The condensation rate, however, is proportional to $\bar{h}/\lambda$, as noted in the previous question. Thus the condensation rate must be modified by a factor $(1 + Ja_V)^{-3/4}$. This produces a reduction of approximately $3Ja_V/4 \simeq 2.5$ per cent.

Q5.8 Show that in forced flow the thickness of a laminar, annular, condensate film is given by

$$\Delta^2 = \int_0^X F(X)\,\mathrm{d}X - \int_0^X G(X)\,\mathrm{d}X,$$

when $\Delta(0) = 0$ and

$$F(X) = \frac{4k_L(T_I - T_0)}{\lambda \rho_L \bar{U}_V},$$

$$G(X) = \frac{2\Delta^2}{\bar{U}_V}\frac{\mathrm{d}\bar{U}_V}{\mathrm{d}X}.$$

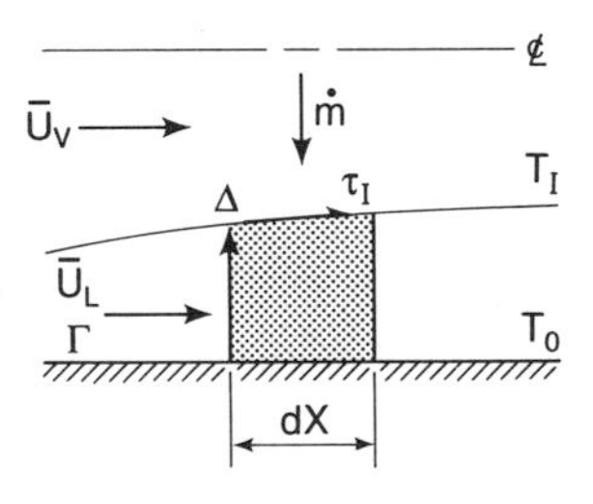

Solution: Film mass balance:

$$\dot{m} = \frac{\mathrm{d}\Gamma}{\mathrm{d}X}.$$

Film heat balance:

$$\dot{q} = \lambda \dot{m} = \frac{k_L(T_I - T_0)}{\Delta}.$$

Hence

$$\frac{k_L(T_I - T_0)}{\lambda} = \frac{\Delta\,\mathrm{d}\Gamma}{\mathrm{d}X}.$$

But

$$\Gamma = \rho_L \bar{U}_L \Delta = \rho_L U_I \Delta/2$$

where, in general,

$$U_I \simeq \bar{U}_V(X),$$

so that

$$\frac{2k_L(T_I - T_0)}{\lambda \rho_L} = \Delta \left(\bar{U}_V \frac{d\Delta}{dX} + \Delta \frac{d\bar{U}_V}{dX} \right)$$

Re-arranging and integrating,

$$\Delta^2 = \int_0^X \frac{4k_L(T_I - T_0)}{\lambda \rho_L \bar{U}_V} dX - \int_0^X \frac{2\Delta^2}{\bar{U}_V} \frac{d\bar{U}_V}{dX} dX$$

if $\Delta(0) = 0$.

The solution to this equation may be obtained iteratively if $\bar{U}_V(X)$ is incorporated using continuity and a film momentum balance along with equations of state.

Q5.9 Use the result of the previous question to find an expression for heat transfer through the forced, laminar condensate film on a flat plate.

Solution: If the vapour stream velocity (U_∞) and temperature (T_∞) are constant, we find that

$$\Delta^2 = \int_0^X \frac{4k_L(T_\infty - T_0)}{\lambda \rho_L U_\infty} dX$$

in which the differences $T_\infty - T_I$ and $U_\infty - U_I$ are neglected. Hence

$$\Delta(X) = \left[\frac{4k_L(T_\infty - T_0)X}{\lambda \rho_L U_\infty} \right]^{1/2}.$$

The Nusselt number is defined by

$$Nu_X = \frac{hX}{k_L} = \frac{\dot{q}X}{k_L(T_\infty - T_0)} = \frac{X}{\Delta}$$

and hence

$$Nu_X = \left[\frac{\lambda \rho_L U_\infty X}{4k_L(T_\infty - T_0)} \right]^{1/2}$$

which may also be written

$$Nu_X = \left(\frac{Pr_L Re_X}{4Ja_L} \right)^{1/2}$$

in which the nondimensional group $Pr_L Re_X/Ja_L$ plays the same role as $PrRe_X$ in single phase forced convection (see Section (1.6)).

Q5.10 Using the empirical relation

$$\overline{Nu}_D = 2.7\left(\frac{Pr_L Re_D}{Ja_L}\right)^{1/2}$$

for forced laminar condensation on a 2 cm diameter horizontal tube, find the average heat transfer coefficient under saturated conditions at 0.1 bar if the vapour (steam) velocity is $1\ \text{m s}^{-1}$ and the tube wall temperature is 36°C.

Solution: At 0.1 bar, T_{sat} i.e., T_∞, is 46°C, from steam tables. Hence $T_\infty - T_w = 10°\text{C}$.
Jakob number $Ja_L = c_{pL}(T_\infty - T_0)/\lambda = 1.75 \times 10^{-2}$, if $c_{pL} = 4.18 \times 10^3\ \text{J kg}^{-1}\text{K}^{-1}$ and $\lambda = 2.39 \times 10^6\ \text{J kg}^{-1}$.
Prandtl number $Pr_L = \mu_L c_{pL}/k_L = 3.88$ if $\mu_L = 5.94 \times 10^{-4}\ \text{kg m}^{-1}\text{s}^{-1}$ and $k_L = 0.64\ \text{W m}^{-1}\text{K}^{-1}$.
Reynolds number $Re_D = U_\infty D/\nu_L = 3.33 \times 10^4$.
Hence

$$\overline{Nu}_D = 2.7\left(\frac{Pr_L Re_D}{Ja_L}\right)^{1/2} = 7.34 \times 10^3$$

and

$$\bar{h} = \frac{k_L \overline{Nu}_D}{D} = 235\ \text{kW m}^{-2}\text{K}^{-1}.$$

Alternatively,

$$\bar{h} = 2.7\left[\frac{\lambda \rho_L U_\infty k_L}{D(T_\infty - T_0)}\right]^{1/2} = 235\ \text{kW m}^{-2}\text{K}^{-1}$$

using only the properties λ, ρ_L and k_L.

Q5.11 Estimate the improvement in heat transfer introduced by the vapour velocity of $1\ \text{m s}^{-1}$ for the same thermal conditions.

Solution: Let

$$\bar{h}_{FC} = 2.7\left[\frac{\lambda \rho_L U_\infty k_L}{D(T_\infty - T_0)}\right]^{1/2},$$

from above, while

$$\bar{h}_{NC} = 0.725\left[\frac{\lambda \rho_L g k_L^3 \Delta\rho}{\mu_L D(T_\infty - T_0)}\right]^{1/4},$$

from Q5.2. Hence

$$\frac{\bar{h}_{FC}}{\bar{h}_{NC}} = 3.72\left[\frac{\lambda\rho_L\mu_L U_\infty^2}{k_L D g\Delta\rho(T_\infty - T_0)}\right]^{1/4}$$

or

$$\frac{\bar{h}_{FC}}{\bar{h}_{NC}} = 3.72\left[\left(\frac{Pr_L Re_D}{Ja_L}\right)^2 \bigg/ \left(\frac{Ar_D Pr_L}{Ja_L}\right)\right]^{1/4}.$$

Now

$$Ar_D = \frac{D^3 g\Delta\rho}{\rho_L \nu_L^2} \simeq \frac{D^3 g}{\nu_L^2} = 218 \times 10^6$$

if $\rho_L \gg \rho_V$. Hence $Ar_D Pr_L / Ja_L = 483 \times 10^8$.

The ratio

$$\left(\frac{Pr_L Re_D}{Ja_L}\right)^2 \bigg/ \left(\frac{Ar_D Pr_L}{Ja_L}\right) = 1.13 \times 10^3$$

is analogous to the single phase ratio Re^2/Gr, where Gr is the Grashof number. These ratios indicate the importance of forced flow relative to natural (buoyant) flow.

From above,

$$\frac{\bar{h}_{FC}}{\bar{h}_{NC}} = 3.72(1.13 \times 10^3)^{1/4} = 21.6.$$

Q5.12 A heavy water recovery unit incorporates a reflux, wetted wall condenser consisting of a 1.2 m diameter vertical cylinder cooled externally. If the internal annular condensate film leaves the base of the unit at a steady rate of 0.188 kg s^{-1}, calculate the (exit) film Reynolds number and the (entry) vapour Reynolds number, and thus determine whether laminar or turbulent conditions exist if pure, saturated water vapour enters at a pressure of 25 kPa.

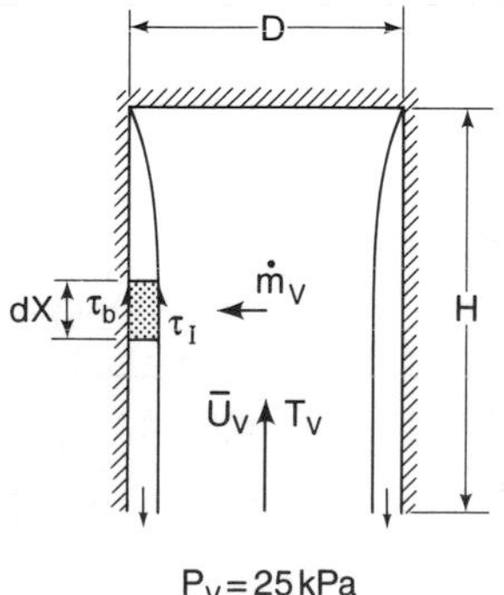

Solution: The condensate mass flow rate is given by

$$\dot{M}_L = \pi D \Gamma = 0.188 \text{ kg s}^{-1}.$$

Therefore

$$\Gamma = \frac{\dot{M}_L}{\pi D} = 49.8 \times 10^{-3} \text{kg m}^{-1}\text{s}^{-1}.$$

The film Reynolds number is

$$Re_f = \frac{\Gamma}{\mu_L} = 100$$

if $\mu_L = 0.5 \times 10^{-3}$ km m^{-1} s^{-1}. This magnitude places the exiting film in the wavy, laminar region (see Fig. 5.7).

The vapour mass flow rate is given by

$$\dot{M}_V = \frac{\pi}{4} D^2 \rho_V \bar{U}_V = 0.188 \text{ kg s}^{-1}$$

under reflux conditions if the film thickness is much less than the cylinder diameter. Hence

$$\bar{U}_V = \frac{4\dot{M}_V}{\pi D^2 \rho_V} = 1.03 \text{ m s}^{-1},$$

taking $\rho_L = 0.161$ kg m^{-3} from steam tables. Therefore the vapour Reynolds number is

$$Re_V = \frac{\rho_V \bar{U}_V D}{\mu_V} = 1.86 \times 10^4,$$

taking $\mu_V = 1.07 \times 10^{-5}$ kg m^{-1} s^{-1}. This suggests that the vapour is turbulent, at least near the inlet.

Q5.13 Assuming the film in Q5.12 is ripple free over most of its height, calculate the shear stresses acting on it and estimate its thickness.

Solution: The inertial shear stress at exit is

$$\tau_I = \frac{1}{2} f \rho_V \bar{U}_V^2$$

in which $f = 0.079/Re_V^{1/4} = 6.76 \times 10^{-3}$. Hence $\tau_I = 5.77 \times 10^{-4}$ N m^{-2}.

This acts upward on the film along with the buoyancy induced drag which may be estimated from the force balance

$$\pi D \tau_b \, dX = \pi D \Delta g (\rho_L - \rho_V) \, dX$$

on an elementary ring dX long. Hence

$$\tau_b = \Delta g(\rho_L - \rho_V).$$

To determine the film thickness it is necessary to use a relation between Γ and Δ. For a falling film

$$\Gamma = \frac{g\Delta^3(\rho_L - \rho_V)}{3\nu_L},$$

and hence

$$\Delta = \left(\frac{3\nu_L\Gamma}{g(\rho_L - \rho_V)}\right)^{1/3} = 1.98 \times 10^{-4}\,\text{m}$$

if $(\rho_L - \rho_V) \simeq \rho_L = 987.8\,\text{kg m}^{-3}$ and $\mu_L = 5 \times 10^{-4}\,\text{kg m}^{-1}\,\text{s}^{-1}$. Hence $\tau_b = 1.92\,\text{N m}^{-2}$ which is much greater than τ_I. This implies that the vapour has little effect on the hydrodynamics of the film.

Q5.14 Estimate the average heat transfer rate for the condenser discussed in Q5.12 and Q5.13 if the wall temperature is maintained at 55°C. Hence determine the cylinder height *H*.

Solution: The heat transfer coefficient may be found from eqn (5.42):

$$\overline{Nu}_f = [(4Re_f)^{-0.44} + 5.82 \times 10^{-6}(4Re_f)^{0.8}Pr_L^{0.33}]^{1/2}$$

in which $Pr_L = \mu_L c_{pL}/k_L = 3.19$ if $c_{pL} = 4.18 \times 10^3\,\text{J kg}^{-1}\,\text{K}^{-1}$ and $k_L = 0.654\,\text{W m}^{-1}\,\text{K}^{-1}$. From Q5.12, $4Re_f = 400$. Hence $\overline{Nu}_f = 0.269$. This result may also be obtained from Fig. 5.7. Thus

$$\bar{h}_H = 0.269 k_L\left(\frac{g}{\nu_L^2}\right)^{1/3} = 5930\,\text{W m}^{-2}\text{K}^{-1}.$$

The average heat flux density is therefore

$$\bar{q} = \bar{h}_H(T_{sat} - T_0) = 5.93 \times 10^4\,\text{W m}^{-2},$$

since $T_{sat} = 65°\text{C}$ at 25 kPa.

A heat balance requires that

$$\bar{q}\pi DH = \pi D\Gamma\lambda,$$

and hence

$$H = \frac{\Gamma\lambda}{\bar{q}} = 1.97\,\text{m},$$

using the mass flow rate found in Q5.12 and taking $\lambda = 2.346 \times 10^6\,\text{J kg}^{-1}$. This result provides the minimum height because T_{sat} decreases with height, thus reducing $\bar{q}$.

Q5.15 If a small amount of air accumulates in the condenser discussed in Q5.12–5.14, thereby creating a bulk vapour mass fraction $m_{V\infty}$, find a relation between the mean interface temperature $\bar{T}_I$ and the suction parameter $B = (m_{V\infty} - m_{VI})/(m_{VI} - 1)$ where m_{VI} is the vapour mass fraction at the interface.

Solution: The average heat flux density is given by

$$\bar{q} = \frac{k_L(\bar{T}_I - T_0)}{\bar{\Delta}},$$

where

$$\bar{\Delta} = \frac{1}{H}\int_0^H \Delta(X)\,\mathrm{d}X \simeq \frac{4}{5}\Delta_H$$

under the laminar film conditions determined in Q5.12. The average heat flux density may also be written

$$\bar{q} = \frac{\Gamma\lambda}{H} \simeq \frac{\lambda g \rho_L^2 \Delta_L^3}{3\mu_L H},$$

while from eqns (5.59) and (5.61) averaged,

$$\bar{q} = -\lambda \bar{h}_M^0 \ln(1 + B),$$

where $\bar{h}_M^0$ is the 'dry' mass transfer coefficient.

Eliminating the film thickness from the above expressions we obtain

$$C(\bar{T}_I - T_0)^{3/4} = -\ln(1 + B),$$

where

$$C = \frac{5k_L}{4\lambda \bar{h}_M^0}\left(\frac{4\lambda g \rho_L^2}{15Hk_L\mu_L}\right)^{1/4}.$$

This provides an explicit relation between $\bar{T}_I$ and B provided that each quantity in the coefficient C may be taken as constant.

Q5.16 Estimate the effect of a 3 per cent bulk concentration of air in the above condenser unit.

Solution: The vapour Reynolds number based on height is given by

$$Re_H = Re_D \cdot \frac{H}{D} = 3.1 \times 10^4.$$

This suggests laminar vapour flow, unlike the conclusion based on the fully developed conditions assumed in Q5.12. The neglect of vapour shear has been confirmed along with the model of a flat plate wrapped into cylindrical shape.

Under these conditions, the 'dry' mass transfer coefficient is determined from the laminar heat transfer analogue. Thus

$$\bar{h}_M^0 = \frac{\bar{\rho}\mathfrak{D}_{Vg}}{H} \times 0.664 Sc_V^{1/3} Re_H^{1/2} = 1.10 \times 10^{-3}\,\mathrm{kg\,m^{-2}\,s^{-1}},$$

taking $Sc_V = 0.6$ and $\mathfrak{D}_{Vg} = 1.16 \times 10^{-4}\,\mathrm{m^2\,s^{-1}}$, and $\bar{\rho} = 0.19\,\mathrm{kg\,m^{-3}}$. Hence, from Q5.15, $C = 3.09\ (\mathrm{K})^{-3/4}$ and therefore

$$3.09(\bar{T}_I - T_0)^{3/4} = -\ln(1 + B)$$

in which

$$B = \frac{m_{V\infty} - m_{VI}}{m_{VI} - 1}.$$

To determine B it is necessary to make use of the following expressions:

$$m_{VI} = \frac{x_{VI}\mathfrak{M}_V}{x_{VI}\mathfrak{M}_V + (1 - x_{VI})\mathfrak{M}_g}$$

and

$$x_{VI} = \frac{P_{VI}}{P} = \exp\left[\frac{\lambda(\bar{T}_I - T_i)}{R_V T_i^2}\right],$$

where x_{VI} is the vapour mole fraction, P_{VI} is the saturated vapour pressure, $T_i = 65°\mathrm{C}$ is the saturation temperature (at P) of pure vapour, and $\mathfrak{M}_V$, $\mathfrak{M}_g$ are the vapour and gas molecular weights, respectively. In these circumstances,

$$m_{VI} = \frac{18x_{VI}}{18x_{VI} + 29(1 - x_{VI})},$$

and

$$x_{VI} = \exp[0.0445(\bar{T}_I - 338)]$$

if $R_V = 461.9\,\mathrm{J\,kg^{-1}\,K^{-1}}$ and $\lambda = 2.346 \times 10^6\,\mathrm{J\,kg^{-1}}$.

The relation betwee $\bar{T}_I$ and B may now be satisfied interatively: $\bar{T}_I$ is guessed to provide x_{VI} and hence m_{VI} which yields B, given that $m_{V\infty} = 0.97$. Proceeding thus we obtain $\bar{T}_I = 55.9°\mathrm{C}$, corresponding to $B = -0.933$ and $m_{VI} = 0.554$. The heat flux density is then

$$\bar{q} = -\lambda \bar{h}_M^0 \ln(1 + B) = 0.70 \times 10^4\,\mathrm{W\,m^{-2}},$$

as compared with $\bar{q} = 5.93 \times 10^4\mathrm{W\,m^{-2}}$ calculated in Q5.14 for a pure vapour. The air concentration has thus reduced the heat transfer rate and condensation rate by about 88 per cent, mainly because of the very low mass (and heat) transfer coefficient. More precisely, it is necessary to accommodate this reduced rate in the axial vapour flux and Re_D, which determines $\bar{h}_M^0$. After adjustment of $\bar{h}_M^0$ and C, the iterative procedure is repeated.

Q5.17 If the water and steam flow rates used in Q3.1–Q3.4 refer to annular condensation, calculate the heat transfer coefficient at atmospheric pressure. Given: $\bar{\Delta} = 2.72\,\text{mm}$ and $\tau_w = 9.1\,\text{N m}^{-2}$.

Solution: Since

$$\bar{\Delta} = \frac{\mu_L}{(\rho_L \tau_w)^{1/2}} \delta^+,$$

we note that $\delta^+ = 911$ if $\mu_L = 2.8 \times 10^{-4}\ \text{kg m}^{-1}\text{s}^{-1}$ and $\rho_L = 958\,\text{kg m}^{-3}$. Using eqn (5.41), this value of δ^+ yields $\theta^+ = 28.6$, if $c_{PL} = 4.2 \times 10^3\,\text{J kg}^{-1}\,\text{K}^{-1}$ and $k_L = 0.68\,\text{W m}^{-1}\text{K}^{-1}$ thus giving $Pr_L = 1.73$.

Equation (5.37),

$$\theta^+ = (T_w - T) c_{PL} \frac{(\tau_w \rho_L)^{1/2}}{\bar{q}},$$

now enables us to define the average heat transfer coefficient as

$$\bar{h} = \frac{\bar{q}}{T_w - T_I} = \frac{c_{PL}(\tau_w \rho_L)^{1/2}}{\theta_I^+}.$$

Hence $\bar{h} = 1.37 \times 10^4\,\text{W m}^{-2}\,\text{K}^{-1}$.

An alternative, empirical method for calculating the heat transfer coefficient in a forced annular flow is described in Section 6.4.4 and illustrated in Q6.12.

Q6.1 A water spray with a liquid water lid content w_φ of $5\,\text{kg m}^{-3}$ and a drop volumetric mean diameter d_m of $15\,\mu\text{m}$ cools a steel strip. Find the average spacing X_d between the drops in the spray. Also find the number per unit area approaching the strip, and determine the diameter D_i of the hemispherical drops produced by impact.

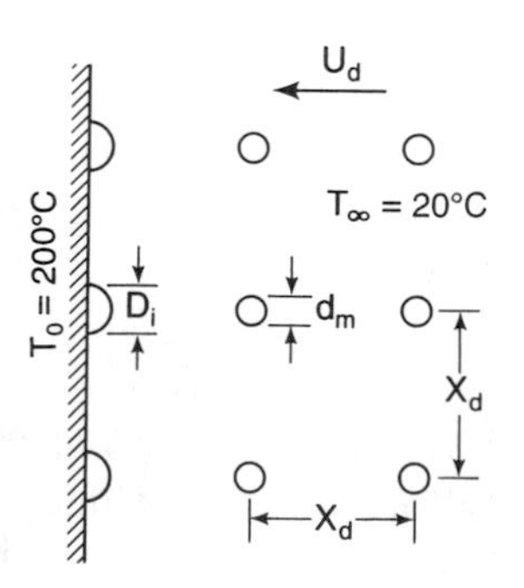

Solution: Liquid water content $w_\varphi = \rho_L n_\varphi v_m$, where n_φ is the number of drops per unit volume and $v_m = \pi d_m^3/6$ is the mean volume. If $\rho_L = 958\,\text{kg m}^{-3}$, this yields

$$n_\varphi = \frac{6w_\varphi}{\rho_L \pi d_m^3} = 2.95 \times 10^{12}\,\text{m}^{-3}.$$

The average drop spacing under isotropic conditions is therefore $X_d = n_\varphi^{1/3} = 69.7\,\mu\text{m}$.

The number approaching the strip surface is $n_A = n_\varphi^{2/3} = 2.06 \times 10^8\,\text{m}^{-2}$. The hemispherical drop diameter formed on impact may be found from the mass balance

$$\frac{\pi d_m^3}{6} = \frac{\pi}{12} D_i^3,$$

and hence

$$D_i = 2^{1/3} d_m = 18.9\,\mu\text{m}.$$

Q6.2 Estimate the conductance K_d and the heat transfer coefficient h_d of a hemispherical drop based on the average path length and the average cross sectional area for the conductive heat flux.

Solution: By definition, $K_d = k_L \bar{A}/\bar{R}$
in which

$$\bar{R} \simeq \left(\frac{D}{2} + 0\right) \Big/ 2,$$

and

$$\bar{A} \simeq \left(\frac{\pi D^2}{2} + \frac{\pi D^2}{4}\right) \Big/ 2.$$

Hence

$$K_d \simeq \frac{3\pi k_L D}{2} = 6 \times 10^{-5}\,\text{W K}^{-1}$$

immediately after impact when $D = D_i$: $k_L = 0.68\,\text{W m}^{-1}\,\text{K}^{-1}$.

The heat transfer coefficient of the drop may be defined by

$$h_d = K_d / A.$$

When $A = A_b = \pi D^2/4$ is the base (contact) area of the drop, this yields

$$h_d = \frac{6k_L}{D_i} = 2.1 \times 10^5\,\text{W m}^{-2}\,\text{K}^{-1}$$

immediately after impact. This magnitude illustrates the ability of small drops to conduct heat very efficiently. The ability increases as the drop evaporates.

Q6.3 Demonstrate that evaporation of a hemispherical surface drop is controlled by drop conductance if an external convective heat transfer coefficient $h_e = 250\ \mathrm{W\,m^{-2}\,K^{-1}}$ is produced by relative motion between spray vapour at 20°C and a steel strip at 200°C.

Solution: The interface equation for the drop may be written

$$A_I \frac{\rho_L}{2} \lambda \frac{dD}{dt} = -K_d(T_0 - T_I) + A_I h_e (T_I - T_\infty),$$

where the interfacial area $A_I = \pi D^2/2$ and T_0 is the surface temperature. Hence

$$A_I \frac{\rho_L}{2} \lambda \frac{dD}{dt} = -K_d(T_o - T_I)\left[1 - \frac{h_e D}{3k_L} \frac{(T_I - T_\infty)}{(T_0 - T_I)}\right].$$

With $h_e = 250\ \mathrm{W\,m^{-2}\,K^{-1}}$, $h_e D/3k_L = 2.3 \times 10^{-3}$ immediately after impact. This Biot number decreases with evaporation.

The second term on the right-hand side of the interface equation may therefore be neglected if $(T_I - T_\infty)/(T_0 - T_I) \leqslant O(1)$. At atmospheric pressure $T_I < 100°\mathrm{C}$ for pure water vapour. The presence of air implies that $T_I < 100°\mathrm{C}$. The conductance of the drop therefore controls the evaporation rate. The interface equation thus reduces to

$$A_I \frac{\rho_L \lambda}{2} \frac{dD}{dt} = -K_d(T_0 - T_I).$$

Q6.4 Estimate the time required for the surface drops mentioned in the previous questions to evaporate completely if the strip surface remains at a temperature of 200°C.

Solution: The heat flux density $\dot{q}_b$ at the drop base is given by the heat balance

$$\frac{\pi}{4} D^2 \dot{q}_b = K_d(T_0 - T_I) = \frac{3\pi}{2} k_L D(T_0 - T_I).$$

Hence

$$\dot{q}_b = \frac{6k_L(T_0 - T_I)}{D}.$$

This may be combined with the drop interface equation

$$\frac{\pi}{4} D^2 \dot{q}_b = -\frac{\pi}{4} D^2 \rho_L \lambda \frac{dD}{dt}$$

to yield

$$\rho_L \lambda \frac{dD}{dt} = -\frac{6k_L(T_0 - T_I)}{D}.$$

Assuming T_I does not change much during evaporation, this equation may be integrated to give

$$D = \left[D_i^2 - \frac{12k_L(T_0 - T_L)t}{\rho_L \lambda} \right]^{1/2},$$

where D_i is the initial diameter. The time for complete evaporation is thus

$$t_f = \frac{\rho_L \lambda D_i^2}{12k_L(T_0 - T_I)} \simeq 0.948 \text{ ms},$$

taking $T_0 - T_I \simeq 100$ K and $\lambda = 2.26 \times 10^6 \text{J kg}^{-1}$. This time interval is only about twice the period required for transient effects to die out. It therefore underestimates the evaporation time but does not alter the order of magnitude, i.e. $t_f = 0$ (1 ms). Dilution of the vapour by the air implies that $T_0 - T_I < 100$ K.

Q6.5 Estimate the average surface heat flux density $\dot{q}_s$ on the steel strip under the above conditions.

Solution: The average heat flux density $\bar{\dot{q}}_{bi}$ beneath each drop may be estimated from the energy balance

$$\frac{\pi}{4} D_i^2 \bar{\dot{q}}_{bi} t_f = \frac{\pi D^3}{12} \rho_L \lambda.$$

Therefore

$$\bar{\dot{q}}_{bi} = \frac{\rho_L \lambda D_i}{3t_f} = 13.6 \text{ MW m}^{-2},$$

based on the evaporation time calculated above. The fraction of the strip surface on which this occurs is equal to $n_A \pi D_i^2/4$, or 5.8 per cent. Since the remaining surface of the strip is exposed to a vapour–air mixture its contribution to heat transfer may be neglected. The average heat flux density of the strip surface is therefore estimated from

$$\bar{\dot{q}}_s = \bar{\dot{q}}_{bi} \frac{n_A \pi D_i^2}{4} = 0.79 \text{ MW m}^{-2},$$

using the value of n_A calculated in Q6.1.

This rate of heat loss occurs only during the evaporation period t_f. For this process to repeat itself without the surface incurring dry intervals, the spray velocity U_d towards the surface is given by $U_d = X_d/t_f = 7 \text{ cm s}^{-1}$, using the spacing X_d calculated in Q6.1.

The effective velocity U_d is a crucial determinant in spray cooling. If it is too small, the surface will experience dry intervals which reduce the cooling rate. If it is too large, the surface will be inundated with a continuous film which, if thick enough, will also reduce the cooling rate. It is also important to note that the drops will be retarded by escaping vapour and must therefore be ejected from the spray nozzle with a higher initial velocity.

Q6.6 Construct an idealized pool boiling curve for water surrounding a 2 cm diameter horizontal tube if the pressure is 1 bar and surface microcavities have a mean diameter of 10 μm.

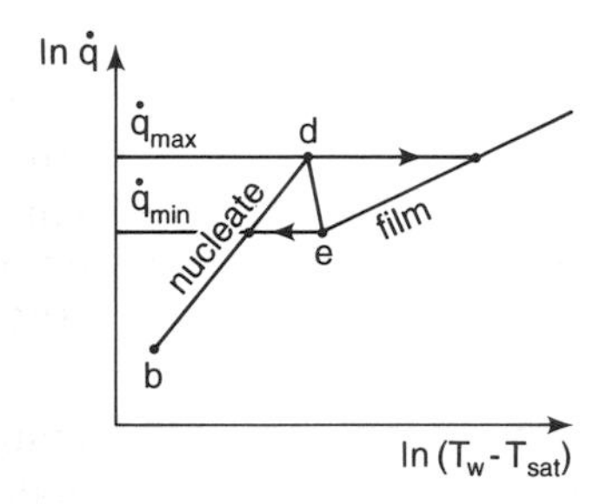

Solution: The main features of pool boiling may be reflected in an ideal representation consisting of two curves: the nucleate boiling curve, extending from the onset of nucleation (b) to the maximum (critical) heat flux (d); and the film boiling curve extending upwards from the minimum heat flux point (e).

The nucleate boiling curve may be represented by eqn (6.13):

$$\dot{q} = K(T_w - T_{sat})^3$$

in which

$$K = 1.89 \times 10^{-14} \frac{g^{1/2}\lambda^{1/8} k_L^{1/2} \rho_L^{17/8} c_{pL}^{19/8} \rho_V^{1/8}}{\sigma^{9/8}(\rho_L - \rho_V)^{5/8} T_{sat}^{1/8}} = 40.37\,\mathrm{W\,m^{-2}\,K^{-3}},$$

given the properties listed at the end of the problem.

The onset of nucleate boiling is given by eqn (6.7):

$$T_w - T_{sat} = \frac{2\sigma R_V T_{sat}^2}{\lambda r_b P_{sat}},$$

where r_b is taken as the effective microcavity radius (5 μm). Hence $T_w - T_{sat} = 6.68$ K at b, thus determining the heat flux density $\dot{q} = 12\,\mathrm{kW\,m^{-2}}$.

The maximum heat flux is given by eqn (6.17):

$$\dot{q}_{max} = 0.15\ \lambda\rho_V^{1/2}(\sigma g \Delta\rho)^{1/4} = 1.27 \times 10^6\,\mathrm{W\,m^{-2}}.$$

The corresponding temperature difference is $T_w - T_{sat} = 31.6$ K at d.

The minimum heat flux is given by eqn (6.19):

$$\dot{q}_{\min} = \left(\frac{\rho_V}{\rho_L}\right)^{1/2} \dot{q}_{\max} = 3.13 \times 10^4 \,\mathrm{W\,m^{-2}}.$$

while the corresponding temperature difference may be found from the film boiling curve, eqn (6.4):

$$\dot{q} = 0.62 \left[\frac{\rho_V \rho_L \lambda k_V^3 g}{\mu_V D}\right]^{1/4} (T_w - T_{sat})^{3/4}.$$

Hence $T_w - T_{sat} = 201$ K at e.

$k_L = 0.680\,\mathrm{W\,m^{-1}\,K^{-1}}$ $\quad k_V = 24.8 \times 10^{-3}\,\mathrm{W\,m^{-1}\,K^{-1}}$
$\rho_L = 958\,\mathrm{kg\,m^{-3}}$ $\quad \rho_V = 0.59\,\mathrm{kg\,m^{-3}}$
$c_{PL} = 4.22 \times 10^3\,\mathrm{J\,kg^{-1}\,K^{-1}}$ $\quad \mu_V = 12 \times 10^{-6}\,\mathrm{kg\,m^{-1}\,s^{-1}}$
$\sigma = 0.0588\,\mathrm{N\,m^{-1}}$ $\quad \lambda = 2.26 \times 10^6\,\mathrm{J\,kg^{-1}}$

Q6.7 For a pressure of 1 bar, find the burnout temperature of the above system if the heat flux density is raised to and maintained at $\dot{q}_{\max}$. Also find the temperature at which nucleate boiling resumes when the heat flux density is lowered slightly beneath $\dot{q}_{\min}$.

Solution: Under these conditions, film boiling is described by

$$\dot{q} = 586(T_w - T_{sat})^{3/4}.$$

Hence when $\dot{q} = \dot{q}_{\max} = 1.27 \times 10^6\,\mathrm{W\,m^{-2}}$, $T_w - T_{sat} = 2.8 \times 10^4$ K! While this magnitude illustrates the dramatic change in substrate temperature it greatly overestimates the rise. Turbulent film boiling augmented by thermal radiation reduces the rise but not enough to avoid serious overheating.

When film boiling collapses into nucleate boiling,

$$\dot{q}_{\min} = 40.4(T_w - T_{sat})^3,$$

and hence $T_w - T_{sat} = 9.18$ K. Therefore $T_w \simeq 382$ K. This temperature is only about 2 K greater than the onset temperature T_b.

Q6.8 Assuming a power law relation ($\dot{q} \propto P^m$) for the above system, calculate the rate of change of the maximum and minimum heat fluxes with pressure near 1 bar.

Solution: The results of Exercise 8 after Chapter 6 are presented below:

$\ln P$	$\ln \dot{q}_{\max}$	$\ln \dot{q}_{\min}$
−2.30	13.06	8.28
0	14.05	10.35
2.30	14.91	12.34

Near 1 bar, these yield

$$\dot{q}_{max} = 1.27 \times 10^6 P^{0.402},$$

and

$$\dot{q}_{min} = 3.13 \times 10^4 P^{0.883},$$

with P measured in bars. If these relations were maintained at higher pressures, $\dot{q}_{min} = \dot{q}_{max}$ when $P = 2200$ bar. Since this is much higher than the critical pressure (221 bar) the transitional régime is not likely to exhibit a positive slope. In fact, $\dot{q}_{max}$ is found to increase only up to $P/P_C \simeq 0.3$, after which it decreases as $P/P_C \rightarrow 1$.

Q6.9 Calculate the length X_{on} at which nucleate boiling begins if a pipe, having a diameter of 4 cm and a wall temperature of 217.4°C, contains water which enters at 2 MPa, 160°C with a flow rate of 10 kg s^{-1}.

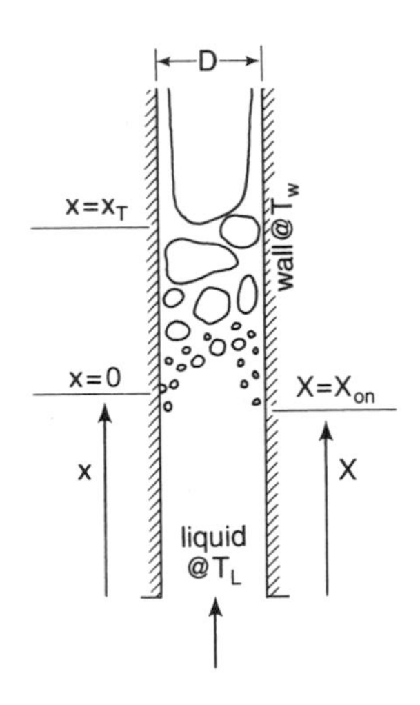

Solution: The heat transfer coefficient may be determined from the correlation

$$Nu_D = 0.023\, Re_D^{0.8} Pr_L^{0.4}$$

for $T_w > T_L$. The Reynolds number is given by

$$Re_D = \frac{4\dot{M}}{\pi \mu_L D} = 2.51 \times 10^6$$

if $\mu_L = 1.27 \times 10^{-4}$ kg m^{-1} s^{-1}. Hence, if $Pr_L = 0.89$,

$$h = \frac{k_L Nu_D}{D} = 4.69 \times 10^4 \text{ W m}^{-2}\text{ K}^{-1},$$

taking $k_L = 0.65$ W m^{-1} K^{-1}.

The onset of nucleate boiling is predicted by eqn (6.22):

$$\dot{q} = \frac{\rho_V k_L \lambda (T_w - T_{sat})^2}{8\sigma T_{sat}} = 2.64 \times 10^6 \text{ W m}^{-2}$$

if $\rho_V = 10 \text{ kg m}^{-3}$, $\lambda = 1.89 \times 10^6 \text{ J kg}^{-1}$, and $\sigma = 0.035 \text{ N m}^{-1}$: $T_{sat} =$ 485.15 K at 2 MPa. Hence

$$\theta = T_w - T_L = \frac{\dot{q}}{h} = 56.24 \text{ K}.$$

Finally, using the result of the exercise after Chapter 6,

$$X_{on} = \frac{\dot{M} c_{pL}}{h\pi D} \ln \frac{\theta_i}{\theta} = 15.8 \text{ cm}$$

if $c_{pL} = 4.56 \times 10^3 \text{ J kg}^{-1} \text{K}^{-1}$.

Q6.10 Estimate the quality x_t at which transition to annular flow occurs in the above pipe if the flow rate is reduced to 0.32 kg s^{-1} and the pressure is reduced to atmospheric under saturated conditions.

Solution: Since $G = 4\dot{M}/\pi D^2 = 255 \text{ kg m}^{-2} \text{s}^{-1}$, the liquid and vapour mass flux densities are given by

$$G_L = G(1 - x) = 255(1 - x)$$

and

$$G_V = Gx = 255x,$$

respectively. The corresponding superficial velocities are therefore

$$U_L^s = G_L/\rho_L = 0.266(1 - x)$$

and

$$U_V^s = G_V/\rho_V = 431x$$

if $\rho_V = 0.59 \text{ kg m}^{-3}$ and $\rho_L = 958 \text{ kg m}^{-3}$.

The locus of these coordinates is a straight line given by

$$U_V^s = 431(1 - 3.76\, U_L^s).$$

This cuts the boundary between the churn and annular regions (see Fig. 3.13 or Fig. 6.17) at $U_V^s \simeq 12.6 \text{ ms}^{-1}$, i.e. at $U_L^s = 0.258 \text{ m s}^{-1}$ and $x_t \simeq 3$ per cent. In practice, the transitional point is not distinct.

Q6.11 For flow boiling in a pipe of diameter D and uniform wall temperature T_w show that

$$\Delta X = \frac{\dot{M} \lambda \Delta x}{\pi D h (T_w - T_{sat})}$$

gives the length interval ΔX for a change in quality Δx if the mass flow rate is $\dot{M}$ and the heat transfer coefficient is h.

Solution: A heat balance on an elementary length $\mathrm{d}X$ gives

$$\pi D\dot{q}_{\mathrm{x}}\,\mathrm{d}X = \dot{M}\lambda\,\mathrm{d}x,$$

where λ is the latent heat of evaporation and $\dot{q}_{\mathrm{w}} = h(T_{\mathrm{w}} - T_{\mathrm{sat}})$ is the wall heat flux density. Hence for any length interval ΔX over which h remains fixed,

$$\Delta X = \frac{\dot{M}\lambda\Delta x}{\pi Dh(T_{\mathrm{w}} - T_{\mathrm{sat}})},$$

if Δx is the corresponding change in quality.

Q6.12 A 2.0 cm diameter horizontal pipe with a wall temperature of 150°C carries boiling water at a pressure of 2.5×10^5 kPa and a flow rate $\dot{M} = 0.06\ \mathrm{kg\,s^{-1}}$. Using the properties listed at the end of the problem, calculate the heat transfer coefficient where the quality $x = 0.1$.

Solution: Following the procedure outlined in the text, we first obtain $\dot{q}_{\mathrm{FC}}$:

$$\chi = \left(\frac{\rho_{\mathrm{L}}}{\rho_{\mathrm{V}}}\right)^{0.5}\left(\frac{\mu_{\mathrm{V}}}{\mu_{\mathrm{L}}}\right)^{0.1}\left(\frac{x}{1-x}\right)^{0.9} = 2.71.$$

Hence

$$F_1 = 0.15(\chi + 2\chi^{0.32}) = 0.8192.$$

Since

$$Re = 4\dot{M}/\mu_{\mathrm{L}}\pi\mathrm{D} = 1.744 \times 10^4, \quad Re_{\mathrm{L}} = Re(1-x) = 1.57 \times 10^4 > 1125.$$

Hence

$$F_2 = 5Pr_{\mathrm{L}} + 5\ln(1 + 5Pr_{\mathrm{L}}) + 2.5\ln(3.1 \times 10^{-3} Re_{\mathrm{L}}^{0.81}) = 22.19 \text{ if } Pr_{\mathrm{L}} = 1.36.$$

Therefore

$$h_{\mathrm{A}} = \frac{k_{\mathrm{L}}}{D} Pr_{\mathrm{L}} Re_{\mathrm{L}}^{0.9} \frac{F_1}{F_2} = 10.27 \times 10^3\ \mathrm{W\,m^{-2}\,K^{-1}},$$

and hence $\dot{q}_{\mathrm{FC}} = h_{\mathrm{A}}(T_{\mathrm{w}} - T_{\mathrm{sat}}) = 2.32 \times 10^5\ \mathrm{W\,m^{-2}}$, since $T_{\mathrm{sat}} = 127.4$°C at a pressure of 2.5×10^5 kPa.

The onset of boiling occurs at

$$T_{\mathrm{wo}} - T_{\mathrm{sat}} = \frac{8h_{\mathrm{A}}\sigma T_{\mathrm{sat}}}{\rho_{\mathrm{V}} k_{\mathrm{L}} \lambda} = 0.67\ \mathrm{K}.$$

Hence $\dot{q}_{NB} = K(T_w - T_{sat})^3 = 6.63 \times 10^5\ \mathrm{W\,m^{-2}}$, where $K = 57.4\ \mathrm{W\,m^{-2}\,K^{-3}}$, and $\dot{q}_N = K(T_{wo} - T_{sat})^3 = 17.3\ \mathrm{W\,m^{-2}}$.

Therefore $\dot{q}_B = \dot{q}_{NB} - \dot{q}_N \simeq 6.63 \times 10^5\ \mathrm{W\,m^{-2}}$, and

$$h = \frac{(\dot{q}_{FC} + \dot{q}_B)}{(T_w - T_{sat})} = 3.96 \times 10^4\ \mathrm{W\,m^{-2}\,K^{-1}}.$$

Properties:

$\rho_L = 937\ \mathrm{kg\,m^{-3}}$ $\qquad$ $\rho_V = 1.39\ \mathrm{kg\,m^{-3}}$

$\mu_L = 2.19 \times 10^{-4}\ \mathrm{kg\,m^{-1}\,s^{-1}}$ $\qquad$ $\mu_V = 0.134 \times 10^{-4}\ \mathrm{kg\,m^{-1}\,s^{-1}}$

$k_L = 0.685\ \mathrm{W\,m^{-1}\,K^{-1}}$ $\qquad$ $\lambda_V = 2.72 \times 10^6\ \mathrm{J\,kg^{-1}}$

$\sigma = 0.053\ \mathrm{N\,m^{-1}}$

Q6.13 Repeat the above procedure at intervals of $\Delta x = 0.1$ in the range $0.05 < x < 0.95$ using the result obtained in Q6.11. Hence estimate the overall length of the pipe.

Solution: With x being the independent variable in Q6.12, the expressions may be reduced to the following:

$$\chi = 19.53\left(\frac{x}{1-x}\right)^{0.9}$$

$$F_1 = 0.15(\chi + 2\chi^{0.32})$$

$$Re_L = 1.744 \times 10^4(1 - x)$$

$$F_2 = 22.4 + 2.02 \ln(1 - x)$$

$$\dot{q}_{FC} = 6.91 \times 10^6(1 - x)^{0.9}\frac{F_1}{F_2}\ \mathrm{W\,m^{-2}}$$

$$\dot{q}_B = 6.63 \times 10^5\ \mathrm{W\,m^{-2}}$$

$$h = \frac{(\dot{q}_{FC} + \dot{q}_B)}{22.6}\ \mathrm{W\,m^{-2}\,K^{-1}}$$

$$\Delta X = \frac{11.5 \times 10^3}{h}\ \mathrm{m}, \quad \text{since} \quad \Delta x = 0.1$$

The overall pipe length $L \simeq \Sigma\Delta X$ over ten intervals from $x = 0.5$ to $x = 0.95$. From the table below we find that $L = 208.3$ cm. Strictly, the calculations assume an annular flow. To accommodate earlier and later heat transfer the pipe length must be increased. Note also that $Re_L > 1125$ except at $x = 0.95$.

x	0.05	0.15	0.25	0.35	0.45	0.55	0.65	0.75	0.85	0.95
h (kW m^{-2} K^{-1})	36.4	42.3	47.2	51.8	56.1	60.4	64.7	69.4	74.8	83.9
ΔX (cm)	31.5	27.2	24.4	22.2	20.5	19.0	17.8	16.6	15.4	13.7

Q6.14 Using the data from Q6.13 estimate the average heat transfer coefficient for the pipe:

$$\bar{h} = \frac{1}{10}\Sigma h.$$

Solution: From the above table, $\Sigma h = 587\,\text{kW m}^{-2}\,\text{K}^{-1}$. Hence $\bar{h} = 58.7\,\text{kW m}^{-2}\,\text{K}^{-1}$.

Q7.1 A steam desuperheater consists of a pipe of diameter $D = 10\,\text{cm}$ into which water is sprayed radially inwards from a collar. If the spray is injected at an angle of $\alpha = 2.866$ degrees from the tube axis, with an absolute velocity of $U = 10\,\text{m s}^{-1}$, calculate the radial velocity U_d of the drops and hence determine the time interval and pipe length if any unevaporated drops are to reach the centreline.

If the steam enters with a superheat of 100 K, determine the minimum water flux so that saturated steam may exit when the average steam velocity is $10\,\text{m s}^{-1}$.

Solution: The radial velocity $U_d = U \sin\alpha = 0.5\,\text{m s}^{-1}$. The time required for an unevaporated drop to reach the centreline is

$$t = \frac{D}{2U_d} = 0.1\,\text{s}.$$

During this period, the axial travel of the drops is

$$L = (U\cos\alpha)t = 1\,\text{m},$$

providing there is no slip relative to the steam.

An energy balance requires that

$$\lambda \dot{M}_L = \dot{M}_V c_{pV}(T_V - T_{sat}),$$

where $\dot{M}_L$ is the minimum water flux, i.e. assuming complete evaporation of the spray.

Hence $\dot{M}_L = \pi D^2 \rho_V U_V Ja_V/4 = 0.0364\,\text{kg s}^{-1}$, where U_V is the average steam velocity, $\rho_V = 5.15\,\text{kg m}^{-3}$, $c_{pV} = 2.5 \times 10^3\,\text{J kg}^{-1}\,\text{K}^{-1}$ and $\lambda = 2.78 \times 10^6\,\text{J kg}^{-1}$.

Q7.2 Calculate the interfacial heat transfer coefficient of the drops in Q7.1 if the vapour pressure is 1 MPa. Hence estimate the time required for evaporation of the drops if their volumetric mean diameter is 50 μm. Compare this time interval with the flight time calculated in Q7.1.

Solution: For a pure vapour, the heat transfer coefficient is given by eqn (5.43):

$$h_I = \frac{2\rho_V \lambda^2}{T_I(2\pi R_V T_I)^{1/2}} = 6.12 \times 10^8 \text{ W m}^{-2}\text{ K}^{-1},$$

taking $R_V = 461.9 \text{ J kg}^{-1}\text{ K}^{-1}$: $T_I = 179.9°\text{C}$.

The interface equation for a drop is given by

$$\rho_L \lambda \frac{dR_I}{dt} = h_I(T_V - T_I)$$

or, in normalized form,

$$\left[\frac{\rho_L \lambda R^c}{h_I(T_V - T_I)t^c}\right]\frac{dr_I}{d\tau} = 1.$$

The time scale for complete evaporation of a drop with a length scale $R^c = O$ (25 μm) is therefore

$$t^c = O\left[\frac{\rho_L \lambda R^c}{h_I(T_V - T_I)}\right] = O(1 \text{ μs})$$

if $\rho_L = 887 \text{ kg m}^{-3}$. This period is very much less than the maximum flight time, indicating that evaporation is essentially instantaneous upon exposure to the superheated vapour.

Q7.3 The above desuperheating process may be modelled as a ring of spray drops converging radially. Evaporation occurs 'instantaneously' at the inner surface of the ring, the last (outer) drops evaporating as they reach the pipe centre line.

Calculate the thickness Δ of the spray ring injected at the collar around the pipe wall if the drop number density $n_{\nu w} = 10^{11} \text{ m}^{-3}$ at the pipe circumference.

Calculate the axial length W of the spray collar.

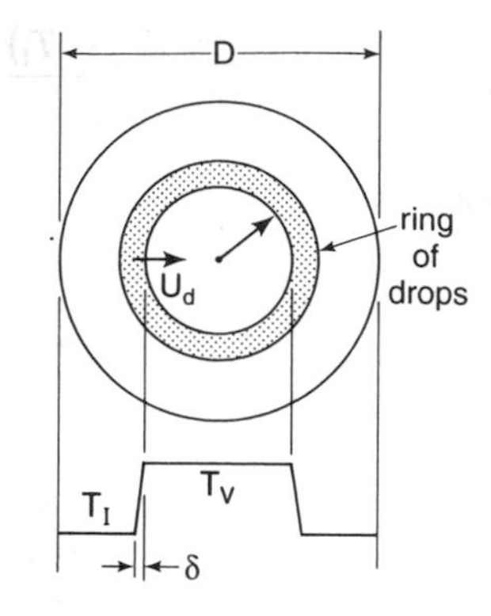

Solution: For complete desuperheating, the latent heat addition to the injected ring must balance the sensible heat loss of the vapour cylinder through which the ring moves. Thus for an elementary ring of axial length $\mathrm{d}X$,

$$(\pi D \Delta \mathrm{d}X) \frac{\pi}{6} d^3 \rho_L \lambda n_{vw} = \left(\frac{\pi D^2}{4} \mathrm{d}X\right) \rho_V c_{pV}(T_V - T_I),$$

where d is the drop diameter. Hence

$$\Delta = \frac{3D\rho_V Ja_V}{2\pi \rho_L n_{vw} d^3} = 2.76\,\text{mm}.$$

A mass balance at the collar gives

$$\dot{M}_L = \pi D W \left(U_d n_{vw} \frac{\pi}{6} d^3 \rho_L\right).$$

Hence

$$W = \frac{6\dot{M}_L}{\pi^2 D U_d n_{vw} d^3 \rho_L} = 5.5\,\text{cm}.$$

Q7.4 Near the inner surface (R_i) of the converging spray ring discussed above, the conversion of latent heat to sensible heat may be described by

$$\mathrm{d}\lambda_v = \rho_V c_{pV} \mathrm{d}T,$$

where

$$\lambda_v = \frac{\pi}{6} d^3 \rho_L \lambda n_v$$

is the spray latent heat per unit volume, and n_v is the number of drops per unit volume; n_v and λ_v are both functions of R. Show that the temperature gradient produced in the spray at R_i is given by

$$\left(\frac{\mathrm{d}T}{\mathrm{d}R}\right)_{R_i} = -\frac{\pi n_v d^2 h_I (T_V - T_I)}{\rho_V c_{pV} U_d},$$

and hence estimate the penetration

$$\delta = -(T_V - T_I) \Big/ \left(\frac{\mathrm{d}T}{\mathrm{d}R}\right)_{R_i},$$

where $R_i = D/2$.

Solution: The gradient of the sensible heat caused by desuperheating of vapour penetrating the spray ring at R_i is given by

$$\rho_V c_{pV} U_d \left(\frac{\mathrm{d}T}{\mathrm{d}R}\right)_{R_i} = U_d \left(\frac{\mathrm{d}\lambda_v}{\mathrm{d}R}\right)_{R_i}.$$

Now

$$\lambda_v = \frac{\pi}{6} d^3 n_v \rho_L \lambda,$$

and therefore

$$\mathrm{d}\lambda_v = \pi d^2 n_v \rho_L \lambda \mathrm{d}(r_I)$$

in which $R_I = d/2$ is the drop radius and the weak variation $n_v(R)$ has been neglected. The interface equation for a single drop is approximated by

$$\rho_L \lambda \frac{\mathrm{d}R_I}{\mathrm{d}t} = h_I (T_V - T_I)/2$$

and hence

$$\mathrm{d}\lambda_v = \pi n_v d^2 h_I (T_V - T_I) \mathrm{d}t/2$$

or, since

$$\mathrm{d}R = U_d \mathrm{d}t,$$

$$\frac{\mathrm{d}\lambda_v}{\mathrm{d}R} = \frac{\pi n_v d^2 h_I (T_V - T_I)}{2U_d}.$$

Therefore

$$\left(\frac{\mathrm{dT}}{\mathrm{d}R}\right)_{R_i} = \frac{\pi n_v d^2 h_I (T_V - T_I)}{2\rho_V c_{pV} U_d}.$$

Normalizing, we find that

$$\left[\frac{\pi n_v d^2 h_I \delta}{2\rho_V c_{pV} U_d}\right] = O(1)$$

where δ is the scale of penetration measured from $R = R_i$. When the spray ring leaves the collar $R_i = D/2$ and $n_v = n_{vw}$. The penetration depth is then

$$\delta = \frac{2\rho_V c_{pV} U_d}{\pi n_{vw} d^2 h_I} = 0.27\ \mu\text{m}.$$

This penetration is very much less than the thickness of the spray ring at the point of injection, thus confirming the model in which instantaneous evaporation of drops at $R = R_i$ creates a surface of discontinuity with superheated vapour on one side and saturated vapour on the other.

Q7.5 Using the Clausius–Clapeyron equation, and treating vapour as an ideal gas, show that if $T_s - T_\infty \ll T_\infty$,

$$\frac{\rho_s - \rho_{s\infty}}{\rho_{s\infty}} = \left(\frac{\lambda}{R_V T_\infty} - 1\right)\left(\frac{T_s - T_\infty}{T_\infty}\right),$$

where ρ_s and T_s are the vapour saturation density and temperature, respectively, and $\rho_{s\infty}$ is the saturation density when $T_s = T_\infty$.

Solution: Since

$$P = P(\rho, T),$$

$$\mathrm{d}P = \left(\frac{\partial P}{\partial \rho}\right)_T \mathrm{d}\rho + \left(\frac{\partial P}{\partial T}\right)_\rho \mathrm{d}T$$

$$= R_V T\,\mathrm{d}\rho + R_V \rho\,\mathrm{d}T,$$

and hence

$$\mathrm{d}\rho = \frac{1}{R_V T}\mathrm{d}P - \frac{\rho}{T}\mathrm{d}T.$$

Therefore, since

$$\mathrm{d}P = \frac{\rho\lambda}{T}\mathrm{d}T,$$

from the Clausius–Clapeyron equation,

$$\frac{\mathrm{d}\rho_s}{\rho_s} = \left(\frac{\lambda}{R_V T_s^2} - \frac{1}{T_s}\right)\mathrm{d}T$$

which integrates to give

$$\ln\frac{\rho_s}{\rho_{s\infty}} = \frac{\lambda(T_s - T_\infty)}{R_V T_s T_\infty} - \ln\frac{T_s}{T_\infty}$$

if $\rho_s = \rho_{s\infty}$ when $T_s = T_\infty$.

Now

$$\frac{T_s}{T_\infty} = 1 + \frac{T_s - T_\infty}{T_\infty},$$

and hence

$$\ln\frac{T_s}{T_\infty} \simeq \frac{T_s - T_\infty}{T_\infty}$$

if $(T_s - T_\infty) \ll T_\infty$.
Similarly,

$$\ln\frac{\rho_s}{\rho_{s\infty}} \simeq \frac{\rho_s - \rho_{s\infty}}{\rho_{s\infty}}.$$

Under these conditions,

$$\frac{\rho_s - \rho_{s\infty}}{\rho_{s\infty}} = \left(\frac{\lambda}{R_V T_\infty} - 1\right)\left(\frac{T_s - T_\infty}{T_\infty}\right).$$

Q7.6 Using energy and mass balances at the interface ($R = R_I$) of a growing water drop in air, derive the following relations:

$$\rho_L \lambda \frac{dR_I}{dt} = k_m\left(\frac{T_I - T_\infty}{R_I}\right) = \lambda \mathfrak{D}\left(\frac{\rho_{V\infty} - \rho_{VI}}{R_I}\right),$$

where ρ_{VI} and $\rho_{V\infty}$ are the vapour densities at T_I and T_∞, respectively. Assume that $m_{V\infty} \ll 1$.

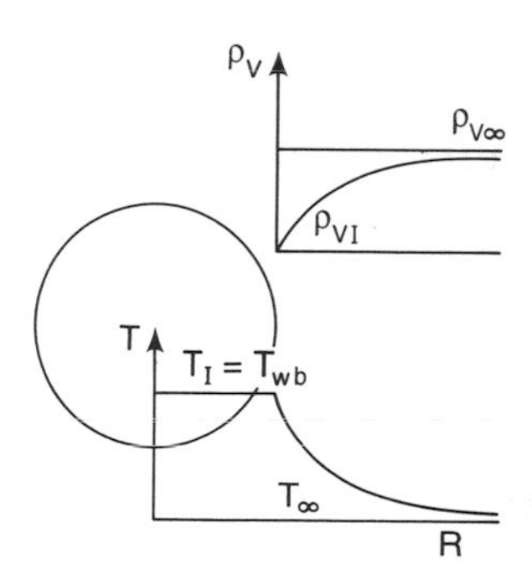

Solution: Continuity at the interface requires

$$\rho_L \frac{dR_I}{dt} = \mathfrak{D}\left(\frac{\partial \rho_V}{\partial R}\right)_I = \frac{\mathfrak{D}(\rho_{V\infty} - \rho_{VI})}{R_I},$$

using the solution from Exercise 4 following Chapter 7. Similarly, if $Ja_V = c_{pV}(T_I - T_\infty)/\lambda \ll 1$, the solution of the heat conduction equation in the vapour enables us to write

$$\rho_L \lambda \frac{dR_I}{dt} = -k_m \left(\frac{\partial T_V}{\partial R}\right)_I = k_m \frac{(T_I - T_\infty)}{R_I}.$$

Hence

$$\rho_L \lambda \frac{dR_I}{dt} = k_m \left(\frac{T_I - T_\infty}{R_I}\right) = \lambda \frac{\mathfrak{D}(\rho_{V\infty} - \rho_{VI})}{R_I}$$

in which continued growth requires $T_I > T_\infty$ and $\rho_{V\infty} > \rho_{VI}$.

Q7.7 Using the results obtained in Q7.5 and Q7.6, show that the growth of a water drop in cool moist air is governed by the equation

$$\frac{d}{dt}(R_I^2) = \frac{2(S-1)}{\dfrac{R_V T_\infty \rho_L}{\mathfrak{D} P_{s\infty}} + \dfrac{\rho_L \lambda}{k_m T_\infty}\left(\dfrac{\lambda}{R_V T_\infty} - 1\right)},$$

where $S = P_{V\infty}/P_{s\infty}$, in which $P_{s\infty}$ is the saturation vapour pressure at T_∞.

Solution: The difficulty with the expressions obtained in Q7.5 and Q7.6 is that the saturation density at the interface ($\rho_s = \rho_{VI}$), where $T_s = T_I = T_{wb}$ is unknown; the wet bulb temperature will be calculated in Q7.13. However, it may be eliminated as follows.

We write

$$\rho_{V\infty} - \rho_{VI} = (\rho_{V\infty} - \rho_{s\infty}) - (\rho_{VI} - \rho_{s\infty})$$

$$= (\rho_{V\infty} - \rho_{s\infty}) - \rho_{s\infty}\left(\frac{\lambda}{R_V T_\infty} - 1\right)\frac{(T_I - T_\infty)}{T_\infty},$$

using the result from Q7.5. The interface equations from Q7.6 then enable us to re-state this expression as

$$\frac{\rho_L R_I}{\mathfrak{D}} \frac{dR_I}{dt} = (\rho_{V\infty} - \rho_{s\infty}) - \rho_{s\infty}\left(\frac{\lambda}{R_V T_\infty} - 1\right)\frac{\rho_L \lambda R_I}{k_m T_\infty}\frac{dR_I}{dt}.$$

Rearranging, we obtain

$$\frac{d}{dt}(R_I^2) = \frac{2(S-1)}{\dfrac{R_V T_\infty \rho_L}{\mathfrak{D} P_{s\infty}} + \dfrac{\rho_L \lambda}{k_m T_\infty}\left(\dfrac{\lambda}{R_V T_\infty} - 1\right)}$$

in which $S = \rho_{V\infty}/\rho_{s\infty} = P_{V\infty}/P_{s\infty}$ is the supersaturation ratio (at T_∞) which is usually obtainable from the prescribed conditions.

Q7.8 A cloud begins to form at an elevation where the ambient pressure P_a is 61 kPa, the vapour pressure $P_{V\infty}$ is 0.711 kPa and the temperature T_∞ is 273.16 K. Beginning at the time when a cluster radius $R_I(0)$ is 1 μm, calculate the distance X travelled by the cloud at 100 kilometres per hour when $R_I(t) = 10\ \mu m$, $100\ \mu m$, and 1 mm. Ignore any relative motion between the drop and the cloud.

Solution: For these conditions, $P_{s\infty} = 0.611$ kPa, the triple point pressure. Hence

$$2(S - 1) = 0.327,$$

and

$$\frac{R_V T_\infty \rho_L}{\mathfrak{D} P_{s\infty}} = 7.68 \times 10^9\ \mathrm{s\ m^{-2}}$$

if $\mathfrak{D} = 0.26 \times 10^{-4}\ \mathrm{m^2\ s^{-1}}$, $R_V = 461.9\ \mathrm{J\ kg^{-1}\ K^{-1}}$ and $\rho_L = 967\ \mathrm{kg\ m^{-3}}$.

Also

$$\frac{\lambda}{R_V T_\infty} - 1 = 15.64$$

if $\lambda = 2.1 \times 10^6\ \mathrm{J\ kg^{-1}}$, and

$$\frac{\rho_L \lambda}{k_m T_\infty} = 0.338 \times 10^9\ \mathrm{s\ m^{-2}}$$

if $k_m = 22 \times 10^{-3}\ \mathrm{W\ m^{-1}\ K^{-1}}$.

Restating the result obtained in Q7.7,

$$\frac{d}{dt}(R_I^2) = A$$

which integrates to yield

$$t = \frac{[R_I^2 - R_I^2(0)]}{A},$$

where $R_I(0) = 1\ \mu m$ and $A = 25.2 \times 10^{-12}\ \mathrm{m^2\ s^{-1}}$, using the data given above. Hence the following results:

$$R_I = 10\ \mu m\text{: } t = 3.92\ s,\ X = 0.11\ km$$
$$R_I = 100\ \mu m\text{: } t = 396\ s,\ X = 11\ km$$
$$R_I = 1\ mm\text{: } T = 3.96 \times 10^4\ s,\ X = 1100\ km$$

Q7.9 A regenerative feed water heater unit consists of a bubble tray condenser in which a layer of water flows over a tray containing an array of orifices from which steam bubbles emerge.

Estimate the initial diameter D_0 of each vapour bubble if it is formed from a vapour slug length equal to one orifice diameter $d = 1$ mm. Also find the vapour exit velocity U_V if it is equal to the bubble terminal velocity U_b. Assume the bubble drag coefficient $C_D = 0.4$.

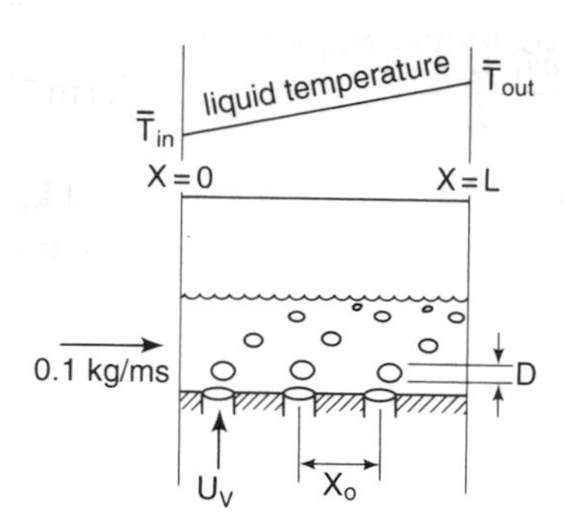

Solution: Equating the bubble and cylindrical slug volumes,

$$\frac{\pi D_0^3}{6} = \left(\frac{\pi}{4} d^2\right) d.$$

Hence

$$D_0 = \left(\frac{3}{2}\right)^{1/3} d = 1.14\,\text{mm}.$$

The terminal velocity U_b of the bubble may be determined from the force balance

$$\frac{1}{2}\rho_L U_b^2 C_D \frac{\pi}{4} D^2 = \frac{\pi D^3}{6} g\Delta\rho,$$

where $\Delta\rho = \rho_L - \rho_V \simeq \rho_L$. Since $C_D = 0.4$, this yields

$$U_V = U_b \simeq \left(\frac{4D_0 g}{3C_D}\right)^{1/2} = \left(\frac{1.52 dg}{C_D}\right)^{1/2}.$$

Hence

$$U_V = U_b = 0.193\,\text{m}\,\text{s}^{-1},$$

corresponding to a bubble Reynolds number of $Re_D = U_V D_0/\nu_L = 756$ if $\nu_L = 0.291 \times 10^{-6}\,\text{m}^2\,\text{s}^{-1}$.

7.10 Determine the vapour mass flux density $\dot{m}$ in a bubble tray condenser if a water flow rate of $0.1\,\text{kg}\,\text{m}^{-1}\,\text{s}^{-1}$ per unit span experiences a temperature rise from 80°C to 99°C over a 1.2 m length. Hence determine the necessary spacing X_0 of the orifices for the situation given in Q7.9.

Solution: A heat balance on unit span of the water layer requires

$$\dot{m}_V \lambda L = \dot{M}_L c_{pL} \Delta T_w,$$

where ΔT_w is the water temperature rise over a length L. Hence

$$\dot{m}_V = \frac{\dot{M}_L c_{pL} \Delta T_w}{L\lambda} = 2.9 \times 10^{-3} \text{ kg m}^{-2} \text{s}^{-1},$$

if $c_{pL} = 4.22 \times 10^3 \text{ J kg}^{-1} \text{K}^{-1}$ and $\lambda = 2.26 \times 10^6 \text{ J kg}^{-1}$.

This vapour mass flux density is created by n_A orifices per unit area of the tray. Hence

$$\dot{m}_V = n_A \frac{\pi}{4} d^2 U_V \rho_V$$

or

$$n_A = \frac{4\dot{m}_V}{\pi d^2 U_V \rho_V} = 3.24 \times 10^4 \text{ m}^{-2}$$

if $\rho_V = 0.59 \text{ kg m}^{-3}$. If the orifices are evenly distributed this value of n_A corresponds to a spacing of

$$X_0 = (n_A)^{-1/2} = 5.55 \text{ mm}.$$

Q7.11 Estimate the collapse times of the bubbles discussed in Q7.9 and Q7.10,

(a) neglecting bubble motion
(b) including bubble motion

Solution:

(a) Equation (7.28) for a stagnant bubble gives the collapse time as

$$t_{cond} = \frac{D_0^2}{4\kappa_L}\left(\frac{\rho_V}{\rho_L Ja_L}\right)^2,$$

where D_0 is the initial bubble diameter, and $Ja_L = c_{pL}(T_{sat} - \bar{T}_L)/\lambda$. Since $T_{sat} - \bar{T}_L$ decreases linearly over the range 20 K to 1 K, the conductive collapse time extends over the range $0.55 \text{ ms} < t_{cond} < 0.22 \text{ s}$, if $\rho_L = 958 \text{ kg m}^{-3}$ and $k_L = 0.670 \text{ W m}^{-1} \text{K}^{-1}$.

(b) The effect of bubble translation may be estimated roughly from the expression

$$t_{conv} = t_{cond}/Pe^{1/2}$$

where $Pe = U_b D/\kappa_L$. See equation (7.42), for example. Using the initial values of $U_b = 0.193 \text{ ms}^{-1}$ and $D = D_0$, we find that $Pe = 1.38 \times 10^3$ and hence 14.8 μs $< t_{conv} <$ 5.94 ms. Given that the bubble must begin and end at rest, these figures greatly underestimate collapse times but they do reveal the significance of bubble motion.

Q7.12 Assuming the above bubbles always move with their terminal velocity, estimate the minimum outlet water depth to prevent vapour from escaping through the free surface of the layer,

(a) assuming conductive heat loss

(b) assuming convective heat loss

Solution: From Q7.9, the terminal velocity of the bubble is given by

$$U_b = \left(\frac{4Dg}{3C_D}\right)^{1/2}.$$

The variation of the bubble diameter D may be found from the interface equation

$$\frac{dD}{dt} = -\frac{4k_L(T_{sat} - \bar{T}_L)}{\lambda\rho_V}\left(\frac{\rho_L Ja_L}{\rho_V D_0}\right),$$

using equation (7.27). Hence

$$D = D_0\left[1 - \frac{4k_L(T_{sat} - \bar{T}_L)}{\lambda\rho_V}\left(\frac{\rho_L Ja_L t}{\rho_V D_0^2}\right)\right]$$

which confirms the earlier finding for t_{cond}, i.e. when $D = 0$. Thus

$$U_b = \left(\frac{4D_0 g}{3C_D}\right)^{1/2}\left[1 - \frac{t}{t_{cond}}\right]^{1/2}$$

assuming C_D remains fixed.

(a) The collapse height of the bubble is given by

$$H = \int_0^{t_f} U_b \, dt$$

For a conductive heat loss, $t_f = t_{cond}$ and

$$H = \frac{2}{3}\left(\frac{4D_0 g}{3C_D}\right)^{1/2} t_{cond}.$$

Using the range of t_{cond} given in Q7.11, we find that 71 μm $< H <$ 2.84 cm, assuming $C_D = 0.4$. Since C_D will gradually rise these figures overestimate the collapse height.

(b) Since $\bar{U}_b = 2\,U_{b0}/3$, and $\bar{D} = D_0/2$, we may improve the estimate of t_{conv}

by writing $t_{conv} = t_{cond}/(Pe)^{1/2}$, where $\bar{P}e = \bar{U}_b \bar{D}/\kappa_L$. Hence 25.6 μs < t_{conv} < 10.3 ms. Substituting these values into the expression for H, we find that 3.3 μm < H < 0.133cm.

At outlet from the tray, H lies between the overestimate of 2.84 cm and the underestimate of 0.133 cm. We thus conclude that the minimum outlet water depth is of the order of 1 cm when the water is well mixed. A more accurate prediction requires a more accurate description of bubble motion in a shallow, agitated film.

Q7.13 Calculate the temperature of the growing drop discussed in Q7.7 and Q7.8, i.e. with an atmospheric pressure of $P_a = 61$kPa, a vapour pressure of $P'_\infty = 0.711$ kPa and a vapour temperature of $T_\infty = 273.16$ K.

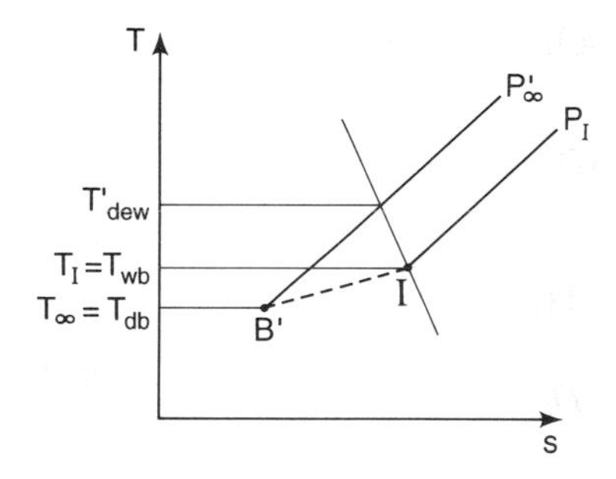

Solution: The drop temperature is $T_I = T_{wb}$, the wet bulb temperature corresponding to a dry bulb temperature $T_{db} = T_\infty$ at B′ in the supersaturated vapour. The dew point temperature $T'_{dew} > T_{wb} > T_{db}$.

From steam tables, T'_{dew}, the saturation temperature corresponding to P'_∞, is 1.92°C. The mass fraction of distant vapour is therefore

$$m_\infty = P'_\infty \left(\frac{R_a}{R_V P_a} \right) = 7.24 \times 10^{-3},$$

while the mass fraction at the interface may be found from the Clausius–Clapeyron equation. Hence

$$m_I = P_I \left(\frac{R_a}{R_V P_a} \right) = m_\infty \exp \frac{\lambda(T_{wb} - T'_{dew})}{R_V (T'_{dew})^2}$$

or, if $\lambda(T_{wb} - T'_{dew}) \ll R_V(T'_{dew})^2$,

$$m_I \simeq m_\infty [1 + \lambda(T_{wb} - T'_{dew})/R_V(T'_{dew})^2].$$

This may be combined with eqn (7.37) to yield

$$T_{wb} - T_\infty = \frac{m_\infty \lambda^2 (T'_{dew} - T_{wb})}{C_{pm} R_V (T'_{dew})^2}$$

from which we find that $T_{wb} = 1.15°C$, if $\lambda = 2.5 \times 10^6$ J kg^{-1}, $c_{pm} = 10^3$ J kg^{-1} K^{-1} and $R_V = 461.9$ J kg^{-1} K^{-1}.

Q7.14 Demonstrate that

$$m_{FOX} = m_{FU} - m_{OX}/r$$

is a conservative property if m_{FU} and m_{OX} are the mass fractions of the fuel and oxidant vapour, respectively, and $r = \dot{s}_{OX}/\dot{s}_{FU}$ is the oxidant/fuel ratio under stoichiometric conditions.

Solution: For the oxidant,

$$\frac{1}{R^2}\frac{d}{dR}(R^2 G m_{OX}) - \frac{1}{R^2}\frac{d}{dR}\left(R^2 \Gamma_{OX}\frac{dm_{OX}}{dR}\right) = -\dot{s}_{OX},$$

while for the fuel,

$$\frac{1}{R^2}\frac{d}{dR}(R^2 G m_{FU}) - \frac{1}{R^2}\frac{d}{dR}\left(R^2 \Gamma_{FU}\frac{dm_{FU}}{dR}\right) = -\dot{s}_{FU}.$$

Given that $\dot{s}_{FU} = \dot{s}_{OX}/r$ and $\Gamma_{OX} = \Gamma_{FU} = \Gamma_{FOX}$ are the diffusion coefficients, these equations may be combined to yield

$$\frac{1}{R^2}\frac{d}{dR}(R^2 G m_{FOX}) - \frac{1}{R^2}\frac{d}{dR}\left(R^2 \Gamma_{FOX}\frac{dm_{FOX}}{dR}\right) = 0,$$

where $m_{FOX} = m_{FU} - m_{OX}/r$. Hence, if we define

$$\dot{\mathbf{m}}_{FOX} = \mathbf{G} m_{FOX} - \Gamma_{FOX} \nabla m_{FOX},$$

$\nabla \cdot \dot{\mathbf{m}}_{FOX} = 0$ and m_{FOX} is a conservative property under the point symmetric conditions considered.

Q7.15 Consider combustion of a liquid hydrocarbon fuel spray at atmospheric pressure in a combustion chamber where the temperature is 1227°C. If the latent heat of evaporation is $\lambda = 36 \times 10^4$ J kg^{-1}, the enthalpy of combustion $\mathcal{H} = 44 \times 10^6$ J kg^{-1}, the mixture specific heat $c_{pm} = 1.15 \times 10^3$ J kg^{-1} K^{-1}, the oxygen/fuel ratio $r = 3.5$ and the boiling temperature of the fuel $T_B = 187°C$, estimate the ratio of the flame radius R_{FL} to the drop radius R_I.

Solution: Using $\psi = m_{FOX}$ in the result stated in Exercise 12 following Chapter 7,

$$\frac{G_{FU}R_I^2}{\Gamma_{FOX}}\left(\frac{1}{R_I}-\frac{1}{R}\right)=\ln\left[1+\frac{m_{FOX}-m_{FOX,I}}{\left(\Gamma_{FOX}\dfrac{dm_{FOX}}{dR}\right)_I\Big/G_{FU}}\right].$$

Hence, at the flame, where $R=R_{FL}$ and $m_{FU}=m_{OX}=m_{FOX}=0$,

$$\frac{G_{FU}R_I^2}{\Gamma_{FOX}}\left(\frac{1}{R_I}-\frac{1}{R_{FL}}\right)=\ln\left[1-\frac{m_{FU,I}}{\left(\Gamma_{FOX}\dfrac{dm_{FOX}}{dR}\right)_I\Big/G_{FU}}\right].$$

since $m_{OX,I}=0$. At $R=\infty$, where $m_{FOX}=-m_{OX,\infty}/r$ and $m_{FU}=0$, we obtain

$$\frac{G_{FU}R_I}{\Gamma_{FOX}}=\ln\left[1-\frac{(m_{FU,I}+m_{OX,\infty}/r)}{\left(\Gamma_{FOX}\dfrac{dm_{FOX}}{dR}\right)_I\Big/G_{FU}}\right].$$

Now using eqn (5.50),

$$\Gamma_{FOX}\left(\frac{dm_{FOX}}{dR}\right)_I=\Gamma_{FOX}\left(\frac{dm_{FU}}{dR}\right)_I=-G_{FU}(1-m_{FU,I}),$$

this second result may be written

$$\frac{G_{FU}R_I}{\Gamma_{FOX}}=\ln\left[1+\frac{m_{FU,I}+m_{OX,\infty}/r}{1-m_{FU,I}}\right]$$

which, by comparison with eqn (7.57),

$$\frac{G_{FU}R_I}{\Gamma_{FOX}}=\ln(1+N),$$

where $N=\theta_\infty/\lambda$, reveals that

$$m_{FU,I}+m_{OX,\infty}/r=N(1-m_{FU,I}).$$

Hence,

$$\Gamma_{FOX}\left(\frac{dm_{FOX}}{dR}\right)_I\Big/G_{FU}=-\frac{(m_{FU,I}+m_{OX,\infty}/r)}{N}.$$

Substituting for this and $G_{FU}R_I/\Gamma_{FOX}$ in the first result above we obtain

$$\ln(1+N)\left(1-\frac{R_I}{R_{FL}}\right)=\ln\left[1+\frac{Nm_{FU,I}}{m_{FU,I}+m_{OX,\infty}/r}\right],$$

and hence

$$\frac{R_I}{R_{FL}}=1-\frac{\ln[1+Nm_{FU,I}/(m_{FU,I}+m_{OX,\infty}/r)]}{\ln(1+N)}$$

or

$$\frac{R_{\mathrm{I}}}{R_{\mathrm{FL}}} \simeq \frac{\ln(1 + m_{\mathrm{OX},\infty}/r m_{\mathrm{FU,I}})}{\ln N}.$$

Since $m_{\mathrm{OX},\infty}/r m_{\mathrm{FU,I}} \ll 1$ typically, this reduces to

$$\frac{R_{\mathrm{FL}}}{R_{\mathrm{I}}} \simeq \frac{m_{\mathrm{FU,I}} r \ln N}{m_{\mathrm{OX},\infty}} \simeq \frac{r \ln N}{0.232}$$

for a fuel burning in atmospheric air if we assume that $m_{\mathrm{FU,I}} \simeq 1$. Using the data given,

$$N = \frac{c_{\mathrm{pm}}(T_\infty - T_{\mathrm{B}})}{\lambda} + m_{\mathrm{OX},\infty}\frac{\mathcal{H}}{r\lambda} = 11.42.$$

Hence

$$\frac{R_{\mathrm{FL}}}{R_{\mathrm{I}}} = 36.7.$$

Q7.16 For the above combustion system, estimate the initial burning rate and the time for complete combustion of a 20 μm diameter drop of fuel with a density $\rho_{\mathrm{L}} = 750\,\mathrm{kg\,m^{-3}}$. Take the mixture thermal conductivity $k_{\mathrm{m}} = 0.064\,\mathrm{W\,m^{-1}\,K^{-1}}$.

Solution: The burning rate is given by eqn (7.57):

$$G_{\mathrm{FU}} = \frac{\Gamma}{R_{\mathrm{I}}} \ln\left(1 + \frac{\theta_\infty}{\lambda}\right),$$

in which $\Gamma \simeq k_{\mathrm{m}}/c_{\mathrm{pm}}$. Hence, when $R_{\mathrm{I}} = 10\,\mu\mathrm{m}$, $G_{\mathrm{FU}} = 14.0\,\mathrm{kg\,m^{-2}\,s^{-1}}$.

The time for complete combustion of the drop is given by eqn (7.60):

$$t_{\mathrm{f}} = \frac{\rho_{\mathrm{L}} R_0^2}{2\Gamma \ln\left(1 + \frac{\theta_\infty}{\lambda}\right)},$$

where $R_0 = 10\,\mu\mathrm{m}$ is the initial radius. Hence $t_{\mathrm{f}} = 0.269\,\mathrm{ms}$.

INDEX

a-axis 210
ablation 107, 109
  dry 109
  shield 13
  wet 109–10
accretion 72–3, 109, 111–12
  dry 109, 111
  wet 109, 111
activation pressure 157
activation temperature 157
adhesion tension 33
alloy 98, 100–2
  magnetic 103
aluminium 104
amplitude-wavelength ratio 62
annealing 214
annular film 56, 60, 63, 140, 142–5, 174–9, 181–4
annular flow 75–7, 145–7
anti-icing 12
anti-wetting agent 34
Archimedean buoyancy 54–6, 119, 122, 152, 159, 162, 166, 179, 184
Archimedes number 55, 121, 124
atmospheric icing, *see* icing
atomised spray 67
avalanche 215

basal face 210
basal liquid layer 168–9
Bernouilli effect 72
binary mixture 38, 40, 42, 131
binary solution 38, 40, 41, 98–9
Biot number 89, 194, 217
bipartioned system 9, 53, 72, 76, 166, 202, 218
Black, Joseph 1–3
Black number 4–5
Blasius flow 20
blowing parameter 135
blown spray 67
body force 27
boiler 13
  marine 14
  riser 178
  sizes 13
Boltzmann constant 115
Boltzmann distribution 115
Bond number 63–4, 69–71, 167
Bondian instability 63, 65, 124, 167
boundary layer
  convection 22, 198
  momentum 22–3
  similarity 123
  thermal 22, 194–6, 201
  thickness 21–2
brass 104
brazing 104
bubble behaviour 69–71, 201
bubble bursting 68, 168, 171, 173
bubble collapse 199–201
bubble condenser 199, 201
bubble-drop 209–10, 213
bubble frequency 161, 163–4
bubble growth 158–9, 161–2, 165, 171, 174
bubble interactions 163–4
bubble point curve 40–1
bubble shape 69–71, 158–9
bubble size 160, 165, 168
bubbly flow 75–6, 146–7, 174–8, 180–1, 184
bulk energy 31
buoyancy force 63, 121, 167
burning rate 221
burnout 183

caloric 1–2
calorimetry 3–4
capillary pressure 218
capillary wave 62–3, 66, 69, 128, 198; *see also* interfacial waves
casting 12, 81, 88, 98, 102
c-axis 210
characteristic scale 18
chemical potential 38, 42, 46, 48
chemical reaction 5–6, 8
  endothermic 5
  exothermic 5–6, 215
chill zone 102–5
chimney effect 218
churn flow 75–6, 175–7
Clapeyron equation 37, 41, 43, 133, 205
Clausius, Rudolf 3
Clausius–Clapeyron equation 43, 135, 157
cluster 48–50, 115, 117, 156, 210
coalescence 69, 72, 75, 118, 163–4, 169, 179, 184, 210
coating, hydrophobic 118
cold condensation 191–2, 202–3
collapse time 201

collision efficiency 74, 112
columnar zone 102–5
combustion 13, 190
condensation domains 190–1
condenser 13
  direct contact 13
  spray 13, 16
  tubular 13, 15
conduction equation 28, 82, 84, 86, 93
conductivity ratio 85
constitutional supercooling 95–7, 102
contact angle 33–4, 64–5, 156, 160
continuity equation 29–30, 52, 55, 119–20
convective boundary condition 87–8, 90
cool evaporation 191, 202, 208
cooling curve 5–6
cooling pond 15, 190
cooling tower 14–15, 17, 190, 201
copper–nickel system 98, 100
copper–zinc system 104
core zone 102, 105
critical heat flux 166–8, 173, 183–4; *see also* heat flux, maximum
critical point 44, 54, 62
critical pressure 168
critical temperature 32
cryopreservation 215
crystal
  discontinuities 214
  equiaxed 102–3, 105
  growth 94, 102, 210
  lattice 210
  size 102
  size distribution 102
crystalline solid 9
crystallizer boiler 190, 213

degree of wetting, *see* wetting coefficient
de-icing 12
dendrites 9, 94, 96, 98–9, 103, 108–9, 116, 211
densification 109
desalination 12
de-wetting 33
dew point 40, 203, 205
diesel oil 221
diffusion
  analogue 133, 204
  coefficient 216, 218
  equation 29, 216
  of heat 29, 31
  layer 132–3, 135–6
  of mass 31, 207
diffusive resistance 133
discovery
  of latent heat 2
  of saturation curves 2
  of sensible heat 1–2
dispersed phase 9
dispersed system 9, 53, 72–3, 166, 202, 210, 217–18
domestic heating 10
drag coefficient 71–2
dripping 126
drop
  behaviour 71–2, 74, 116, 196, 202
  combustion 218
  flight 194
  internal circulation 70–1, 194, 196, 209
  median diameter 72
  mode diameter 72
  trajectory 73–4
  travel 194
dry-out 176–7, 179, 182–3

eddy diffusivity 59, 129
embryo 48–50, 66, 68
energy 5, 27
  geothermal 10
  of hydrogen bond 6–7
  of lattice vibration 6–7
  levels 7
  of melting 6
  of valence bond 7–8
  of vaporization 6
energy conservation 14
energy equation 27–8, 35, 119–20, 122, 200, 217
enthalpy 5
  of formation 7–8, 217
  of melting 7
  ratio 4–5, 221
  of reaction 5, 8
  of sublimation 7
  of vaporization 7
entrainment 6, 8, 139, 143, 176, 181, 184, 198
  splash-induced 68, 168–9
entropy
  change 37
  of fusion 7–8
  generation 37
  of sublimation 7
  of vaporization 7–8
equation of motion 29, 53–6, 58, 119, 121–2, 124
eutectic
  mixture 136
  solution 98
  temperature 40, 96, 98, 104
evaporation domains 190–1
evaporative cooling 14, 74, 206
exchange coefficient, *see* diffusion coefficient
experiment, historic 1

Fahrenheit, Gabriel 5
failure regimes 183–4
falling film 54, 63, 128, 151, 171
Fick's law 30
film flow rate 58, 60
film thickness 57–8, 64, 125, 150
first order model 23
first order phase change 42
flame 215–19
flooding 143, 156, 173
flow boiling 174–84
flow condensation 140–7, 152
fogging 115
force balance 57, 158, 209
Fourier's law 29
Fourier number 20, 84
frazil 95
freeze index, *see* thaw index
freeze shrinkage 88, 103
freezing
  of aqueous solution 81
  of biological tissue 214–15
  point 1, 5, 38, 87, 95–6
  rain 111
  range 98, 100, 102, 104
  time 90
  zone 98, 101, 109
friction factor 60, 142
frost 11, 106, 108
Froude number 62, 137, 139, 198
Froudian instability 62, 65, 68, 75, 77
fuel consumption 217
fuel spray 13, 218
fully developed boiling 180–1

Gibbs equation 38, 48
Gibbs function 37, 42–3, 48–50, 115
glacier 11, 106
glaciology 190
glaze 111
global climate 11
Graetz number 141, 197–9, 201, 206
grain boundaries 214
grain distribution, *see* crystal size distribution
gravity drainage 118
gravity wave 61–2; *see also* interfacial waves

hail 110–11, 211
hanging waves 68
hardening, *see* quenching
heat capacity 3
heat of combustion 217
heat engine 213, 222
heat exchanger 14, 140, 153, 174, 176–9
heat flux
  density 27, 36, 81
  maximum 166–8, 173, 183
  minimum 166, 168, 173
heating curve 6
heat of melting 2
hoarfrost 108
holdup 143
hot evaporation 191, 202–3
humidifier, domestic 14
hydroelectric intake 95
hysteresis 33, 50, 65, 161–2, 166

ice
  bridges 11
  cover 83–5
  formation 11, 95, 108, 210
  matrix 108–9, 111, 215
  nuclei 94, 210
  sheets 11
  in soil 11
iceberg 11
ice-water transition 43–4
icing
  of aircraft 106, 111
  atmospheric 11–12, 81, 211, 214
  of cables 112
  rate 112
  of ships 12, 106, 111
incipient flow boiling 174–5, 180–81
incondensable gas 131–2, 134, 191, 196, 198, 201–3, 207
industrial revolution 3
inertial force 21, 63, 167
ingot 102–3
interfacial concentration 96
interfacial energy 31–2, 34, 50
interfacial forces
  of buoyancy 9, 139
  of inertia 9, 61, 69, 139
  of surface tension 9, 62–3, 136, 139
interfacial geometry 9, 32, 65
interfacial heat transfer coefficient 37, 130
interfacial instability 3, 9, 60–5, 67, 69, 76, 108–9
interfacial motion 34–5, 96
interfacial resistance 36–7, 130
interfacial ripples 9; *see also* interfacial waves
interfacial shear 5–7, 136, 139, 142, 145
interfacial surface 31
interfacial waves 60, 62, 110, 129, 138, 142, 145
  capillary 63, 66, 128, 137
  gravity 61, 137
interfacial work 31
intermolecular bonds 5–7, 9, 50, 116, 118, 210, 214

intermolecular energy 31–2
intermolecular ordering 6–7
internal combustion engine 222
internal energy 27
intramolecular bonds 7–9
inverted boiling 173, 177
iron 104

Jakob number 4, 121–4, 128, 130, 141, 166, 192, 196, 202, 206
jet behaviour 66–7, 69, 218
jet condenser, *see* spray condenser
jet, cone 198
jet, fan 198–9, 218
Joule, James 3

Kelvin, Lord 3
kinetic energy 27, 36–7
kinetic theory 37
Kutateladze number 75, 167, 170

Leidenfrost, Johann 2
  temperature 155, 166–7, 169–70, 172–3, 177, 183
Lewis number 96, 204, 207, 220
liquefaction 13
liquid lenses 34
liquidus 38, 98, 100–2, 104, 106
liquid water content 112

macrolayer 163–170, 179, 184–5
mass diffusivity 96; *see also* diffusion
mass flux 55, 72
  of component 30
  density 29–30, 36, 131
  diffusive 131; *see also* diffusion
mass fraction 30, 74, 106, 110, 115–17, 153
melting point 1
meniscus 218–19
metastable equilibrium 44–8, 50, 111
meteor 106, 110
meteorology 190
microcavity 117–18, 156–7, 174
  density 156, 158–61, 163
  geometry 34, 163
microfilm 118
microlayer 158, 163–4
mist flow 176
mixing length 59
molecular accommodation 32
molecular adhesion 34
molecular cohesion 34
molten lava 10
momentum diffusivity 59, 121, 129
Morton number 69
mould conductance 102, 104
mushy zone 101–2, 104, 106, 109

needle crystals 211–12
Neumann solution 90, 92, 98, 101
Newtonian cooling 18
Newtonian fluid 28, 53
Newtonian heating 217
normalization 16, 20, 23, 132–3, 153, 194, 196, 201, 204
normalized temperature 18
normalized time 18–19
normalized variable 17–21, 23, 55, 58, 82, 85, 119–20, 125; *see also* scale
nucleation 46, 94
  curve 174
  heterogeneous 50, 102, 116–17, 156, 210
  homogeneous 47, 49–50, 156
  promoter 46, 50
  sites 162–3
Nukiyama, Shiro 3
Nusselt number, travel 194
Nusselt problem 85, 119, 121–2, 123, 125, 126, 130, 137, 144, 151–2, 155
Nusselt solution 122, 126, 128–9, 137, 151
Nusselt, Wilhelm 3, 122

oceanography 190
order of magnitude 17, 23, 128
OTEC (ocean thermal energy conversion) 12–13, 202, 213
oxidant 215–16

Peclet number 22, 24, 205
  travel 194
penetration depth 20
permafrost 11
phase diagram 40–1, 96–8, 105
pinch-off process 159–60, 165
plate crystals 211–12
plug flow 75–7, 146, 174–5, 184
Plumbers' solder 104
potential energy 3, 64
Prandtl number 23, 121–3, 182
precipitation 10
precursory cooling 173
pre-transitional boiling 161–5, 184
prism face 210–11
product of combustion 216, 218
psychrometry 205

quenching 172–3, 178, 207

Rayleigh, Lord 3
reactant 216
re-crystallization 214–15
reflooding 160, 162, 165
reflux condensation 142–3
renucleation 106
response time 19
re-wetting 155

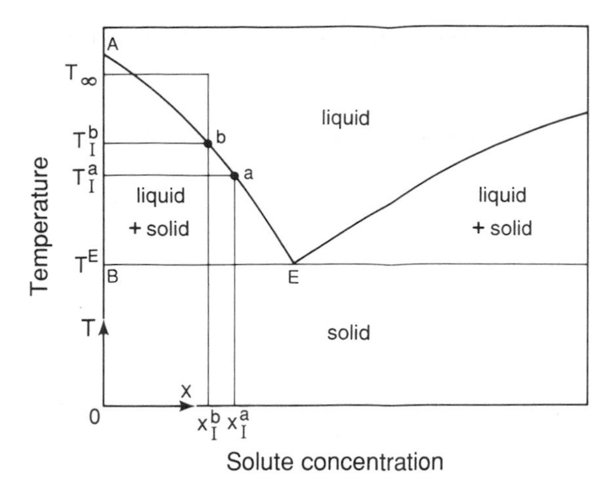

Fig. 4.8 Binary phase diagram for a solute insoluble in the solid phase.

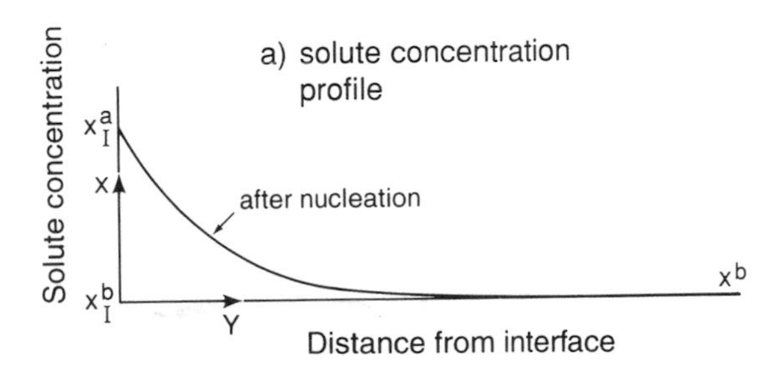

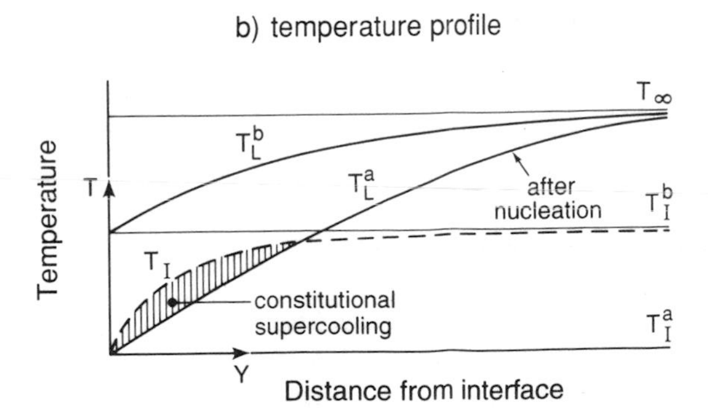

Fig. 4.9 Temperature and concentration profiles near the interface in a freezing binary solution.

AE in Fig. 4.8. As the cooled substrate temperature T_s is lowered, freezing continues and the interfacial temperature T_I^a follows the liquidus downward, the neighbouring solute concentration x increasing until the eutectic point is reached at E. Under these conditions, dendritic crystals form a freezing zone between the pure solid and the unfrozen solution, as illustrated in Fig. 10(a). However, when the highest temperature in the *pure* solid, T_I^a, falls beneath the eutectic temperature, no further increase in solute concentration may occur at this boundary. Solid eutectic with composition x^E is then deposited in a layer separating the pure solid from the dendritic freezing zone. This is illustrated in Fig. 4.10(b). It is worth noting that a eutectic solution, cooled slowly, freezes without dendritic growth. Since it possesses a unique freezing temperature T^E it is amenable to the analyses presented in Sections 4.1–4.3, and may thus yield Stefan and Neumann solutions. In general, however, the solid appears in three separate zones. Of these, the freezing zone is the most complex. It lies between the fixed values of the initial freezing temperature T_I^b and the eutectic temperature T^E; in the absence of the solid eutectic, the lower temperature limit becomes the unknown T_I^a. Latent heat is then no longer absorbed or released at a unique temperature, but is distributed throughout the freezing zone over a temperature range. This fundamental change in conditions is discussed more fully below.

4.5 Thermal fabrication: casting and welding

4.5.1 Freezing of alloys

Figure 4.8 is the equilibrium phase diagram of a binary solution for which the solute is soluble only in the liquid phase. This restriction seldom applies to metal alloys for which the components are usually soluble in both phases. The corresponding equilibrium phase diagram is well illustrated by the copper–nickel system described in Fig. 4.11. By comparison with Fig. 4.8, it is evident that this alloy may not exist as a eutectic. Only the pure components have a unique melting (and solidification) temperature. Any other composition melts (and freezes) over a range of temperature. This feature is common to many materials used in the fabrication of engineering components.

Using Fig. 4.11 to illustrate, solidification of a 45 per cent copper melt begins (b) at the liquidus temperature T_I^b and ends (e) at the solidus temperature T_I^e; conversely, melting begins at T_I^e and ends at T_I^b. This indicates that the first metal frozen (c) is poorer in copper than the last metal which has the composition of the original melt (e); likewise, the last fraction to solidify has a copper composition (d) greater than the first at (b). Under stable equilibrium conditions, these variations of composition within the freezing range would be evened out by diffusion but in practice

Reynolds number 55, 60, 69–74, 127–128, 141, 144, 182
Reynolds stress 58–9
Richards' rule 7–8
rime 111–12, 211
rivulet
formation 64, 74, 136, 144, 172–3, 183
stability 65, 74, 152

scale
of length 19–20, 55, 81, 83–4, 86, 119, 150, 201
of temperature 18–19, 81, 92, 119
of time 18–20, 81, 83–4, 196–7, 201, 222
of variable 17–18
of velocity 55, 81, 119, 150
Schmidt number 133
seam weld 105–6
secondary nucleation 68, 168, 171, 176, 211
sensible heat 2–5, 7
shear stress 57, 60
Sherwood number 133
silver solder 104
similarity variable 123
sintering 211
sliding bubble 171, 185
slug flow 76–7, 145
snow 11, 215
crystals 215
roads 11
snowflake 211
solar still 12–13, 202
soldering 13, 104
solidus 38, 98, 100, 104
specific entropy 7, 42–3
specific heat 2, 4, 5, 7
specific volume 42–3
splashing 127, 155
spray
condenser 16, 192
cooling 74, 152, 154, 170; *see also* evaporative cooling
generation 68, 76, 139, 143, 145
size distribution 71–2, 74
spreading coefficient 34, 193, 196
sputtering 173, 177
stable equilibrium 37–42, 44–6, 50
steam engine 2
steam generator, *see* boiler
steel 104
Stefan, Josef 3
Stefan number 4, 84, 86, 90, 92–3, 100–1, 110, 121
Stefan problem 3, 81–2, 85, 87–9, 118, 122
Stefan solution 86, 88, 90–2, 98
Stokes flow 73
stratified flow 76–7, 145–6
superficial velocity 74
superheat ratio 85
supersaturation ratio 115–16
Suratman number 198
surface degradation 118, 155, 157
surface geometry, effect of 124, 126, 130, 151
surface ripples 77, 110, 137; *see also* interfacial waves
surface roughness 157
swarm 72, 210
sweat cooling 219

terminal velocity 70
thaw index 87
thermal conductivity 129, 209
thermal diffusivity 19, 96, 101, 121; *see also* diffusion
thermal energy balance 28
thermal equilibrium 9, 195
thermal flux density 29, 131
thermal penetration 96, 214
thermocapillarity 160
thermodynamics
first law of 27, 37
second law of 37
thermosyphon 11–13, 16, 142–3
three-phase system 213
time constant, *see* response time
tin–lead system 104–5
Tollmien–Schlichting waves 60, 128
transitional energy 31
transitional flow 127–9, 146, 151, 176, 184
transitional zone 31–2
trapped air pockets 34, 156
triangular relation 58, 60, 141–2
triple point 44, 106, 110, 116
Trouton's rule 7–8
tube bundle 74, 126–7, 174, 184
tubular condenser 67, 116, 127; *see also* heat exchanger
turbine blade 103
turbulence 58, 60, 69, 95, 123, 126–9, 142, 151, 198, 207, 211
turbulent fluctuations 58–9
turbulent pipe flow 59, 144–5
two-phase flow
regimes 74–8
relation 181

universal velocity profile 59, 141
unstable equilibrium 37, 44–6

vapour mixture 131, 136, 191–2
viscous dissipation 28
viscous force 21, 55, 58, 61, 121, 125
viscous sublayer 58–9, 128, 174
vitrification 214

void fraction 145, 176, 181, 184
volcano 10
volumetric source 30–1

waiting period 162
wake effect 72, 209
warm condensation 191–3
water conservation 14, 190
water pollution 190
Watt, James 2
wavy flow 77, 146, 179
Weber number 63–4, 74, 137, 139, 142, 167
Weberian instability 63, 65–6, 68, 72, 151–2, 218
welding 13, 98, 104
wet bulb temperature 203, 205, 217–18
wetting 31–2, 34, 156–7, 160–1
  agent 34
  coefficient 33–4, 50, 156
  tension 33
whisky distillers 2
whitecap 67
wick 218–19
wispy-annular flow 76
work
  of compression 35, 38, 48
  flux density 35
  transfer 27

Young's equation 32